Hepatic De Novo Lipogenesis and Regulation of Metabolism

James M. Ntambi
Editor

Hepatic De Novo Lipogenesis and Regulation of Metabolism

Springer

Editor
James M. Ntambi
Department of Biochemistry
University of Wisconsin-Madison
Madison, WI, USA

ISBN 978-3-319-25063-2 ISBN 978-3-319-25065-6 (eBook)
DOI 10.1007/978-3-319-25065-6

Library of Congress Control Number: 2015956534

Springer Cham Heidelberg New York Dordrecht London

Printed on acid-free paper

Springer International Publishing AG Switzerland is part of Springer Science+Business Media (www.springer.com)

Preface

The liver is the largest solid vital organ in mammals that supports other organs in the body in some facet. Without a healthy liver, a person cannot survive. In the human it is located in the upper right quadrant of the abdomen, below the diaphragm. This book constitutes an effort to sum-up all the primary and relevant metabolic information that one needs to review to understand the complex and diverse role of the liver in metabolism.

With the current epidemic of metabolic diseases, it is of immediate importance to understand the contribution of the liver in health and its role in the development of impaired metabolic regulation. Many studies have unmasked important roles that proteins expressed in the liver play in the development of or protection from metabolic diseases.

Obesity is currently a worldwide public health burden that increases the risk for developing insulin resistance and several metabolic diseases such as diabetes, cardiovascular disease, inflammation and non-alcoholic fatty liver disease. Dietary approaches to limit fat intake are commonly prescribed to achieve the hypocaloric conditions necessary for weight loss. But dietary fat restriction is often accompanied by increased carbohydrate intake, which can dramatically increase endogenous fatty acid synthesis depending upon carbohydrate composition. It should be pointed out that although western societies consume high fat diets, the consumption of high carbohydrate diets in developing countries is on the rise. Since both dietary and endogenously synthesized fatty acids contribute to the whole-body fatty acid pool, obesity can therefore result from excessive fat or carbohydrate consumption.

One of the major metabolic functions of the liver is to carry out de novo lipogenesis, which is the metabolic pathway that allows the conversion of excess carbohydrates into fatty acids. The process of de novo lipogenesis utilizes several enzymes that convert the two-carbon acetyl-CoA into 16- and 18-carbon long chain saturated fatty acids (palmitate and stearate, respectively). Stearoyl-CoA desaturase (SCD) enzymes catalyze the conversion of palmitate and stearate into the monounsaturated fatty acids palmitoleate and oleate, which upon esterification into triglycerides are transported by very low-density lipoprotein (VLDL) to white adipose tissue (WAT) for storage. There is also extensive crosstalk between the liver and WAT via signaling factors such as adipokines, lipokines and hepatokines that are involved in regulation of metabolic homeostasis.

There is a vast literature and ongoing research on liver function. Our contributions in this book are focused on some aspects of metabolism. I am therefore extremely grateful to many people who have helped me make possible the current series on the role of the liver in de novo lipogenesis and metabolism. I would like to express my sincere appreciation to all contributors of the diverse chapters for their cooperation and excellent work. Without their participation this edition would not have been possible. I would like to thank Cori Praska, Mackenzie Carlson, and Andrew Denu who have helped in the editing of some chapters of the book. I would also like to thank Springer for their assistance and support during the course of this project. Finally I would like to thank my family for their support, encouragement and patience over the years.

Madison, WI, USA James M. Ntambi, PhD

Contents

1 **Transcriptional Regulation of De Novo Lipogenesis in Liver** 1
Deqiang Zhang and Lei Yin

2 **The Molecular Basis of Hepatic De Novo Lipogenesis in Insulin Resistance** 33
Mengwei Zang

3 **Regulation of Hepatic Metabolism by the Mechanistic Target of Rapamycin** 59
Dudley W. Lamming

4 **FoxO1: A Conductor of Insulin Signaling to Glucose and Lipid Metabolism** 79
Ting Zhang and H. Henry Dong

5 **Molecular Mechanisms of Lipotoxicity in Nonalcoholic Fatty Liver Disease** 101
Sophie C. Cazanave and Arun J. Sanyal

6 **Role of the Endocannabinoid System in Hepatic Lipogenesis** 131
Joseph Tam

7 **Role of Lipogenesis and Lipid Desaturases in Non-alcoholic Fatty Liver Disease** 143
Mohamed Amine Lounis, Sabri Rial, James M. Ntambi, and Catherine Mounier

8 **Lipid Droplet Proteins and Hepatic Lipid Metabolism** 165
Yumi Imai, Michelle B. Trevino, and Rexford S. Ahima

9 **Influence of Gut Microbiota on Hepatic Lipogenesis and Disease Pathogenesis** 189
Vishal Singh, Beng San Yeoh, James M. Ntambi, and Matam Vijay-Kumar

10 Hepatic Lipogenesis: Nutritional Control and Pathophysiological Relevance 211
Shih-Lung Woo, Ting Guo, and Chaodong Wu

11 Metabolic Rhythm of Hepatic Lipogenesis: Regulation and Roles in Metabolism 235
David Jacobi, Hyunjii J. Cho, Ryan K. Alexander, and Chih-Hao Lee

12 Monounsaturated Fatty Acid Mediated Liver-Adipose Tissue Crosstalk and Metabolic Regulation 255
Maggie S. Burhans and James M. Ntambi

13 Adipose Tissue DNL and Its Role in Metabolic Homeostasis 267
Ludger Scheja and Christoph Buettner

14 The Lipogenic Effect of Insulin Revisited 285
Sonia M. Najjar

Index 297

Contributors

Rexford S. Ahima, M.D., Ph.D. Division of Endocrinology, Diabetes and Metabolism, Perelman School of Medicine, University of Pennsylvania, Philadelphia, PA, USA

Ryan K. Alexander Division of Biological Sciences, Department of Genetics and Complex Diseases, Harvard T.H. Chan School of Public Health, Boston, MA, USA

Christoph Buettner Department of Medicine and Department of Neuroscience, Diabetes, Obesity and Metabolism Institute, Icahn School of Medicine at Mount Sinai, New York, NY, USA

Maggie S. Burhans Department of Medicine, University of Wisconsin-Madison, Madison, WI, USA

Sophie C. Cazanave, Ph.D. Division of Gastroenterology, Hepatology and Nutrition, Department of Internal Medicine, Virginia Commonwealth University, Richmond, VA, USA

Hyunjii J. Cho Division of Biological Sciences, Department of Genetics and Complex Diseases, Harvard T.H. Chan School of Public Health, Boston, MA, USA

H. Henry Dong, Ph.D. Division of Pediatric Endocrinology, Department of Pediatrics, Childern's Hospital of Pittsburgh of UPMC, University of Pittsburgh School of Medicine, Pittsburgh, PA, USA

Ting Guo Department of Nutrition and Food Science, Texas A&M University, College Station, TX, USA

Yumi Imai, M.D. Division of Endocrinology, Diabetes and Metabolism, Perelman School of Medicine, University of Pennsylvania, Philadelphia, PA, USA

David Jacobi Division of Biological Sciences, Department of Genetics and Complex Diseases, Harvard T.H. Chan School of Public Health, Boston, MA, USA

Dudley W. Lamming Division of Endocrinology, Department of Medicine, University of Wisconsin-Madison and William S. Middleton Memorial Veterans Hospital, Madison, WI, USA

Chih-Hao Lee Division of Biological Sciences, Department of Genetics and Complex Diseases, Harvard T.H. Chan School of Public Health, Boston, MA, USA

Mohamed Amine Lounis Département des Sciences Biologiques, Centre de recherche BioMed, Université du Québec à Montréal, Montréal, QC, Canada

Catherine Mounier Département des Sciences Biologiques, Centre de recherche BioMed, Université du Québec à Montréal, Montréal, QC, Canada

Sonia M. Najjar, Ph.D. Center for Diabetes and Endocrine Research, University of Toledo College of Medicine and Life Sciences, Toledo, OH, USA

Department of Physiology and Pharmacology, University of Toledo College of Medicine and Life Sciences, Toledo, OH, USA

James M. Ntambi Department of Biochemistry, University of Wisconsin-Madison, Madison, WI, USA

Department of Nutritional Sciences, University of Wisconsin-Madison, Madison, WI, USA

Sabri Rial Département des Sciences Biologiques, Centre de recherche BioMed, Université du Québec à Montréal, Montréal, QC, Canada

Arun J. Sanyal, M.D. Division of Gastroenterology, Hepatology and Nutrition, Department of Internal Medicine, Virginia Commonwealth University, Richmond, VA, USA

Ludger Scheja Department of Biochemistry and Molecular Cell Biology, University Medical Center Hamburg-Eppendorf, Hamburg, Germany

Vishal Singh Department of Nutritional Sciences, The Pennsylvania State University, University Park, PA, USA

Joseph Tam Obesity and Metabolism Laboratory, Faculty of Medicine, The Institute for Drug Research, School of Pharmacy, The Hebrew University of Jerusalem, Jerusalem, Israel

Michelle B. Trevino Department of Internal Medicine, The Strelitz Diabetes Center, Eastern Virginia Medical School, Norfolk, VA, USA

Matam Vijay-Kumar, Ph.D. Department of Nutritional Sciences, The Pennsylvania State University, University Park, PA, USA

Shih-Lung Woo Department of Nutrition and Food Science, Texas A&M University, College Station, TX, USA

Chaodong Wu Department of Nutrition and Food Science, Texas A&M University, College Station, TX, USA

Beng San Yeoh Department of Nutritional Sciences, The Pennsylvania State University, University Park, PA, USA

Lei Yin, Ph.D. Department of Molecular and Integrative Physiology, University of Michigan Medical School, Ann Arbor, MI, USA

Mengwei Zang, M.D., Ph.D. Department of Medicine, Whitaker Cardiovascular Institute, Boston University School of Medicine, Boston, MA, USA

Deqiang Zhang, Ph.D. Department of Molecular and Integrative Physiology, University of Michigan Medical School, Ann Arbor, MI, USA

Ting Zhang Division of Pediatric Endocrinology, Department of Pediatrics, Childern's Hospital of Pittsburgh of UPMC, University of Pittsburgh School of Medicine, Pittsburgh, PA, USA

Chapter 1
Transcriptional Regulation of De Novo Lipogenesis in Liver

Deqiang Zhang and Lei Yin

Abstract The liver is one of the main metabolic organs where de novo lipogenesis (DNL) takes place. DNL in the liver is tightly controlled by insulin and other metabolic hormones, as well as the availability of substrate, i.e., glucose level. In the positive energy balance, high concentration of glucose stimulates insulin secretion, activates ChREBP, and at the same time, provides substrate, which facilitate DNL in the liver. Accurate regulation of lipid levels is critical for cellular and organismal homeostasis, while deregulations in the lipogenic pathway are associated with diverse metabolic syndromes. This chapter will first introduce our current understanding about the DNL regulation by insulin-SREBP-1c pathway and ChREBP pathway, as well as other factors, including PPARγ, AMPK, SIRT1, circadian clock, ER stress and inflammation. At last, we will discuss the deregulation of DNL in metabolic syndromes and cancer, and potential targets for therapy.

Keywords De novo lipogenesis • ChREBP • Insulin • SREBP-1c • Metabolic syndromes

1 Introduction

All metazoans need to eat. This defining characteristic poses a challenge because nutrient sources are often scarce and caloric demands constantly change. Animals solve this problem by developing highly integrated mechanisms to promote anabolism efficiently when calorie supply is abundant.

Storing energy is crucial for survival during starvation. Mammals have a limited ability to store energy in forms of carbohydrates but a seemingly unlimited capacity to store calories as fat (Lodhi et al. 2011). In conditions of nutrients are present in excess, carbohydrates can be converted to fatty acids via the process of de novo lipogenesis. De novo lipogenesis (DNL) or de novofatty acid (FA) synthesis is the

D. Zhang, Ph.D. • L. Yin, Ph.D. (✉)
Department of Molecular and Integrative Physiology, University of Michigan Medical School, 7711 Med Sci II, 1337 Catherine street, Ann Arbor, MI 48103, USA
e-mail: deqiangz@umich.edu; Leiyin@umich.edu

J.M. Ntambi (ed.), *Hepatic De Novo Lipogenesis and Regulation of Metabolism*,
DOI 10.1007/978-3-319-25065-6_1

metabolic pathway that synthesizes fatty acids from acetyl-CoA, derived from excess carbohydrates (Ameer et al. 2014). These fatty acids can then be incorporated into triglycerides (TGs) for energy storage through subsequent triglyceride synthesis. During the postprandial phase, DNL mainly takes place in liver and adipose tissue, and liver is thought to be quantitatively more important than adipose tissue (Lodhi et al. 2011). In normal physiological conditions, DNL is considered to contribute a minority of total serum TG. Most of the serum TG content is uptaken from dietary sources (Björntorp and Sjöström 1978). However, under some physiological conditions, such as in individuals on high carbohydrate diet, hepatic DNL can make significant contribution to serum lipid content (Nomura and Yamanouchi 2012) (Fig. 1.1).

1.1 Lipogenic Program in Liver

De novo lipogenesis mainly take place in the adipose tissue and the liver (Pearce 1983). Synthesizing fats from simple precursors involves the generation of saturated fatty acids and the processes of elongating as well as desaturating these molecules.

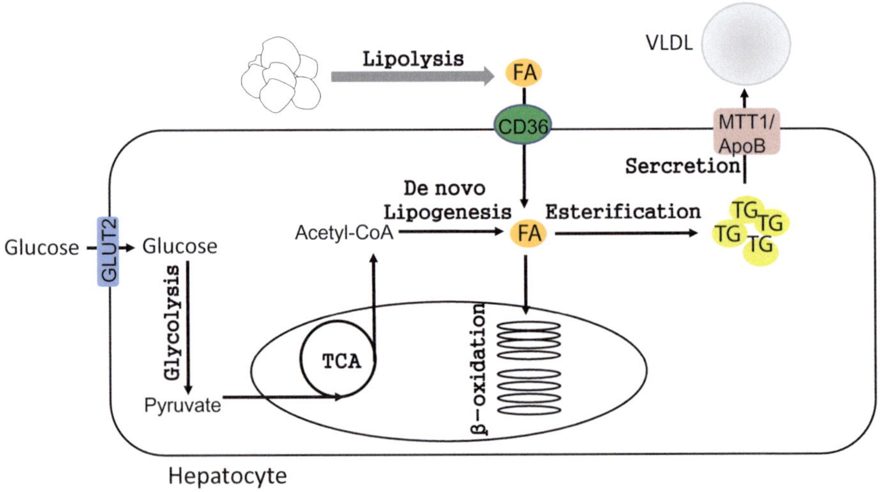

Fig. 1.1 TG metabolism in hepatocytes. In hepatocytes, there are mainly two sources of fatty acid: (a) In condition feeding with high carbohydrate diet, glucose is transported into hepatocytes by GLUT-2. After Glycolysis and TCA, glucose is transformed as acetyl-CoA, providing substrate of de novo lipogenesis to produce fatty acids. (b) In fasting condition, the liver uptakes fatty acids from serum, production of lipolysis in white adipose tissue. There are mainly two destinations for fatty acids: (a) In condition of positive energy balance, fatty acids are synthesized into triglyceride and secreted into serum, which are delivered to the peripheral tissues and used as energy source or energy storage. (b) In condition of fasting, fatty acids are transported into mitochondria and be oxidized to produce energy

The lipogenic pathway includes a coordinated series of enzymatic reactions. The synthesis pathway occurs in the cytoplasm. Glucose is first converted into pyruvate by glycolysis. Then pyruvate enters mitochondrion and is converted into citrate in the tricarboxylic acid (TCA) cycle. After exiting mitochondrion, citrate is used to form acetyl-CoA, the start point of de novo lipogenesis, by ATP-citrate lyase (ACLY) (Ameer et al. 2014). Acetyl-CoA is carboxylated to malonyl-CoA by acetyl-CoA carboxylase (ACC). Then malonyl-CoA is used by fatty acid synthase (FASN) to produce a 16-carbon saturated fatty acid, palmitate. In animals, ACC1 (Mr = 265 kDa) and ACC2 (Mr = 280 kDa) are the two isoforms of acetyl-CoA carboxylase, and they display distinct tissue distributions. ACC1 is abundant in lipogenic tissues, such as liver and adipose tissue, whereas ACC2 is highly expressed in heart, skeletal muscle, and liver (Abu-Elheiga et al. 1997, 2005; Ha et al. 1996). Liver-specific deletion of ACC1 reduces hepatic de novo fatty acid synthesis and triglyceride accumulation (Mao et al. 2006). However, it was reported that in the liver, ACC2 also have a compensatory effect to ACC1 (Harada et al. 2007). FASN is the key rate-limiting enzyme of fatty acid synthesis. It is organized into a multienzyme complex, which is composed of two identical 272 kDa multifunctional polypeptides. The substrates are handed from one functional domain to the next (Stoops et al. 1975; Smith et al. 2003).

The main product of DNL is palmitate, but stearate and shorter fatty acids are also generated. Palmitate is further elongated into very long chain fatty acids (VLCFA) in a four-step reaction cycle by membrane-bound enzymes predominantly located in the endoplasmic reticulum. Elovl (elongation-of-very-long-chain-fatty acids) family members are believed to perform the first, regulatory, step (condensation) in the elongation cycle in mammals. This family of enzymes consists of at least six members in mouse and human, carrying out substrate-specific elongation with fatty acids of different lengths and degrees of unsaturation (Jakobsson et al. 2006). For example, Elovl6 is involved in the elongation of saturated and monounsaturated FAs with 12, 14, and 16 carbons and is reported to be responsible to liver steatosis (Moon et al. 2001; Matsuzaka et al. 2007, 2012). VLCFAs are desaturated by stearoyl-CoA desaturase (SCD), which introduces double bonds at specific positions, to form mono- and poly-unsaturated fatty acids (Hodson and Fielding 2013). Desaturation has profound effects on metabolic fate of fatty acids. There are four isoforms of SCDs. SCD1 is the predominant isoform in adipose tissue and liver (Miyazaki et al. 2007). SCD1 is upregulated in animals fed a diet rich in carbohydrates. It has been shown that oleate production by SCD1 is required for carbohydrate-induced adiposity (Miyazaki et al. 2007).

1.2 Triglyceride Secretion and β-Oxidation

After fatty acid synthesis, the resulting fatty acids are esterified with glycerol to form triglyceride, which are assembled with Apolipoprotein B (APOB) to form very low density lipoprotein (VLDL) particles and secreted into the blood for delivery to adipose tissues and other peripheral tissues. Blocking triglyceride (TG) secretion results in hepatosteatosis (Sun and Lazar 2013). Each VLDL has one apoB100 molecule, making apoB100 availability a key determinant of the number of TG-rich VLDL particles secreted (Fisher 2012). Mice deficient in the ApoB gene develop hepatosteatosis, independent of glucose intolerance or insulin resistance (Schonfeld et al. 2008). The pool of apoB100 in the liver is typically regulated not by its level of synthesis, which is relatively constant, but by its level of degradation (Fisher 2012).

Except apoB100 and triglycerides, nascent VLDL released from the liver also contains apolipoprotein C1 (apoC1), apolipoprotein E (apoE), cholesterol and cholesteryl esters. And it also picks up apolipoprotein C-II (apoC-II) and additional apoE from high-density lipoprotein (HDL) in circulation. Once in circulation, VLDL will come in contact with lipoprotein lipase (LPL) in tissues like adipose tissue, heart, and skeletal muscle. LPL will remove triglycerides from VLDL for storage or energy production. Then apoC-II of VLDL will be picked up back by high-density lipoprotein (HDL) (Shelness and Sellers 2001). As the composition of the molecule changes, VLDL becomes intermediate-density lipoprotein (IDL). Fifty percentage of the IDL are recycled by apoB100 and apoE receptors on surface of the liver cells. The other 50 % IDL lose apoE and become low-density lipoprotein (LDL), which is up-taken by the LDL receptor via endocytosis. The contents are either stored, used for cell membrane structure, or converted into other products such as steroid hormones or bile acids (Nimpf and Schneider 2000).

Except being secreted into circulation, the TG can also provide energy via oxidation. In the liver, fatty acids are oxidized by β-oxidation in mitochondria and peroxisomes and by ω-oxidation in microsomes (Yu et al. 2003; Reddy and Rao 2006). Mitochondrial β-oxidation is responsible for the oxidation of short, medium and long chain fatty acids, which constitutes the major process by which fatty acids are oxidized to generate energy, producing ATP. Fatty acids are completely oxidized to acetyl-CoA by mitochondrial β-oxidation, and the acetyl-CoA then either enters the Krebs cycle for further oxidation or condenses to ketone bodies (acetoacetate, acetone, and β-hydroxybutyrate) in liver to serve as oxidizable fuels for extrahepatic tissues (Bartlett and Eaton 2004; Gibbons et al. 2000; Vluggens and Reddy 2012). Peroxisomal β-oxidation is responsible for the oxidation of very long chain fatty acids (VLCFAs, >C20). VLCFAs, are almost exclusively β-oxidized in peroxisomes because mitochondria are devoid of very-long-chain acyl-CoA synthetase (Reddy and Mannaerts 1994; Uchiyama et al. 1996). VLCFAs are also oxidized by the cytochrome P450 CYP4A ω-oxidation system in microsome to long-chain dicarboxylyl-CoAs that serve as substrates for peroxisomal β-oxidation (Mannaerts et al. 2000). The genes encoding peroxisomal, microsomal and some mitochondrial fatty acid oxidizing enzymes in the liver are transcriptionally regulated by peroxisome proliferator-activated receptor α (PPARα).

2 Regulation of De Novo Lipogenesis in Liver

2.1 *Nutritional Control of Lipogenesis*

In the liver, lipogenesis is mainly driven by carbohydrate availability. It has been reported that high-carbohydrate diet induces an increase in hepatic DNL which leads to hypertriglyceridemia, and simple sugars are more effective than complex carbohydrates in stimulating hepatic DNL (Schwarz et al. 2003; Hudgins et al. 1998; Parks et al. 1999). Plasma glucose derived from dietary directly stimulates hepatic lipogenesis. Glucose metabolism via the glycolytic pathway provides a carbon source for de novo lipogenesis. Pyruvate, the main glycolytic product, links glycolysis to lipogenesis. At the same time, glucose induces expression of enzymes involved in the lipogenic pathway, which is primarily mediated by carbohydrate responsive element binding protein (ChREBP). Plasma glucose also induces expression of lipogenic enzymes by stimulating the release of insulin and inhibiting the release of glucagon from the pancreas (Kersten 2001). In hepatocytes, insulin activates the protein kinase AKT2 (Strable and Ntambi 2010). AKT2 in turn regulates two downstream transcription factors, FOXO1 and SREBP1c, which control the expression of the metabolic enzymes underlying these processes. Both SREBP-1c and SREBP-2 are abundantly expressed in the liver. SREBP1c is the dominant insulin-stimulated isoform in the liver, responsible for inducing lipogenic gene expression and promoting fatty acid synthesis, while SREBP-2 activates the genes that control cholesterol biosynthesis (Horton et al. 2002).

2.2 *Insulin Pathway*

In response to increased blood glucose levels, β cells in the islets of pancreas synthesize and release insulin (Sonksen and Sonksen 2000). The main targets of insulin are liver, adipose tissue and muscle. Insulin increases glucose uptake in muscle and adipose tissue, inhibits hepatic gluconeogenesis and stimulates hepatic lipogenesis to promote the storage of energy as fat (Saltiel and Kahn 2001). In the liver, insulin promotes lipogenesis through various mechanisms, including induction of glucose uptake and glycolysis, which provides the carbon intermediates required for lipid synthesis (Manning and Cantley 2007). At the same time, insulin also promotes the transcription of key lipogenic genes. These effects of insulin are mediated by Akt2, which phosphorylate two targets, FOXO1 (Forkhead box protein O1) and SREBP1c (sterol-regulatory element binding protein 1c) (Yecies et al. 2011; Leavens et al. 2009). FOXO1 stimulates gluconeogenic gene expression in the liver and is directly phosphorylated and inhibited by Akt2 (Gross et al. 2009). Srebp1c is responsible for inducing lipogenic gene expression and promoting fatty acid synthesis (Horton et al. 2002).

2.2.1 SREBPs

There are three isoforms of SREBPs (SREBP-1a, -1c and -2), of which both SREBP-1c and SREBP-2 are abundantly expressed in the liver. SREBP-1a and SREBP-1c are encoded by a single gene, but transcribed by different promoters (Shimomura et al. 1997). SREBP1-c activates the genes that control fatty acid and TAG synthesis, and SREBP-2 activates the genes that control cholesterol biosynthesis (Horton et al. 2002). SREBPs belong to the basic helix–loop–helix leucine zipper family of transcription factors (Goldstein et al. 2006). The SREBPs are synthesized as inactive precursors residing in the endoplasmic reticulum (ER) (Brown and Goldstein 1997; Horton et al. 2002). SREBP precursor consist of about 1150 amino acids, including three domains: (a) an NH_2-terminal domain which contains the bHLH-Zip region for binding DNA; (b) two hydrophobic transmembrane-spanning segments, which projects into the lumen of the ER; and (c) a COOH-terminal domain which performs the essential regulatory function (Horton et al. 2002). In conditions of high ER cholesterol concentration, SREBP precursors binds to SREBP-cleavage activating protein (SCAP) and ER retention protein called Insig, which sequesters SREBP activity (Sun et al. 2005, 2007; Radhakrishnan et al. 2008).

ER cholesterol level decrease induces conformational change of SCAP, facilitating SCAP-SREBP complex binding to COPII proteins and hence translocation from the ER to the Golgi apparatus, where Site-1 protease (SIP1) and Site-2 protease (S2P) reside. After cleaved sequentially by SIP1 and SIP2 proteases, The NH2-terminal bHLH-Zip domain of SREBP1c, also designated nuclear SREBP (nSREBP), released released from the membrane. nSREBP translocates to the nucleus, where it activates transcription by binding to sterol response elements (SREs) in the promoter/enhancer regions of multiple target genes (Horton et al. 2002).

The insulin-Akt signaling pathway also plays a critical role in SREBP-1c proteolytic processing (Hegarty et al. 2005; Yellaturu et al. 2009; Owen et al. 2012; Shimomura et al. 1999). It was reported that Akt promotes SREBP-1c phosphorylation, which increases the affinity of SCAP-SREBP-1c complex for COPII-coated vesicles and facilitates the Golgi transportation of SREBP-1c (Yellaturu et al. 2009). Another mechanism which insulin utilized to stimulate SREBP1c processing is suppressesing the expression of Insig-2a, which interacts with SCAP-SREBP-1c complex and prevents their binding to COPII-coated vesicles (Yecies et al. 2011; Yellaturu et al. 2009). Hence, insulin stimulates SREBP-1c processing through two distinct mechanisms, involving the suppression of Insig-2a, and the induction of SREBP-1c phosphorylation which facilitates the association of SCAP-SREBP-1c complex with COPII-coated vesicles.

Insulin also transcriptionally regulates srebp-1c in livers and cultured hepatocytes (Shimomura et al. 1997, 1999; Horton et al. 1998). Interaction of insulin with its cell surface receptor induces the phosphorylation of the insulin-receptor substrates (IRS), which initiates a signaling cascade that finally activates AKT. Active AKT phosphorylates a wide range of downstream targets involved in cell metabolism, such as forkhead box protein O1 (Foxo1), glycogen synthase kinase 3 (GSK3), and tuberous sclerosis 2 (TSC2) within TSC1–TSC2 complex (Hemmings and Restuccia 2012). TSC1–TSC2 complex acts as a GTPase-activating protein (GAP), i.e. the suppressor of a Ras-related small G protein called Rheb, by enhancing its conversion to the GDP-bound off state. GTP-bound Rheb stimulates mTORC1 kinase activity. Therefore, Akt-mediated inhibition of the TSC1–TSC2 complex serves to activate Rheb and mTORC1 (Huang and Manning 2009). Increased activation of mTORC1 has been found to activate SREBP isoforms and promote an SREBP-dependent increase in de novo lipid synthesis (Porstmann et al. 2008; Düvel et al. 2010), while mTORC1-specific inhibitor rapamycin blocked insulin induced srebp-1c expression (Li et al. 2010). The details of the molecular mechanism how insulin activate SREBP remains to be determined (Yecies 2013).

mTORC1 protein kinase directly phosphorylates two downstream targets, initiation factor 4E-binding protein (4E-BP) and p70 ribosomal S6 kinase (p70S6K), which affects upstream regulators of Akt, acting as a negative feedback mechanism (Dibble et al. 2009; Um et al. 2004). Acute inhibition of mTORC1 activity suppresses insulin-induced SREBP-1c mRNA expression. However, liver-specific deletion of TSC1 resulted constitutive activation of mTORC1 protects mice from age- and diet-induced hepatic steatosis and leads to hepatocyte-intrinsic defects in SREBP1c activation and de novo lipogenesis (Yecies et al. 2011).

2.2.2 Regulation of SREBP

An important question remains to be answered is how does insulin-Akt pathway activate SREBP. Liver X receptors (LXRs) are proposed to mediate it (Chen et al. 2004; Repa et al. 2000). LXRs are members of the nuclear hormone receptor superfamily that are bound and activated by oxysterols. These receptors serve as sterol sensors to regulate the transcription of gene products that control intracellular cholesterol homeostasis (Calkin and Tontonoz 2010). It was reported that insulin treatment increased LXRα mRNA level via stabilization of LXRα transcripts in cultured primary hepatocytes (Tobin et al. 2002). Mangelsdorf group revealed an RXR/LXR DNA-binding site in mouse SREBP-1c gene promoter. Dietary cholesterol and synthetic agonists markedly increased SREBP-1c expression and lipogenesis in mouse tissues in an LXR-dependent manner (Repa et al. 2000), while the lipogenic activity of LXRα was abrogated in SREBP-1c deficient mice (Liang et al. 2002), and LXR inhibitor or LXR deficiency blocked SREBP-1c response to insulin (Chen et al. 2004; Tobin et al. 2002), indicating that LXRα is responsible for SREBP-1c induction in response to insulin. The role of LXRα in insulin signaling cascade appears to be specific to SREBP-1c, since insulin does not induce other LXRα target genes (Xu et al. 2013).

A positive feedback loop was also proposed, that SREBP-1c autoregulates SREBPs promoter through binding to the SRE complex, possibly leading to the overshoot in induction of SREBP-1c and its downstream genes. The relative contribution of LXR and SREBP-1c autoregulation in the insulin-induced transcriptional activation of SREBP-1c remains to be further investigated (Amemiya-Kudo et al. 2000).

Peterson et al. found that mTORC1 regulates SREBP by directly phosphorylating and inhibiting the nuclear entry of lipin 1, a phosphatidic acid phosphatase which represses SREBP-dependent transcription and reduces nuclear SREBP protein abundance (Peterson et al. 2011).

2.2.3 Regulation of SREBP Protein Stability

The active SREBPs in the nucleus are degraded very rapidly via ubiquitin-dependent proteasomal degradation. Treatment with proteasome inhibitors elevates the amount of the endogenous nuclear SREBPs (Hirano et al. 2001).

F-box and WD repeat domain-containing 7 (Fbw7), a cullin-RING type E3 ubiquitin ligase has emerged as the major ubiquitin ligase for SREBPs (Sundqvist et al. 2005). Fbxw7, also known as Fbw7, SEL-10, hCdc4, and hAgo, is a member of the F-box protein family, which targets for degradation of various mammalian proteins (Nakayama and Nakayama 2006; Welcker and Clurman 2008; Mao et al. 2008). Liver-specific ablation of *Fbxw7* resulted in hepatomegaly and steatohepatitis, with massive deposition of triglyceride, indicating Fbxw7 plays negative roles in regulating lipogenesis (Onoyama et al. 2011; Bengoechea-Alonso and Ericsson 2009).

Phosphorylation of SREBP is critically required for its ubiquitination (Xu et al. 2013). In response to SREBP-1a specific DNA binding, glycogen synthase kinase (GSK)-3β is recruited to the promoters of SREBP target genes, facilitating phosphorylation of SREBP-1a at T426 and S430, which resemble Cdc4 phosphodegron (CPD) motif, a recognition site for Fbw7 (Kim et al. 2004; Sundqvist et al. 2005; Punga et al. 2006). Fbw7 also similarly ubiquitinates SREBP-1c and SREBP-2. Insulin induced AKT activation induces phosphorylates GSK3 at Ser-9 and suppresses its kinase activity, which consequently promotes SREBP activity by suppressing its phosphorylation and ubiquitination (Cross et al. 1995; Krycer et al. 2010). Cyclin-dependent kinase 8 (CDK8) and its regulatory partner cyclin C (CycC) also phosphorylate SREBP-1c in *Drosophila*, mammalian hepatocytes, and mouse liver, acting as a negative regulators of the lipogenic pathway. Consistent with the physiologic regulation of lipogenesis, CDK8 and CycC proteins are rapidly downregulated by feeding and insulin, resulting in decreased SREBP-1c phosphorylation (Zhao et al. 2012). Thus, Insulin-Akt pathway might influence SREBP at numerous levels, from mRNA transcription to protein degradation, by parallel or interacted mechanisms.

On the other hand, acetylation stabilizes SREBPs. SREBPs are reported to be acetylated by the acetyltransferases p300 and CBP. The acetylated Lys-289 and Lys-309 residues are also targets of ubiquitination. Thus acetylation by p300 and CBP increases SREBPs stability. Coexpression with p300 dramatically increases the expression of both SREBP1a and SREBP2 (Giandomenico et al. 2003; Ponugoti et al. 2010). On the other hand, SIRT1 decreases SREBP-1c stability and inhibits transactivation by deacetylating SREBP-1c, which facilitates ubiquitination. Abnormally elevated SREBP-1c acetylation increases SREBP-1c lipogenic activity in obese mice (Ponugoti et al. 2010; Walker et al. 2010).

2.2.4 Bmal1, a Novel Regular of Insulin-AKT Pathway

BMAL1, a basic helix-loop-helix (bHLH) transcription factor, is the master regular of the mammalian circadian clock (Bunger et al. 2000). It has been observed that *Bmal1* knockout mice show insulin resistance and impaired lipid homeostasis (Rudic et al. 2004; Shi et al. 2013; Shimba et al. 2011). Our lab recently reported that BMAL1 is both necessary and sufficient to promote the expression of lipid synthesis enzymes in the mouse liver and de novo lipogenesis in mouse hepatocytes. Both feeding and insulin treatment promote the expression of DNL genes in a BMAL1-dependent manner. Mechanistically, BMAL1 is required for stabilization of RICTOR, the component of mTORC2, which activates AKT upon feeding or insulin stimulation in hepatocytes (Zhang et al. 2014). It is intriguing to speculate that BMAL1 may positively regulate synthesis and activity of SREBP1C during a circadian cycle via maintaining cellular mTORC2-AKT2 (Fig. 1.2).

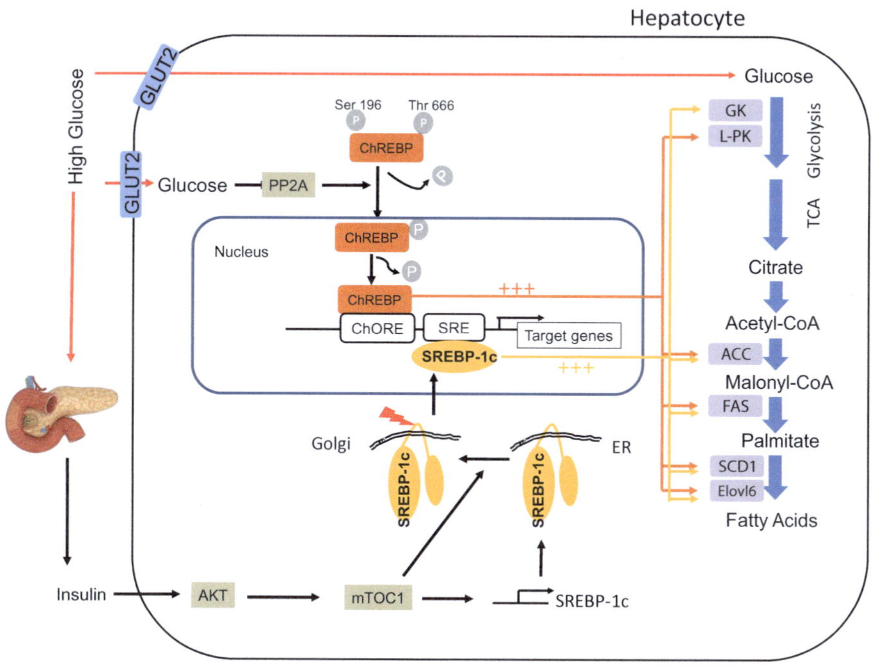

Fig. 1.2 De novo lipogenesis are regulated by SREBP-1c and ChREBP. In high glucose condition, ChREBP is dephosphorylated by PP2A, which facilitates nucleus entry and ChORE binding of ChREBP, resulting in transcription of glycolytic and lipogenic enzymes, including L-PK, ACC, FAS, SCD1 and Elovl6. At the same time, high glucose level in the serum stimulates insulin secretion of pancreas. In hepatocytes, insulin activated AKT-mTOC1 pathway enhances both transcription and cleavage of SREBP-1c. Activated SREBP-1c enters nucleus and promotes transcription of GK, ACC, FAS, SCD1 and Elovl6

2.2.5 Discrete Insulin Response and SREBP Activity in Insulin Resistance

Constitutively active hepatic lipogenesis and steatosis is present in insulin-resistant obese rodents. In *ob/ob* mice and lipodystrophic mice, increased food intake due to leptin deficiency increases insulin secretion, resulting in hyperinsulinemia. Chronic hyperinsulinemia downregulates the mRNA for IRS-2, an essential component of the insulin-signaling pathway in liver, thereby produces insulin resistance. Despite fails to supress gluconeogenesis, insulin continues to stimulate production of SREBP-1c (Shimomura et al. 2000). The paradoxical combination of insulin resistance (elevated gluconeogenesis) and insulin sensitivity (elevated lipogenesis) is called selective insulin resistance, which establishes a vicious cycle that aggravates hyperinsulinemia and insulin resistance (Shimomura et al. 2000; Brown and Goldstein 2008). Elevated triglycerides from hepatic lipogenesis raises plasma triglyceride levels, worsening the insulin-resistant state in muscle and adipose tissue, resulting in hyperglycemia, hyperinsulinemia, and hypertriglyceridemia, the classic triad of type 2 diabetic (Brown and Goldstein 2008).

3 Glucose-ChREBP Pathway

3.1 Introduction

Insulin had long been thought to be the main inductor of lipogenic gene transcription. Transgenic mice that overexpress SREBP-1c, the mediator of lipogenesis induction effect of insulin, in the liver have increased mRNA of most lipogenic genes and develop liver steatosis (Shimano et al. 1997; Ferre and Foufelle 2010). However, Srebp-1c gene deletion in mice only reduced fatty acid synthesis by 50 % (Liang et al. 2002). In addition, in hepatocytes overexpressing a constitutive active form of SREBP-1c while lacking Glucokinase (GK), glucose fails to induce lipogenic gene expression (Dentin et al. 2004). Hepatic GK catalyzes the phosphorylation of glucose to glucose 6-phosphate (G6P), a step which is essential for glucose entering glycolysis and lipogenesis (Hariharan et al. 1997; Niswender et al. 1997; Ferre et al. 1996). Therefore, it seems that glucose not only serves as an energetic metabolism substrate, but also acts as a signaling molecule in adipose tissue and liver. The postprandial increasing of glucose level stimulates the transcription of enzymes involved in glycolysis and lipogenesis, including liver-pyruvate kinase (L-PK), acetyl-CoA carboxylase (ACC), fatty acid synthase (FAS), and stearoyl-CoA desaturase (SCD1), thereby promoting long-term energy storage in forms of triglycerides (Girard et al. 1997). Among these lipogenic enzymes, some are activated by insulin. However, elevated carbohydrate level is required for the optimal transcription of most lipogenic genes. At the same time, the transcription of other lipogenic enzymes, such as L-PK, is stimulated by glucose alone, independent of insulin (Doiron et al. 1994). Then, with promoter-mapping analysis, it was found that glucose-regulated genes share a conserved consensus sequence, named the carbohydrate response element (ChoRE), which is composed of two E-boxes (Enhancer Box), or E-box-like sequences, separated by five base pairs (5′-CACGTGnnnnnCACGTG-3′) (Bergot et al. 1992; Cuif et al. 1993; Thompson and Towle 1991; Rufo et al. 2001; Shih et al. 1995). The discovery of a glucose-responsive bHLH/LZ transcription factor, carbohydrate responsive element-binding protein (ChREBP), has shed light on mechanism by which glucose promotes lipogenic gene transcription.

3.2 Structure and Binding Basis

In 2001, using affinity chromatography and mass spectrometry, K Uyeda's group identified a ChoRE binding protein, composing of 864 amino acids and with a relative molecular weight of 94.6 kDa, which is named ChREBP (Yamashita et al. 2001). The DNA-binding activity of ChREBP) in rat livers is specifically induced by a high carbohydrate diet, and forced ChREBP overexpression in primary hepatocytes activates transcription of the L-type Pyruvate kinase in response to high glucose levels, indicating that ChREBP was the long-sought glucose-responsive transcription factor.

The structure of the ChREBP gene is highly conserved among species. It contains several other key domains, including a nuclear localization signal (NLS) near the N terminus, polyproline domains, a bHLH/LZ domain, and a leucine zipper-like (Zip-like) domain (Yamashita et al. 2001). Structure-function analysis revealed a glucose-sensing module (GSM), which is conserved among Mondo family members. Thus the NH2-terminal region of 300 amino acids of ChREBP is also called the MondoA conserved region (MCR) I–V. The GSM mediates glucose responsiveness of ChREBP (Li et al. 2006; Davies et al. 2010). GSM is composed of a low-glucose inhibitory domain (LID, residues 37–192) and a glucose-response activation conserved element (GRACE, residues 197–298). Under low glucose concentration, LID inhibits the activity of ChREBP conferred by GRACE. The resulting conformation is incapable of binding to DNA and stimulating transcription. While under high glucose conditions, this inhibition is relieved. One possibility is that a glucose metabolite binds directly to the MCR1–4 region, resulting in a de-repressive conformation (Davies et al. 2010; Filhoulaud et al. 2013). Therefore, deletion of the LID domain leads to a glucose-independently activated form of ChREBP (Davies et al. 2010).

There are two nuclear export signals (NES1, residues 5–15, and NES2, residues 86–95, respectively) near the N terminus of ChREBP (Fukasawa et al. 2010; Merla et al. 2004). These two NESs coordinate to interact together with CRM1 (exportin) for nuclear export of ChREBP. Deletion or mutation of NES1 and NES2 leads to increased ChREBP nuclear localization (Fukasawa et al. 2010).

In 2004, with a yeast two-hybrid screen, HC Towle's group reported that ChREBP requires the heterodimeric partner Max-like factor X (Mlx)to bind to ChoRE sequences (Stoeckman et al. 2004). Mlx is a member of the Myc/Max/Mad family of transcription factors, and similar to Max, it can serve as a common interaction partner of a transcription factor network (Meroni et al. 2000). The inhibition of Mlx with expressing a dominant negative form of Mlx directly interferes with the endogenous ChREBP/Mlx complex and abrogates the glucose response of the ACC reporter gene in primary cultures of hepatocytes (Ma et al. 2005). Mlx interacts with the bHLH/LZ domain of ChREBP in liver. The most recent model proposed that two ChREBP/Mlx heterodimers interact to stabilize binding to the tandem E box-like motifs in the ChoRE. Mlx has a significantly longer loop domain than that of most other bHLH/LZ proteins, facilitating its interaction across the inter-face between heterodimer pairs. Intermolecular interactions between residues within the Mlx loop regions of adjacent heterodimers are responsible for stabilizing the complex. Three critical residues (F164, I166, and K170) within the Mlx loop that play a crucial role in the binding of the ChREBP/Mlx complex to the ChoRE, as Mlx variants in which the loop region was replaced with that of another family member or mutated at the key residues no longer bound to the ChoRE nor supported glucose responsive activity (Ma et al. 2006, 2007). Therefore, it appears that it is the Mlx loop region, instead of the loop region of ChREBP, determinant for responsiveness to glucose (Postic et al. 2007).

3.3 Regulation of ChREBP

The DNA-binding activity of ChREBP is induced by glucose in both liver and in cultured primary hepatocytes through multiple post-translational modifications and translocation (Yamashita et al. 2001; Havula and Hietakangas 2012; Filhoulaud et al. 2013). The phosphorylation/dephosphorylation and cytosol/nucleus translocation of ChREBP have been proposed as the classic mechanism for regulating ChREBP transactivity. Under starvation conditions, ChREBP is in an inactivated status, with phosphorylation at Ser196, Ser626 and Thr666 by cAMP-dependent protein kinase (PKA), and at Ser568 by AMP-activated protein kinase (AMPK), locating in cytosol (Kawaguchi et al. 2001). In high glucose conditions, elevated xylulose 5-phosphate (X5P) level activates protein phosphorylase 2A (PP2A), which sequentially activates ChREBP by dephosphorylation at Ser196, locating near the NLS, and leading to its nucleus translocation. Additionally, heterodimerization of ChREBP with Mlx is also required for their nuclear entry (Havula and Hietakangas 2012; Stoeckman et al. 2004; Kabashima et al. 2003). In nucleus, ChREBP undergoes a second dephosphorylation by PP2A on residue Thr666 which permits its binding to ChoRE (Postic et al. 2007).

Although this is the most popular model of ChREBP activity regulation, there is evidence challenging it. Analysis of mutant forms of ChREBP that lack one or more of the proposed PKA sites found that these forms of ChREBP still require glucose for activation, indicating that in addition to its repression by PKA, glucose activation of ChREBP involves a second mechanism that is independent of PKA phosphorylation (Tsatsos and Towle 2006). Chan group found a mechanism of intramolecular inhibition and its release by glucose, i.e., specific inhibition of the transactivation activity of glucose-response activation conserved element (GRACE) by low-glucose inhibitory domain (LID) under low glucose concentration and reversal of this inhibition by glucose (Li et al. 2006). Mutant of LID created a constitutive activated isoform of ChREBP, while mutant of the PKA targeting sites did not, which rise more challenge to the phosphorylation/dephosphorylation model. As mentioned above, ChREBP dimerizes with Mlx to activate their target gene transcription. However, nothing is known about the regulation of Mlx by glucose (Postic et al. 2007).

Besides phosphorylation, ChREBP activity is regulated through several other post-translational modifications, such as acetylation and O-GlcNAcylation. O-GlcNAc is proposed as a nutrient sensor and metabolic regulator (Wells et al. 2003; Ruan et al. 2013). In response to high glucose, there is elevated level of UDP-GlcNAc, the donor substrate for O-GlcNAcylation, which is derived from extracellular glucose through the hexosamine biosynthetic pathway. ChREBP interacts with *O*-linked GlcNAc transferase (OGT), which transfers the monosaccharide N-acetylglucosamine to serine/threonine residues on target proteins, and is subjected to *O*-GlcNAcylation in liver cells. *O*-GlcNAcylation stabilizes the ChREBP protein and increases its transcriptional activity toward its target glycolytic and lipogenic genes (Guinez et al. 2011; Sakiyama et al. 2010). OGT overexpression induces lipogenesis by increasing ChREBP levels, whereas *O*-GlcNAc hydrolase (OGA) overexpression prevents hepatic steatosis in db/db mice (Guinez et al. 2011).

Histone acetyltransferase (HAT) co-activator p300 was reported as key upstream regulators of ChREBP activity. In mouse hepatocytes, glucose-activated p300 acetylated ChREBP on Lys672 and increased its transcriptional activity by enhancing its recruitment to its target gene promoters. While serine/threonine kinase salt-inducible kinase 2 (SIK2) inhibited p300 HAT activity by phosphorylation on Ser89 (Bricambert et al. 2010).

Neither acetylation nor O-GlcNAcylation influences ChREBP nuclear shuttling, but instead increase its transcriptional activity within the nucleus by favoring its recruitment to its DNA binding sites (Bricambert et al. 2010; Guinez et al. 2011; Filhoulaud et al. 2013).

It is known that O-GlcNAcylation interplays with other kinds of post-translational modifications. O-GlcNAcylation provides a docking site for de-ubiquitin enzymes that remove ubiquitination moiety from modified substrates (Ruan et al. 2013). Forkhead box O1 (FoxO1) was showed to reduce activity of ChREBP by suppressing O-linked glycosylation and increasing ubiquitination, which reduces the protein stability (Ido-Kitamura et al. 2012). Till now, the enzymes that either promote or remove ChREBP ubiquitination remain to be identified.

3.4 ChREBPβ

Recently, Herman et al. discovered a novel isoform of ChREBP, ChREBPβ in adipose tissue (Herman et al. 2012). To understand the transcriptional regulation of ChREBP, the authors searched the ChREBP genomic region for carbohydrate response elements (ChoREs) and found two ChoREs, located in two separate promoters, corresponding to two variants of ChREBP. Translation from the start-site in exon 1a produces the full-length protein of 864 amino acids (ChREBP-α), whereas translation from the start-site located in exon 4 produces a shorter form of 687 amino acids (ChREBP-β). *Chrebpβ* mRNA is less abundant than *Chrebpα*, however, ChREBP-β is a more potent transactivator at both low and high glucose concentrations. The ChREBP-β protein, missing the first 177 amino acids (most of the LID), shares common features with a constitutively activated ChREBP which lacks the first 196 amino acids. Thus, glucose activates lipogenesis by a two-step mechanism. First glucose activates ChREBPα, which in turn stimulates transcription of ChREBP-β, a much more potent transcriptional activator (Herman et al. 2012; Dentin et al. 2012). Furthermore, the mRNA level of *Chrebpα* is not affected by high-fat or high-carbohydrate diets, while that of *Chrebpβ* is elevated in all tissues. Therefore ChREBPβ appears to be responsible for carbohydrate-induced lipogenic gene induction whereas ChREBPα is likely responsible for inducing ChREBPβ (Paton et al. 2014).

3.5 The Effect of High-Fat Diet and High-Fructose Diet on ChREBP

High-fat diet was reported to suppress ChREBP activity. According to Kawaguchi's study, ChREBP is phosphorylated at Ser568 by AMPK, which is activated by the increased AMP that was generated by the fatty acid activation (Kawaguchi et al. 2002). While Dentin et al. reported that dietary polyunsaturated fatty acids (PUFAs), which are potent inhibitors of hepatic glycolysis and lipogenesis, suppressed ChREBP activity by accelerating ChREBP mRNA decay and by decreasing xylulose 5-phosphate concentrations, sequentially blocking ChREBP nucleus entry, independently of an activation of the AMPK, while saturated and monounsaturated fatty acids have no effect (Dentin et al. 2005).

High-fructose diet has been shown to increase liver steatosis by increasing protein level of ChREBP (Iizuka et al. 2004; Koo et al. 2008), while ChREBP deficiency attenuated high-fructose diet feeding induced liver steatosis (Erion et al. 2012). Our lab also confirmed that ChREBP is required for high-fructose diet induced liver steatosis (unpublished data).

3.6 ChREBP in Liver Steatosis

The relationship of ChREBP with insulin resistance is complex: Liver-specific inhibition of ChREBP improves liver steatosis and insulin resistance in ob/ob mice (Dentin et al. 2006). ChREBP overexpression promotes hepatic steatosis which is dissociated from insulin resistance by modifying hepatic fatty acid composition (Benhamed et al. 2012).

In human, gene expression levels of ChREBP are positively correlated with hepatosteatosis and negatively related to insulin resistance. Conversely, ChREBP knockout mice fed high starch diet display mitigated hepatosteatosis, but develop severe insulin resistance along with hyperglycemia, hyperinsulinemia, and increased hepatic glycogen store, without changes in body weight (Iizuka et al. 2004; Benhamed et al. 2012).

4 PPARγ and PPARδ Regulation

Peroxisome proliferator-activated receptor (PPAR) γ is a nuclear receptor central to glucose and lipid homeostasis (Barak and Kim 2007). It is activated by a variety of ligands, such as fatty acids and eicosanoids synthetic ligands like thiazolidinediones (TZDs) and antidiabetic agents increasing insulin sensitivity (Kersten et al. 2000; Olefsky 2000). PPARγ expression increases in mice with obesity. Hepatic PPARγ overexpression has been found to exacerbated steatosis by activating lipogenic genes and de novo lipogenesis and increasing hepatic triglyceride concentrations (Matsusue et al. 2003). Hepatocyte specific deletion of PPARγ leads to down-regulated expression of genes involved in lipogenesis (SCD1, SREBP-1c, and ACC) and protects against hepatic steatosis in mice fed a HFD, suggesting PPARγ as a lipogenesis inducer in hepatocytes (Morán-Salvador et al. 2011; Matsusue et al. 2003).

5 Negative Regulation on DNL

5.1 5′ Adenosine Monophosphate (AMP)-Activated Protein Kinase

The 5′ adenosine monophosphate (AMP)-activated protein kinase (AMPK) is a heterotrimeric enzyme complex consisting of a catalytic subunit AMPKα and two regulatory subunits AMPKβ and AMPKγ (Hardie et al. 2012). Activation of AMPK inhibits hepatic lipogenesis (Viollet et al. 2006). One of the mechanisms is suppression of SREBP-1c (Zhou et al. 2001; Yang et al. 2008; Li et al. 2011). In response to polyphenols and metformin, AMPK stimulates Ser372 phosphorylation of SREBP-1c, suppressing SREBP-1c cleavage and nuclear translocation, and repressing SREBP-1c target gene expression in hepatocytes exposed to high glucose, leading to reduced lipogenesis and lipid accumulation (Li et al. 2011). At the same time, AMPK inhibits LXR-dependent SREBP-1c transcription and processing by suppressing endogenous LXR ligand production as well as ligand-induced LXR activity (Yang et al. 2009; Yap et al. 2011). AMPK was also reported to phosphorylate ChREBP at Ser^{568}, which inactivated the DNA binding activity (Kawaguchi et al. 2002), and to phosphorylate ACC1 (Munday et al. 1988). Thus, AMPK regulates lipogenesis by targeting at the master lipogenic regulators ChREBP and SREBP-1c at various levels.

5.2 *SIRT1*

In mammals the sirtuin family comprises seven proteins (SIRT1–SIRT7), among which SIRT1 is mostly studied. SIRT1 is mainly localized in the nucleus but is also present in the cytosol. It deacetylates a variety of protein targets (Houtkooper et al. 2012). SIRT1 is a positive regulator of liver X receptor (LXR) proteins by promoting its deacetylation and subsequent ubiquitination (Li et al. 2007). On the other hand, SIRT1 deacetylates SREBP-1c to facilitate the latter's ubiquitination, thus suppress lipogenesis. Therefore, theoretically, SIRT1 positively and negatively regulates lipogenesis at the same time. In fact, hepatic Sirt1 deficiency in mice impairs mTORC2/AKT signaling and results in hepatic steatosis, hyperglycemia and insulin resistance (Wang et al. 2011; Purushotham et al. 2009). Treatment with SRT1720, a SIRT1 activator, ameliorates fatty liver with reduced expression of lipogenic enzymes in mice exhibiting obesity and insulin resistance (Yamazaki et al. 2009). Thus, SIRT1-mediated LXR activation seems to specifically target cholesterol metabolism rather than triglyceride synthesis (Houtkooper et al. 2012).

SIRT1 positively regulates expression of SIRT6, which has been shown to play a critical role in fat metabolism. Liver-specific deletion of SIRT6 in mice leads to increased lipogenic gene expression and increased triglyceride synthesis and fatty liver formation. Human fatty liver samples exhibited significantly lower levels of SIRT6 (Kim et al. 2010), indicating that loss of SIRT6 expression or function may contribute to the development of liver steatosis in human.

6 Circadian Regulation of DNL

In mammals, most metabolic processes are under the control of a circadian rhythm (Asher and Schibler 2011). Feeding behavior, in terms of its timing and periodicity, has a significant impact on circadian rhythms and metabolism. De novo lipid synthesis increases during the time period of taking meals and then drops during the rest period (Hems et al. 1975). Disruption of the circadian clock exacerbates metabolic diseases, including obesity and diabetes. *Bmal1*-knockout mice are locked into the trough of insulin action and lack rhythmicity in insulin action and activity (Shi et al. 2013). Emerging evidences also suggest that genetic variation in circadian genes also influences fat accumulation in humans. Polymorphisms in the *CLOCK* gene have been associated with predisposition to human obesity (Scott et al. 2008; Sookoian et al. 2008).

The mechanisms that couple lipid metabolism to circadian oscillators are just emerging. Deficiency of Rev-erbα and Rev-erbβ causes marked hepatic steatosis (Bugge et al. 2012; Cho et al. 2012). Genomic recruitment of HDAC3 by Rev-erbα directs a circadian rhythm of histone acetylation and gene expression required for normal hepatic lipid homeostasis (Feng et al. 2011). Liver-specific deletion of histone deacetylase 3 disrupts metabolic transcriptional networks (Knutson et al. 2008). REV-ERBα is also reported to participate in the circadian modulation of SREBP activity, and thereby in the daily expression of SREBP target genes involved in cholesterol and lipid metabolism (Le Martelot et al. 2009).

7 ER Stress and UPR Regulate DNL

The endoplasmic reticulum (ER) is the cellular organelle responsible for protein folding and assembly, lipid and sterol biosynthesis, and calcium storage. UPR is an adaptive intracellular stress response to accumulation of unfolded or misfolded proteins in the ER (Bravo et al. 2013). ER stress and UPR had been proposed as the main contributor for pathogenesis of insulin resistance (Samuel and Shulman 2012). A significant amount of work has focused on the role of ER stress and unfolded protein response (UPR) on lipogenesis. All three arms of the UPR are correlated with lipogenesis regulation (Zeng et al. 2004; Lee et al. 2008; Zhang et al. 2011; Bobrovnikova-Marjon et al. 2008). The UPR regulates lipogenesis to expand the ER membrane and increase the capacity of the ER to handle proteins. ER stress promotes proteolytic cleavage and activation of SREBP-1c in the liver, increasing lipogenesis, while overexpression of GRP78 can suppress activation of SREBP1c and ChREBP, as well as the UPR and reduce liver steatosis in ob/ob mice (Kammoun et al. 2009). Deficiency of a single arm of the UPR leads to increased signaling via the unaffected arms and results in an more drastically increased lipogenesis and a greater degree liver steatosis (Samuel and Shulman 2012).

8 Impact of Inflammation on DNL in the Liver

In liver of obese and high-fat diet fed mice, inflammatory cytokine expression level increase (Cai et al. 2005), which may contribute to progress of liver steatosis into steatohepatitis, the inflammatory state of non-alcoholic fatty liver disease (NAFLD). Conversely, increased inflammation also elevated hepatic lipogenesis. It has been shown that TNF-α stimulates hepatic lipogenesis in the rat in vivo. TNF-α and IL-6 stimulates hepatic lipogenesis by increasing hepatic citrate concentrations (Grunfeld et al. 1990). IL-6 treatment increased hepatic triglyceride secretion and serum triglyceride level, by increase both hepatic lipogenesis and lipolysis (Nonogaki et al. 1995). It was reported that ethanol treatment elevates Kupffer cell-derived prostaglandin E_2

(PGE_2), which is responsible for alcohol-induced fat accumulation in rat liver (Enomoto et al. 2000; Neyrinck et al. 2004). Till now, the effect of inflammation on hepatic lipogenesis is an understudied area (Gregor and Hotamisligil 2011).

9 Dysregulation of DNL in Metabolic Syndromes and Cancer

9.1 Dysregulated Lipogenesis in Metabolic Syndromes

Dysregulated DNL processes are commonly observed in metabolic syndromes, including insulin resistance, non-alcoholic fatty liver disease and obesity (Ameer et al. 2014). Insulin resistance is a complex metabolic disorder in which cells fail to respond to the normal actions of insulin and that precedes the development of type 2 diabetes. The mechanisms of pathogenesis of insulin resistance involve ectopic lipid metabolites, activation of the unfolded protein response (UPR) pathway, and innate immune pathways (Samuel and Shulman 2012). Nonalcoholic fatty liver disease (NAFLD) refers to a large spectrum of diseases ranging from fatty liver (hepatic steatosis) to steatosis with inflammation, necrosis and cirrhosis (Postic and Girard 2008). Liver steatosis is mainly a result from increased delivery of NEFA from adipose tissue to the liver and enhanced hepatic de novo lipogenesis, while lipid disposal via β-oxidation and well VLDL export only contribute moderately (Lewis et al. 2002). In normal physiological conditions, DNL contributes a minority of total hepatic FFAs, while for NAFLD patients, 26.1 % hepatic FFAs are derived from hepatic de novo lipogenesis (Donnelly et al. 2005). Lipogenic gene expression substantially upregulated in the liver of obese subjects (Eissing et al. 2013). SREBP-1c knockout markedly attenuated fatty livers in ob/ob mice, while systemic insulin resistance was not prevented (Yahagi et al. 2002).

In diabetic *ob/ob* mice, liver specific ChREBP knockdown by adenoviral delivered short hairpin ChREBP–RNA suppresses lipogenic gene expression, leading to a 50 % reduction in hepatic and circulating TG concentrations (Dentin et al. 2006). As a consequence, insulin signaling was improved in liver, skeletal muscles, and white adipose tissue, and overall glucose tolerance and insulin sensitivity were restored (Iizuka et al. 2006).

The pathogenesis of NAFLD is thought to be related to insulin resistance and oxidant stress (Abdelmalek and Diehl 2007). Insulin resistance increases lipolysis of adipose tissue, resulting in increased free fatty acids influx into the liver. Furthermore, hyperinsulinemia promotes de novo lipogenesis through up-regulation of lipogenic transcription factors in the liver and inhibits free fatty acid oxidation (Lewis et al. 2002). The resultant hepatic TG accumulation ultimately leads to liver steatosis.

On the other hand, numerous studies in humans and animals have shown that insulin-resistant states are often accompanied by hepatic steatosis (Yki-Järvinen 2005; Petersen and Shulman 2006), and decreasing hepatic TG pools improves insulin sensitivity (Savage et al. 2006; Dentin et al. 2006). Thus, hepatic steatosis was widely believed to result in insulin resistance. However, in fact, it remains uncertain whether liver steatosis is a causal factor of insulin resistance. Hepatic steatosis is not necessarily associated with insulin resistance (Monetti et al. 2007; Buettner et al. 2004; Postic and Girard 2008). Although liver specific overexpression of GPAT2, the enzyme catalyzes the final step of triacylglycerol (TG) biosynthesis, promotes hepatic steatosis in mice, the insulin intolerance remains normal (Monetti et al. 2007). In fact, cellular TG accumulation is not initially toxic, and converting excess fatty acids into TG protects liver from lipotoxicity, which is caused by saturated fatty acids, such as palmitate (Listenberger et al. 2003). Inhibiting triglyceride synthesis via inhibiting DGAT2 improves hepatic steatosis but exacerbates liver damage and fibrosis in obese mice (Yamaguchi et al. 2007).

Despite the existing correlation between fatty liver and insulin resistance, it remains unclear whether insulin resistance causes the excessive accumulation of TG in liver, or whether the increase in TG itself or of metabolite intermediates may play a causal role in the development of hepatic or systemic insulin resistance (Ameer et al. 2014; Postic and Girard 2008; Marchesini et al. 1999).

9.2 *Increased Lipogenesis in Cancer*

Elevated lipogenesis has been observed in a variety of cancer cells, including prostate, ovarian, colon and liver cancer. It is hypothesized that unconstrained lipogenesis is necessary to maintain a constant supply of lipids and lipid precursors as building blocks for membranes and lipid-based post-translational modification of proteins for elevated proliferation (Calvisi et al. 2011). The lipid building blocks come from either exogenous sources or from de novo fatty acid synthesis. While most normal human cells prefer exogenous sources, tumors synthesize FA de novo (Medes et al. 1953). Fatty-acid synthesis is now associated with clinically aggressive tumor behavior and tumor-cell growth and survival and has become a novel target pathway for chemotherapy development (Swinnen et al. 2006; Currie et al. 2013).

Activation of the PI3K/AKT pathway is reported to be responsible for elevated lipogenesis in cancer (Calvisi et al. 2011; Swinnen et al. 2006). This signaling pathway stimulates lipogenic gene transcription through activation of SREBP1c and directly activates lipogenic enzymes such as ATP-citrate lyase (Ettinger et al. 2004). Steroid hormones, overexpression of the ubiquitin-specific protease-2a and mutations in breast cancer susceptibility gene 1 may further enhance lipid synthesis (Swinnen et al. 2006).

Besides transcriptional activation of lipogenic enzymes in cancer cells, lipogenesis in cancer cells may also be upregulated by stabilization of the lipogenic enzymes. In prostate cancer as well as human hepatocellular carcinoma, FAS was shown to be stabilized by USP2a, an ubiquitin-specific protease (Graner et al. 2004; Calvisi et al. 2011).

10 Therapeutic Targeting of Hepatic DNL

Since de novo lipogenesis is reported to be dysregulated in metabolic syndrome as well as cancer, inhibiting the DNL pathway may have substantial therapeutic significance. However, several mouse genetic studies which directly target the specific DNL enzymes have not reached a general consensus. For example, liver-specific Acc1-knockout failed to protect mice from high-fructose and high- carbohydrate diet induced obesity, fatty liver or insulin-resistance, because of compensation of ACC2 (Mao et al. 2006; Harada et al. 2007). However, ACC1 and ACC2 synergistically knockdown with antisense oligonucleotide (ASO) showed lower hepatic lipids and improve hepatic insulin sensitivity (Savage et al. 2006). Surprisingly, liver-specific FAS deficiency does not protected mice from fatty liver but rather exacerbated it under either normal chow or high-fat diet feeding conditions. The underline mechanism is that FAS is required to activate some nuclear receptors involved in β-oxidation, such as PPARα (Chakravarthy et al. 2005). Mice with a global knock-out of *Scd1* are protected from diet-induced obesity and insulin resistance due to decreased lipogenesis and increased β-oxidation (Cohen et al. 2002; Ntambi et al. 2002; Miyazaki et al. 2007).

Since fatty acids are essential for cancer cell proliferation, limiting their availability could provide a therapeutic strategy, which could be achieved by blocking fatty acids synthesis and release from storage or increasing fatty acids oxidation and storage (Currie et al. 2013). Chemical or RNAi-mediated inhibition of key lipogenic enzymes, such as ACLY (Hatzivassiliou et al. 2005; Migita et al. 2008), ACC1 and ACC2 (Beckers et al. 2007), FASN (Lupu and Menendez 2006), SCD1 (Ariyama et al. 2010; Fritz et al. 2010) and SREBPs (Guo et al. 2009).

Inhibition of SREBP by 25-HC, fatostatin, and FGH10019, or knockdown SREBP by shRNA all significantly reduce cellular growth in a variety of cancer cell lines. SREBP deficient cancer cells maintain long-chain saturated fatty acid synthesis, while losing fatty acid desaturation capacity (Williams et al. 2013; Griffiths et al. 2013).

In summary, abnormality in hepatic DNL is often observed in cancers and various metabolic disorders and may significantly contribute to the pathogenesis of these diseases. A better understanding of the regulation of DNL in specific pathological condition and identify disease-specific regulators of DNL will lead to more effective therapeutic avenue to correct abnormal DNL process.

References

Abdelmalek MF, Diehl AM (2007) Nonalcoholic fatty liver disease as a complication of insulin resistance. Med Clin North Am 91(6):1125–1149

Abu-Elheiga L, Almarza-Ortega DB, Baldini A, Wakil SJ (1997) Human acetyl-CoA carboxylase 2 molecular cloning, characterization, chromosomal mapping, and evidence for two isoforms. J Biol Chem 272(16):10669–10677

Abu-Elheiga L, Matzuk MM, Kordari P, Oh W, Shaikenov T, Gu Z, Wakil SJ (2005) Mutant mice lacking acetyl-CoA carboxylase 1 are embryonically lethal. Proc Natl Acad Sci U S A 102(34):12011–12016

Ameer F, Scandiuzzi L, Hasnain S, Kalbacher H, Zaidi N (2014) *De novo* lipogenesis in health and disease. Metabolism 63(7):895–902

Amemiya-Kudo M, Shimano H, Yoshikawa T, Yahagi N, Hasty AH, Okazaki H, Tamura Y, Shionoiri F, Iizuka Y, Ohashi K (2000) Promoter analysis of the mouse sterol regulatory element-binding protein-1c gene. J Biol Chem 275(40):31078–31085

Ariyama H, Kono N, Matsuda S, Inoue T, Arai H (2010) Decrease in membrane phospholipid unsaturation induces unfolded protein response. J Biol Chem 285(29):22027–22035

Asher G, Schibler U (2011) Crosstalk between components of circadian and metabolic cycles in mammals. Cell Metab 13(2):125–137

Barak Y, Kim S (2007) Genetic manipulations of PPARs: effects on obesity and metabolic disease. PPAR Res 2007:12781

Bartlett K, Eaton S (2004) Mitochondrial β-oxidation. Eur J Biochem 271(3):462–469

Beckers A, Organe S, Timmermans L, Scheys K, Peeters A, Brusselmans K, Verhoeven G, Swinnen JV (2007) Chemical inhibition of acetyl-CoA carboxylase induces growth arrest and cytotoxicity selectively in cancer cells. Cancer Res 67(17):8180–8187

Bengoechea-Alonso MT, Ericsson J (2009) A phosphorylation cascade controls the degradation of active SREBP1. J Biol Chem 284(9):5885–5895

Benhamed F, Denechaud P-D, Lemoine M, Robichon C, Moldes M, Bertrand-Michel J, Ratziu V, Serfaty L, Housset C, Capeau J (2012) The lipogenic transcription factor ChREBP dissociates hepatic steatosis from insulin resistance in mice and humans. J Clin Invest 122(6):2176–2194

Bergot M-O, Diaz-Guerra M-JM, Puzenat N, Raymondjean M, Kahn A (1992) Cis-regulation of the L-type pyruvate kinase gene promoter by glucose, insulin and cyclic AMP. Nucleic Acids Res 20(8):1871–1878

Björntorp P, Sjöström L (1978) Carbohydrate storage in man: speculations and some quantitative considerations. Metabolism 27(12):1853–1865

Bobrovnikova-Marjon E, Hatzivassiliou G, Grigoriadou C, Romero M, Cavener DR, Thompson CB, Diehl JA (2008) PERK-dependent regulation of lipogenesis during mouse mammary gland development and adipocyte differentiation. Proc Natl Acad Sci U S A 105(42):16314–16319

Bravo R, Parra V, Gatica D, Rodriguez AE, Torrealba N, Paredes F, Wang ZV, Zorzano A, Hill JA, Jaimovich E (2013) Endoplasmic reticulum and the unfolded protein response: dynamics and metabolic integration. Int Rev Cell Mol Biol 301:215

Bricambert J, Miranda J, Benhamed F, Girard J, Postic C, Dentin R (2010) Salt-inducible kinase 2 links transcriptional coactivator p300 phosphorylation to the prevention of ChREBP-dependent hepatic steatosis in mice. J Clin Invest 120(12):4316–4331

Brown MS, Goldstein JL (1997) The SREBP pathway: regulation of cholesterol metabolism by proteolysis of a membrane-bound transcription factor. Cell 89(3):331–340

Brown MS, Goldstein JL (2008) Selective versus total insulin resistance: a pathogenic paradox. Cell Metab 7(2):95–96

Buettner R, Ottinger I, Schölmerich J, Bollheimer LC (2004) Preserved direct hepatic insulin action in rats with diet-induced hepatic steatosis. Am J Physiol Endocrinol Metab 286(5):E828–E833

Bugge A, Feng D, Everett LJ, Briggs ER, Mullican SE, Wang F, Jager J, Lazar MA (2012) Rev-erbα and Rev-erbβ coordinately protect the circadian clock and normal metabolic function. Genes Dev 26(7):657–667

Bunger MK, Wilsbacher LD, Moran SM, Clendenin C, Radcliffe LA, Hogenesch JB, Simon MC, Takahashi JS, Bradfield CA (2000) Mop3 is an essential component of the master circadian pacemaker in mammals. Cell 103(7):1009–1017

Cai D, Yuan M, Frantz DF, Melendez PA, Hansen L, Lee J, Shoelson SE (2005) Local and systemic insulin resistance resulting from hepatic activation of IKK-β and NF-κB. Nat Med 11(2):183–190

Calkin AC, Tontonoz P (2010) Liver x receptor signaling pathways and atherosclerosis. Arterioscler Thromb Vasc Biol 30(8):1513–1518

Calvisi DF, Wang C, Ho C, Ladu S, Lee SA, Mattu S, Destefanis G, Delogu S, Zimmermann A, Ericsson J (2011) Increased lipogenesis, induced by AKT-mTORC1-RPS6 signaling, promotes development of human hepatocellular carcinoma. Gastroenterology 140(3):1071–1083.e1075

Chakravarthy MV, Pan Z, Zhu Y, Tordjman K, Schneider JG, Coleman T, Turk J, Semenkovich CF (2005) "New" hepatic fat activates PPARα to maintain glucose, lipid, and cholesterol homeostasis. Cell Metab 1(5):309–322

Chen G, Liang G, Ou J, Goldstein JL, Brown MS (2004) Central role for liver X receptor in insulin-mediated activation of Srebp-1c transcription and stimulation of fatty acid synthesis in liver. Proc Natl Acad Sci U S A 101(31):11245–11250

Cho H, Zhao X, Hatori M, Ruth TY, Barish GD, Lam MT, Chong L-W, DiTacchio L, Atkins AR, Glass CK (2012) Regulation of circadian behaviour and metabolism by REV-ERB-[agr] and REV-ERB-[bgr]. Nature 485(7396):123–127

Cohen P, Miyazaki M, Socci ND, Hagge-Greenberg A, Liedtke W, Soukas AA, Sharma R, Hudgins LC, Ntambi JM, Friedman JM (2002) Role for stearoyl-CoA desaturase-1 in leptin-mediated weight loss. Science 297(5579):240–243

Cross DA, Alessi DR, Cohen P, Andjelkovich M, Hemmings BA (1995) Inhibition of glycogen synthase kinase-3 by insulin mediated by protein kinase B. Nature 378(6559):785–789

Cuif M, Porteu A, Kahn A, Vaulont S (1993) Exploration of a liver-specific, glucose/insulin-responsive promoter in transgenic mice. J Biol Chem 268(19):13769–13772

Currie E, Schulze A, Zechner R, Walther TC, Farese RV (2013) Cellular fatty acid metabolism and cancer. Cell Metab 18(2):153–161

Davies MN, O'Callaghan BL, Towle HC (2010) Activation and repression of glucose-stimulated ChREBP requires the concerted action of multiple domains within the MondoA conserved region. Am J Physiol Endocrinol Metab 299(4):E665–E674

Dentin R, Pégorier J-P, Benhamed F, Foufelle F, Ferré P, Fauveau V, Magnuson MA, Girard J, Postic C (2004) Hepatic glucokinase is required for the synergistic action of ChREBP and SREBP-1c on glycolytic and lipogenic gene expression. J Biol Chem 279(19):20314–20326

Dentin R, Benhamed F, Pégorier J-P, Foufelle F, Viollet B, Vaulont S, Girard J, Postic C (2005) Polyunsaturated fatty acids suppress glycolytic and lipogenic genes through the inhibition of ChREBP nuclear protein translocation. J Clin Invest 115(10):2843–2854

Dentin R, Benhamed F, Hainault I, Fauveau V, Foufelle F, Dyck JR, Girard J, Postic C (2006) Liver-specific inhibition of ChREBP improves hepatic steatosis and insulin resistance in ob/ob mice. Diabetes 55(8):2159–2170

Dentin R, Langin D, Postic C (2012) Hidden variant of ChREBP in fat links lipogenesis to insulin sensitivity. Cell Metab 15(6):795–797

Dibble CC, Asara JM, Manning BD (2009) Characterization of Rictor phosphorylation sites reveals direct regulation of mTOR complex 2 by S6K1. Mol Cell Biol 29(21):5657–5670

Doiron B, Cuif M-H, Kahn A, Diaz-Guerra M (1994) Respective roles of glucose, fructose, and insulin in the regulation of the liver-specific pyruvate kinase gene promoter. J Biol Chem 269(14):10213–10216

Donnelly KL, Smith CI, Schwarzenberg SJ, Jessurun J, Boldt MD, Parks EJ (2005) Sources of fatty acids stored in liver and secreted via lipoproteins in patients with nonalcoholic fatty liver disease. J Clin Invest 115(5):1343

Düvel K, Yecies JL, Menon S, Raman P, Lipovsky AI, Souza AL, Triantafellow E, Ma Q, Gorski R, Cleaver S (2010) Activation of a metabolic gene regulatory network downstream of mTOR complex 1. Mol Cell 39(2):171–183

Eissing L, Scherer T, Tödter K, Knippschild U, Greve JW, Buurman WA, Pinnschmidt HO, Rensen SS, Wolf AM, Bartelt A (2013) De novo lipogenesis in human fat and liver is linked to ChREBP-β and metabolic health. Nat Commun 4:1528

Enomoto N, Ikejima K, Yamashina S, Enomoto A, Nishiura T, Nishimura T, Brenner DA, Schemmer P, Bradford BU, Rivera CA (2000) Kupffer cell-derived prostaglandin E2 is involved

in alcohol-induced fat accumulation in rat liver. Am J Physiol Gastrointest Liver Physiol 279(1):G100–G106
Erion DM, Popov V, Hsiao JJ, Vatner D, Mitchell K, Yonemitsu S, Nagai Y, Kahn M, Gillum MP, Dong J (2012) The role of the carbohydrate response element-binding protein in male fructose-fed rats. Endocrinology 154(1):36–44
Ettinger SL, Sobel R, Whitmore TG, Akbari M, Bradley DR, Gleave ME, Nelson CC (2004) Dysregulation of sterol response element-binding proteins and downstream effectors in prostate cancer during progression to androgen independence. Cancer Res 64(6):2212–2221
Feng D, Liu T, Sun Z, Bugge A, Mullican SE, Alenghat T, Liu XS, Lazar MA (2011) A circadian rhythm orchestrated by histone deacetylase 3 controls hepatic lipid metabolism. Science 331(6022):1315–1319
Ferre P, Foufelle F (2010) Hepatic steatosis: a role for de novo lipogenesis and the transcription factor SREBP-1c. Diabetes Obes Metab 12(s2):83–92
Ferre T, Riu E, Bosch F, Valera A (1996) Evidence from transgenic mice that glucokinase is rate limiting for glucose utilization in the liver. FASEB J 10(10):1213–1218
Filhoulaud G, Guilmeau S, Dentin R, Girard J, Postic C (2013) Novel insights into ChREBP regulation and function. Trends Endocrinol Metab 24(5):257–268
Fisher EA (2012) The degradation of apolipoprotein B100: multiple opportunities to regulate VLDL triglyceride production by different proteolytic pathways. Biochim Biophys Acta 1821(5):778–781
Fritz V, Benfodda Z, Rodier G, Henriquet C, Iborra F, Avancès C, Allory Y, De La Taille A, Culine S, Blancou H (2010) Abrogation of de novo lipogenesis by stearoyl-CoA desaturase 1 inhibition interferes with oncogenic signaling and blocks prostate cancer progression in mice. Mol Cancer Ther 9(6):1740–1754
Fukasawa M, Ge Q, Wynn RM, Ishii S, Uyeda K (2010) Coordinate regulation/localization of the carbohydrate responsive binding protein (ChREBP) by two nuclear export signal sites: discovery of a new leucine-rich nuclear export signal site. Biochem Biophys Res Commun 391(2):1166–1169
Giandomenico V, Simonsson M, Grönroos E, Ericsson J (2003) Coactivator-dependent acetylation stabilizes members of the SREBP family of transcription factors. Mol Cell Biol 23(7):2587–2599
Gibbons GF, Islam K, Pease RJ (2000) Mobilisation of triacylglycerol stores. Biochim Biophys Acta 1483(1):37–57
Girard J, Ferré P, Foufelle F (1997) Mechanisms by which carbohydrates regulate expression of genes for glycolytic and lipogenic enzymes. Annu Rev Nutr 17(1):325–352
Goldstein JL, DeBose-Boyd RA, Brown MS (2006) Protein sensors for membrane sterols. Cell 124(1):35–46
Graner E, Tang D, Rossi S, Baron A, Migita T, Weinstein LJ, Lechpammer M, Huesken D, Zimmermann J, Signoretti S (2004) The isopeptidase USP2a regulates the stability of fatty acid synthase in prostate cancer. Cancer Cell 5(3):253–261
Gregor MF, Hotamisligil GS (2011) Inflammatory mechanisms in obesity. Annu Rev Immunol 29:415–445
Griffiths B, Lewis CA, Bensaad K, Ros S, Zhang Q, Ferber EC, Konisti S, Peck B, Miess H, East P (2013) Sterol regulatory element binding protein-dependent regulation of lipid synthesis supports cell survival and tumor growth. Cancer Metab 1(3)
Gross DN, Wan M, Birnbaum MJ (2009) The role of FOXO in the regulation of metabolism. Curr Diab Rep 9(3):208–214
Grunfeld C, Adi S, Soued M, Moser A, Fiers W, Feingold KR (1990) Search for mediators of the lipogenic effects of tumor necrosis factor: potential role for interleukin 6. Cancer Res 50(14):4233–4238
Guinez C, Filhoulaud G, Rayah-Benhamed F, Marmier S, Dubuquoy C, Dentin R, Moldes M, Burnol A-F, Yang X, Lefebvre T (2011) O-GlcNAcylation increases ChREBP protein content and transcriptional activity in the liver. Diabetes 60(5):1399–1413

Guo D, Prins RM, Dang J, Kuga D, Iwanami A, Soto H, Lin KY, Huang TT, Akhavan D, Hock MB (2009) EGFR signaling through an Akt-SREBP-1-dependent, rapamycin-resistant pathway sensitizes glioblastomas to anti-lipogenic therapy. Sci Signal 2(101):ra82

Ha J, Lee J-K, Kim K-S, Witters LA, Kim K-H (1996) Cloning of human acetyl-CoA carboxylase-beta and its unique features. Proc Natl Acad Sci U S A 93(21):11466–11470

Harada N, Oda Z, Hara Y, Fujinami K, Okawa M, Ohbuchi K, Yonemoto M, Ikeda Y, Ohwaki K, Aragane K (2007) Hepatic de novo lipogenesis is present in liver-specific ACC1-deficient mice. Mol Cell Biol 27(5):1881–1888

Hardie DG, Ross FA, Hawley SA (2012) AMPK: a nutrient and energy sensor that maintains energy homeostasis. Nat Rev Mol Cell Biol 13(4):251–262

Hariharan N, Farrelly D, Hagan D, Hillyer D, Arbeeny C, Sabrah T, Treloar A, Brown K, Kalinowski S, Mookhtiar K (1997) Expression of human hepatic glucokinase in transgenic mice liver results in decreased glucose levels and reduced body weight. Diabetes 46(1):11–16

Hatzivassiliou G, Zhao F, Bauer DE, Andreadis C, Shaw AN, Dhanak D, Hingorani SR, Tuveson DA, Thompson CB (2005) ATP citrate lyase inhibition can suppress tumor cell growth. Cancer Cell 8(4):311–321

Havula E, Hietakangas V (2012) Glucose sensing by ChREBP/MondoA–Mlx transcription factors. Semin Cell Dev Biol 23(6):640–647

Hegarty BD, Bobard A, Hainault I, Ferré P, Bossard P, Foufelle F (2005) Distinct roles of insulin and liver X receptor in the induction and cleavage of sterol regulatory element binding protein-1c. Proc Natl Acad Sci U S A 102(3):791–796

Hemmings BA, Restuccia DF (2012) Pi3k-pkb/akt pathway. Cold Spring Harb Perspect Biol 4(9):a011189

Hems DA, Rath EA, Verrinder TR (1975) Fatty acid synthesis in liver and adipose tissue of normal and genetically obese (ob/ob) mice during the 24-hour cycle. Biochem J 150:167–173

Herman MA, Peroni OD, Villoria J, Schön MR, Abumrad NA, Blüher M, Klein S, Kahn BB (2012) A novel ChREBP isoform in adipose tissue regulates systemic glucose metabolism. Nature 484(7394):333–338

Hirano Y, Yoshida M, Shimizu M, Sato R (2001) Direct demonstration of rapid degradation of nuclear sterol regulatory element-binding proteins by the ubiquitin-proteasome pathway. J Biol Chem 276(39):36431–36437

Hodson L, Fielding BA (2013) Stearoyl-CoA desaturase: rogue or innocent bystander? Prog Lipid Res 52(1):15–42

Horton JD, Bashmakov Y, Shimomura I, Shimano H (1998) Regulation of sterol regulatory element binding proteins in livers of fasted and refed mice. Proc Natl Acad Sci U S A 95(11):5987–5992

Horton JD, Goldstein JL, Brown MS (2002) SREBPs: activators of the complete program of cholesterol and fatty acid synthesis in the liver. J Clin Invest 109(9):1125–1131

Houtkooper RH, Pirinen E, Auwerx J (2012) Sirtuins as regulators of metabolism and healthspan. Nat Rev Mol Cell Biol 13(4):225–238

Huang J, Manning BD (2009) A complex interplay between Akt, TSC2, and the two mTOR complexes. Biochem Soc Trans 37(Pt 1):217

Hudgins LC, Seidman CE, Diakun J, Hirsch J (1998) Human fatty acid synthesis is reduced after the substitution of dietary starch for sugar. Am J Clin Nutr 67(4):631–639

Ido-Kitamura Y, Sasaki T, Kobayashi M, Kim H-J, Lee Y-S, Kikuchi O, Yokota-Hashimoto H, Iizuka K, Accili D, Kitamura T (2012) Hepatic FoxO1 integrates glucose utilization and lipid synthesis through regulation of Chrebp O-glycosylation. PLoS One 7(10):e47231

Iizuka K, Bruick RK, Liang G, Horton JD, Uyeda K (2004) Deficiency of carbohydrate response element-binding protein (ChREBP) reduces lipogenesis as well as glycolysis. Proc Natl Acad Sci U S A 101(19):7281–7286

Iizuka K, Miller B, Uyeda K (2006) Deficiency of carbohydrate-activated transcription factor ChREBP prevents obesity and improves plasma glucose control in leptin-deficient (ob/ob) mice. Am J Physiol Endocrinol Metab 291(2):E358–E364

Jakobsson A, Westerberg R, Jacobsson A (2006) Fatty acid elongases in mammals: their regulation and roles in metabolism. Prog Lipid Res 45(3):237–249
Kabashima T, Kawaguchi T, Wadzinski BE, Uyeda K (2003) Xylulose 5-phosphate mediates glucose-induced lipogenesis by xylulose 5-phosphate-activated protein phosphatase in rat liver. Proc Natl Acad Sci U S A 100(9):5107–5112
Kammoun HL, Chabanon H, Hainault I, Luquet S, Magnan C, Koike T, Ferré P, Foufelle F (2009) GRP78 expression inhibits insulin and ER stress-induced SREBP-1c activation and reduces hepatic steatosis in mice. J Clin Invest 119(5):1201–1215
Kawaguchi T, Takenoshita M, Kabashima T, Uyeda K (2001) Glucose and cAMP regulate the L-type pyruvate kinase gene by phosphorylation/dephosphorylation of the carbohydrate response element binding protein. Proc Natl Acad Sci U S A 98(24):13710–13715
Kawaguchi T, Osatomi K, Yamashita H, Kabashima T, Uyeda K (2002) Mechanism for fatty acid "sparing" effect on glucose-induced transcription regulation of carbohydrate-responsive element-binding protein by AMP-activated protein kinase. J Biol Chem 277(6):3829–3835
Kersten S (2001) Mechanisms of nutritional and hormonal regulation of lipogenesis. EMBO Rep 2(4):282–286
Kersten S, Desvergne B, Wahli W (2000) Roles of PPARs in health and disease. Nature 405(6785):421–424
Kim KH, Song MJ, Yoo EJ, Choe SS, Dai Park S, Kim JB (2004) Regulatory role of glycogen synthase kinase 3 for transcriptional activity of ADD1/SREBP1c. J Biol Chem 279(50):51999–52006
Kim H-S, Xiao C, Wang R-H, Lahusen T, Xu X, Vassilopoulos A, Vazquez-Ortiz G, Jeong W-I, Park O, Ki SH (2010) Hepatic-specific disruption of SIRT6 in mice results in fatty liver formation due to enhanced glycolysis and triglyceride synthesis. Cell Metab 12(3):224–236
Knutson SK, Chyla BJ, Amann JM, Bhaskara S, Huppert SS, Hiebert SW (2008) Liver-specific deletion of histone deacetylase 3 disrupts metabolic transcriptional networks. EMBO J 27(7):1017–1028
Koo H-Y, Wallig MA, Chung BH, Nara TY, Cho B, Nakamura MT (2008) Dietary fructose induces a wide range of genes with distinct shift in carbohydrate and lipid metabolism in fed and fasted rat liver. Biochim Biophys Acta 1782(5):341–348
Krycer JR, Sharpe LJ, Luu W, Brown AJ (2010) The Akt–SREBP nexus: cell signaling meets lipid metabolism. Trends Endocrinol Metab 21(5):268–276
Le Martelot G, Claudel T, Gatfield D, Schaad O, Kornmann B, Sasso GL, Moschetta A, Schibler U (2009) REV-ERBα participates in circadian SREBP signaling and bile acid homeostasis. PLoS Biol 7(9):e1000181
Leavens KF, Easton RM, Shulman GI, Previs SF, Birnbaum MJ (2009) Akt2 is required for hepatic lipid accumulation in models of insulin resistance. Cell Metab 10(5):405–418
Lee A-H, Scapa EF, Cohen DE, Glimcher LH (2008) Regulation of hepatic lipogenesis by the transcription factor XBP1. Science 320(5882):1492–1496
Lewis GF, Carpentier A, Adeli K, Giacca A (2002) Disordered fat storage and mobilization in the pathogenesis of insulin resistance and type 2 diabetes. Endocr Rev 23(2):201–229
Li MV, Chang B, Imamura M, Poungvarin N, Chan L (2006) Glucose-dependent transcriptional regulation by an evolutionarily conserved glucose-sensing module. Diabetes 55(5):1179–1189
Li X, Zhang S, Blander G, Jeanette GT, Krieger M, Guarente L (2007) SIRT1 deacetylates and positively regulates the nuclear receptor LXR. Mol Cell 28(1):91–106
Li S, Brown MS, Goldstein JL (2010) Bifurcation of insulin signaling pathway in rat liver: mTORC1 required for stimulation of lipogenesis, but not inhibition of gluconeogenesis. Proc Natl Acad Sci U S A 107(8):3441–3446
Li Y, Xu S, Mihaylova MM, Zheng B, Hou X, Jiang B, Park O, Luo Z, Lefai E, Shyy JY-J (2011) AMPK phosphorylates and inhibits SREBP activity to attenuate hepatic steatosis and atherosclerosis in diet-induced insulin-resistant mice. Cell Metab 13(4):376–388
Liang G, Yang J, Horton JD, Hammer RE, Goldstein JL, Brown MS (2002) Diminished hepatic response to fasting/refeeding and liver X receptor agonists in mice with selective deficiency of sterol regulatory element-binding protein-1c. J Biol Chem 277(11):9520–9528

Listenberger LL, Han X, Lewis SE, Cases S, Farese RV, Ory DS, Schaffer JE (2003) Triglyceride accumulation protects against fatty acid-induced lipotoxicity. Proc Natl Acad Sci U S A 100(6):3077–3082

Lodhi IJ, Wei X, Semenkovich CF (2011) Lipoexpediency: *de novo* lipogenesis as a metabolic signal transmitter. Trends Endocrinol Metab 22(1):1–8

Lupu R, Menendez JA (2006) Pharmacological inhibitors of fatty acid synthase (FASN)-catalyzed endogenous fatty acid biogenesis: a new family of anti-cancer agents? Curr Pharm Biotechnol 7(6):483–494

Ma L, Tsatsos NG, Towle HC (2005) Direct role of ChREBP· Mlx in regulating hepatic glucose-responsive genes. J Biol Chem 280(12):12019–12027

Ma L, Robinson LN, Towle HC (2006) ChREBP* Mlx is the principal mediator of glucose-induced gene expression in the liver. J Biol Chem 281(39):28721–28730

Ma L, Sham YY, Walters KJ, Towle HC (2007) A critical role for the loop region of the basic helix–loop–helix/leucine zipper protein Mlx in DNA binding and glucose-regulated transcription. Nucleic Acids Res 35(1):35–44

Mannaerts GP, Van Veldhoven PP, Casteels M (2000) Peroxisomal lipid degradation via β-and α-oxidation in mammals. Cell Biochem Biophys 32(1–3):73–87

Manning BD, Cantley LC (2007) AKT/PKB signaling: navigating downstream. Cell 129(7):1261–1274

Mao J, DeMayo FJ, Li H, Abu-Elheiga L, Gu Z, Shaikenov TE, Kordari P, Chirala SS, Heird WC, Wakil SJ (2006) Liver-specific deletion of acetyl-CoA carboxylase 1 reduces hepatic triglyceride accumulation without affecting glucose homeostasis. Proc Natl Acad Sci U S A 103(22):8552–8557

Mao J-H, Kim I-J, Wu D, Climent J, Kang HC, DelRosario R, Balmain A (2008) FBXW7 targets mTOR for degradation and cooperates with PTEN in tumor suppression. Science 321(5895):1499–1502

Marchesini G, Brizi M, Morselli-Labate AM, Bianchi G, Bugianesi E, McCullough AJ, Forlani G, Melchionda N (1999) Association of nonalcoholic fatty liver disease with insulin resistance. Am J Med 107(5):450–455

Matsusue K, Haluzik M, Lambert G, Yim S-H, Gavrilova O, Ward JM, Brewer B Jr, Reitman ML, Gonzalez FJ (2003) Liver-specific disruption of PPARγ in leptin-deficient mice improves fatty liver but aggravates diabetic phenotypes. J Clin Invest 111(5):737

Matsuzaka T, Shimano H, Yahagi N, Kato T, Atsumi A, Yamamoto T, Inoue N, Ishikawa M, Okada S, Ishigaki N (2007) Crucial role of a long-chain fatty acid elongase, Elovl6, in obesity-induced insulin resistance. Nat Med 13(10):1193–1202

Matsuzaka T, Atsumi A, Matsumori R, Nie T, Shinozaki H, Suzuki-Kemuriyama N, Kuba M, Nakagawa Y, Ishii K, Shimada M (2012) Elovl6 promotes nonalcoholic steatohepatitis. Hepatology 56(6):2199–2208

Medes G, Thomas A, Weinhouse S (1953) Metabolism of neoplastic tissue. IV. A study of lipid synthesis in neoplastic tissue slices in vitro. Cancer Res 13(1):27–29

Merla G, Howald C, Antonarakis SE, Reymond A (2004) The subcellular localization of the ChoRE-binding protein, encoded by the Williams–Beuren syndrome critical region gene 14, is regulated by 14-3-3. Hum Mol Genet 13(14):1505–1514

Meroni G, Cairo S, Merla G, Messali S, Brent R, Ballabio A, Reymond A (2000) Mlx, a new Max-like bHLHZip family member: the center stage of a novel transcription factors regulatory pathway? Oncogene 19(29):3266–3277

Migita T, Narita T, Nomura K, Miyagi E, Inazuka F, Matsuura M, Ushijima M, Mashima T, Seimiya H, Satoh Y (2008) ATP citrate lyase: activation and therapeutic implications in non-small cell lung cancer. Cancer Res 68(20):8547–8554

Miyazaki M, Flowers MT, Sampath H, Chu K, Otzelberger C, Liu X, Ntambi JM (2007) Hepatic stearoyl-CoA desaturase-1 deficiency protects mice from carbohydrate-induced adiposity and hepatic steatosis. Cell Metab 6(6):484–496

Monetti M, Levin MC, Watt MJ, Sajan MP, Marmor S, Hubbard BK, Stevens RD, Bain JR, Newgard CB, Farese RV (2007) Dissociation of hepatic steatosis and insulin resistance in mice overexpressing DGAT in the liver. Cell Metab 6(1):69–78

Moon Y-A, Shah NA, Mohapatra S, Warrington JA, Horton JD (2001) Identification of a mammalian long chain fatty acyl elongase regulated by sterol regulatory element-binding proteins. J Biol Chem 276(48):45358–45366
Morán-Salvador E, López-Parra M, García-Alonso V, Titos E, Martínez-Clemente M, González-Périz A, López-Vicario C, Barak Y, Arroyo V, Clària J (2011) Role for PPARγ in obesity-induced hepatic steatosis as determined by hepatocyte-and macrophage-specific conditional knockouts. FASEB J 25(8):2538–2550
Munday MR, Campbell DG, Carling D, Hardie DG (1988) Identification by amino acid sequencing of three major regulatory phosphorylation sites on rat acetyl-CoA carboxylase. Eur J Biochem 175(2):331–338
Nakayama KI, Nakayama K (2006) Ubiquitin ligases: cell-cycle control and cancer. Nat Rev Cancer 6(5):369–381
Neyrinck AM, Margagliotti S, Gomez C, Delzenne NM (2004) Kupffer cell-derived prostaglandin E2 is involved in regulation of lipid synthesis in rat liver tissue. Cell Biochem Funct 22(5):327–332
Nimpf J, Schneider WJ (2000) From cholesterol transport to signal transduction: low density lipoprotein receptor, very low density lipoprotein receptor, and apolipoprotein E receptor-2. Biochim Biophys Acta 1529(1):287–298
Niswender KD, Shiota M, Postic C, Cherrington AD, Magnuson MA (1997) Effects of increased glucokinase gene copy number on glucose homeostasis and hepatic glucose metabolism. J Biol Chem 272(36):22570–22575
Nomura K, Yamanouchi T (2012) The role of fructose-enriched diets in mechanisms of nonalcoholic fatty liver disease. J Nutr Biochem 23(3):203–208
Nonogaki K, Fuller GM, Fuentes NL, Moser AH, Staprans I, Grunfeld C, Feingold KR (1995) Interleukin-6 stimulates hepatic triglyceride secretion in rats. Endocrinology 136(5):2143–2149
Ntambi JM, Miyazaki M, Stoehr JP, Lan H, Kendziorski CM, Yandell BS, Song Y, Cohen P, Friedman JM, Attie AD (2002) Loss of stearoyl-CoA desaturase-1 function protects mice against adiposity. Proc Natl Acad Sci U S A 99(17):11482–11486
Olefsky JM (2000) Treatment of insulin resistance with peroxisome proliferator-activated receptor γ agonists. J Clin Invest 106(4):467
Onoyama I, Suzuki A, Matsumoto A, Tomita K, Katagiri H, Oike Y, Nakayama K, Nakayama KI (2011) Fbxw7 regulates lipid metabolism and cell fate decisions in the mouse liver. J Clin Invest 121(1):342
Owen JL, Zhang Y, Bae S-H, Farooqi MS, Liang G, Hammer RE, Goldstein JL, Brown MS (2012) Insulin stimulation of SREBP-1c processing in transgenic rat hepatocytes requires p70 S6-kinase. Proc Natl Acad Sci U S A 109(40):16184–16189
Parks EJ, Krauss RM, Christiansen MP, Neese RA, Hellerstein MK (1999) Effects of a low-fat, high-carbohydrate diet on VLDL-triglyceride assembly, production, and clearance. J Clin Invest 104(8):1087–1096
Paton C, Stamatikos A, Rogowski M (2014) Tissue specific induction of ChREBP isoforms in carbohydrate refed mice and their impact on lipogenesis (642.8). FASEB J 28(1 Suppl):642–648
Pearce J (1983) Fatty acid synthesis in liver and adipose tissue. Proc Nutr Soc 42(02):263–271
Petersen KF, Shulman GI (2006) Etiology of insulin resistance. Am J Med 119(5):S10–S16
Peterson TR, Sengupta SS, Harris TE, Carmack AE, Kang SA, Balderas E, Guertin DA, Madden KL, Carpenter AE, Finck BN (2011) mTOR complex 1 regulates lipin 1 localization to control the SREBP pathway. Cell 146(3):408–420
Ponugoti B, Kim D-H, Xiao Z, Smith Z, Miao J, Zang M, Wu S-Y, Chiang C-M, Veenstra TD, Kemper JK (2010) SIRT1 deacetylates and inhibits SREBP-1C activity in regulation of hepatic lipid metabolism. J Biol Chem 285(44):33959–33970
Porstmann T, Santos CR, Griffiths B, Cully M, Wu M, Leevers S, Griffiths JR, Chung Y-L, Schulze A (2008) SREBP activity is regulated by mTORC1 and contributes to Akt-dependent cell growth. Cell Metab 8(3):224–236

Postic C, Girard J (2008) Contribution of de novo fatty acid synthesis to hepatic steatosis and insulin resistance: lessons from genetically engineered mice. J Clin Invest 118(3):829
Postic C, Dentin R, Denechaud P-D, Girard J (2007) ChREBP, a transcriptional regulator of glucose and lipid metabolism. Annu Rev Nutr 27:179–192
Punga T, Bengoechea-Alonso MT, Ericsson J (2006) Phosphorylation and ubiquitination of the transcription factor sterol regulatory element-binding protein-1 in response to DNA binding. J Biol Chem 281(35):25278–25286
Purushotham A, Schug TT, Xu Q, Surapureddi S, Guo X, Li X (2009) Hepatocyte-specific deletion of SIRT1 alters fatty acid metabolism and results in hepatic steatosis and inflammation. Cell Metab 9(4):327–338
Radhakrishnan A, Goldstein JL, McDonald JG, Brown MS (2008) Switch-like control of SREBP-2 transport triggered by small changes in ER cholesterol: a delicate balance. Cell Metab 8(6):512–521
Reddy JK, Mannaerts GP (1994) Peroxisomal lipid metabolism. Annu Rev Nutr 14(1):343–370
Reddy JK, Rao MS (2006) Lipid metabolism and liver inflammation. II. Fatty liver disease and fatty acid oxidation. Am J Physiol Gastrointest Liver Physiol 290(5):G852–G858
Repa JJ, Liang G, Ou J, Bashmakov Y, Lobaccaro J-MA, Shimomura I, Shan B, Brown MS, Goldstein JL, Mangelsdorf DJ (2000) Regulation of mouse sterol regulatory element-binding protein-1c gene (SREBP-1c) by oxysterol receptors, LXRα and LXRβ. Genes Dev 14(22):2819–2830
Ruan H-B, Nie Y, Yang X (2013) Regulation of protein degradation by O-GlcNAcylation: crosstalk with ubiquitination. Mol Cell Proteomics 12(12):3489–3497
Rudic RD, McNamara P, Curtis A-M, Boston RC, Panda S, Hogenesch JB, FitzGerald GA (2004) BMAL1 and CLOCK, two essential components of the circadian clock, are involved in glucose homeostasis. PLoS Biol 2(11):e377
Rufo C, Teran-Garcia M, Nakamura MT, Koo S-H, Towle HC, Clarke SD (2001) Involvement of a unique carbohydrate-responsive factor in the glucose regulation of rat liver fatty-acid synthase gene transcription. J Biol Chem 276(24):21969–21975
Sakiyama H, Fujiwara N, Noguchi T, Eguchi H, Yoshihara D, Uyeda K, Suzuki K (2010) The role of *O*-linked GlcNAc modification on the glucose response of ChREBP. Biochem Biophys Res Commun 402(4):784–789
Saltiel AR, Kahn CR (2001) Insulin signalling and the regulation of glucose and lipid metabolism. Nature 414(6865):799–806
Samuel VT, Shulman GI (2012) Mechanisms for insulin resistance: common threads and missing links. Cell 148(5):852–871
Savage DB, Choi CS, Samuel VT, Liu Z-X, Zhang D, Wang A, Zhang X-M, Cline GW, Yu XX, Geisler JG (2006) Reversal of diet-induced hepatic steatosis and hepatic insulin resistance by antisense oligonucleotide inhibitors of acetyl-CoA carboxylases 1 and 2. J Clin Invest 116(3):817
Schonfeld G, Yue P, Lin X, Chen Z (2008) Fatty liver and insulin resistance: not always linked. Trans Am Clin Climatol Assoc 119:217
Schwarz J-M, Linfoot P, Dare D, Aghajanian K (2003) Hepatic de novo lipogenesis in normoinsulinemic and hyperinsulinemic subjects consuming high-fat, low-carbohydrate and low-fat, high-carbohydrate isoenergetic diets. Am J Clin Nutr 77(1):43–50
Scott E, Carter A, Grant P (2008) Association between polymorphisms in the Clock gene, obesity and the metabolic syndrome in man. Int J Obes (Lond) 32(4):658–662
Shelness GS, Sellers JA (2001) Very-low-density lipoprotein assembly and secretion. Curr Opin Lipidol 12(2):151–157
Shih H-M, Liu Z, Towle HC (1995) Two CACGTG motifs with proper spacing dictate the carbohydrate regulation of hepatic gene transcription. J Biol Chem 270(37):21991–21997
Shimano H, Horton JD, Shimomura I, Hammer RE, Brown MS, Goldstein JL (1997) Isoform 1c of sterol regulatory element binding protein is less active than isoform 1a in livers of transgenic mice and in cultured cells. J Clin Invest 99(5):846

Shimba S, Ogawa T, Hitosugi S, Ichihashi Y, Nakadaira Y, Kobayashi M, Tezuka M, Kosuge Y, Ishige K, Ito Y (2011) Deficient of a clock gene, brain and muscle Arnt-like protein-1 (BMAL1), induces dyslipidemia and ectopic fat formation. PLoS One 6(9):e25231

Shimomura I, Shimano H, Horton JD, Goldstein JL, Brown MS (1997) Differential expression of exons 1a and 1c in mRNAs for sterol regulatory element binding protein-1 in human and mouse organs and cultured cells. J Clin Invest 99(5):838

Shimomura I, Bashmakov Y, Ikemoto S, Horton JD, Brown MS, Goldstein JL (1999) Insulin selectively increases SREBP-1c mRNA in the livers of rats with streptozotocin-induced diabetes. Proc Natl Acad Sci U S A 96(24):13656–13661

Shimomura I, Matsuda M, Hammer RE, Bashmakov Y, Brown MS, Goldstein JL (2000) Decreased IRS-2 and increased SREBP-1c lead to mixed insulin resistance and sensitivity in livers of lipodystrophic and ob/ob mice. Mol Cell 6(1):77–86

Smith S, Witkowski A, Joshi AK (2003) Structural and functional organization of the animal fatty acid synthase. Prog Lipid Res 42(4):289–317

Sonksen P, Sonksen J (2000) Insulin: understanding its action in health and disease. Br J Anaesth 85(1):69–79

Sookoian S, Gemma C, Gianotti TF, Burgueño A, Castaño G, Pirola CJ (2008) Genetic variants of Clock transcription factor are associated with individual susceptibility to obesity. Am J Clin Nutr 87(6):1606–1615

Shi S-Q, Ansari TS, McGuinness OP, Wasserman DH, Johnson CH (2013) Circadian disruption leads to insulin resistance and obesity. Curr Biol 23(5):372–381

Stoeckman AK, Ma L, Towle HC (2004) Mlx is the functional heteromeric partner of the carbohydrate response element-binding protein in glucose regulation of lipogenic enzyme genes. J Biol Chem 279(15):15662–15669

Stoops JK, Arslanian MJ, Oh YH, Aune KC, Vanaman TC, Wakil SJ (1975) Presence of two polypeptide chains comprising fatty acid synthetase. Proc Natl Acad Sci U S A 72(5):1940–1944

Strable MS, Ntambi JM (2010) Genetic control of de novo lipogenesis: role in diet-induced obesity. Crit Rev Biochem Mol Biol 45(3):199–214

Sun Z, Lazar MA (2013) Dissociating fatty liver and diabetes. Trends Endocrinol Metab 24(1):4–12

Sun L-P, Li L, Goldstein JL, Brown MS (2005) Insig required for sterol-mediated inhibition of Scap/SREBP binding to COPII proteins in vitro. J Biol Chem 280(28):26483–26490

Sun L-P, Seemann J, Goldstein JL, Brown MS (2007) Sterol-regulated transport of SREBPs from endoplasmic reticulum to Golgi: Insig renders sorting signal in Scap inaccessible to COPII proteins. Proc Natl Acad Sci U S A 104(16):6519–6526

Sundqvist A, Bengoechea-Alonso MT, Ye X, Lukiyanchuk V, Jin J, Harper JW, Ericsson J (2005) Control of lipid metabolism by phosphorylation-dependent degradation of the SREBP family of transcription factors by SCF Fbw7. Cell Metab 1(6):379–391

Swinnen JV, Brusselmans K, Verhoeven G (2006) Increased lipogenesis in cancer cells: new players, novel targets. Curr Opin Clin Nutr Metab Care 9(4):358–365

Thompson KS, Towle H (1991) Localization of the carbohydrate response element of the rat L-type pyruvate kinase gene. J Biol Chem 266(14):8679–8682

Tobin KAR, Ulven SM, Schuster GU, Steineger HH, Andresen SM, Gustafsson J-Å, Nebb HI (2002) Liver X receptors as insulin-mediating factors in fatty acid and cholesterol biosynthesis. J Biol Chem 277(12):10691–10697

Tsatsos NG, Towle HC (2006) Glucose activation of ChREBP in hepatocytes occurs via a two-step mechanism. Biochem Biophys Res Commun 340(2):449–456

Uchiyama A, Aoyama T, Kamijo K, Uchida Y, Kondo N, Orii T, Hashimoto T (1996) Molecular cloning of cDNA encoding rat very long-chain acyl-CoA synthetase. J Biol Chem 271(48):30360–30365

Um SH, Frigerio F, Watanabe M, Picard F, Joaquin M, Sticker M, Fumagalli S, Allegrini PR, Kozma SC, Auwerx J (2004) Absence of S6K1 protects against age-and diet-induced obesity while enhancing insulin sensitivity. Nature 431(7005):200–205

Viollet B, Foretz M, Guigas B, Horman S, Dentin R, Bertrand L, Hue L, Andreelli F (2006) Activation of AMP-activated protein kinase in the liver: a new strategy for the management of metabolic hepatic disorders. J Physiol 574(1):41–53

Vluggens A, Reddy JK (2012) Nuclear receptors and transcription factors in the development of fatty liver disease. Curr Drug Metab 13(10):1422–1435

Walker AK, Yang F, Jiang K, Ji J-Y, Watts JL, Purushotham A, Boss O, Hirsch ML, Ribich S, Smith JJ (2010) Conserved role of SIRT1 orthologs in fasting-dependent inhibition of the lipid/cholesterol regulator SREBP. Genes Dev 24(13):1403–1417

Wang R-H, Kim H-S, Xiao C, Xu X, Gavrilova O, Deng C-X (2011) Hepatic Sirt1 deficiency in mice impairs mTorc2/Akt signaling and results in hyperglycemia, oxidative damage, and insulin resistance. J Clin Invest 121(11):4477

Welcker M, Clurman BE (2008) FBW7 ubiquitin ligase: a tumour suppressor at the crossroads of cell division, growth and differentiation. Nat Rev Cancer 8(2):83–93

Wells L, Vosseller K, Hart G (2003) A role for N-acetylglucosamine as a nutrient sensor and mediator of insulin resistance. Cell Mol Life Sci 60(2):222–228

Williams KJ, Argus JP, Zhu Y, Wilks MQ, Marbois BN, York AG, Kidani Y, Pourzia AL, Akhavan D, Lisiero DN (2013) An essential requirement for the SCAP/SREBP signaling axis to protect cancer cells from lipotoxicity. Cancer Res 73(9):2850–2862

Xu X, So J-S, Park J-G, Lee A-H (2013) Transcriptional control of hepatic lipid metabolism by SREBP and ChREBP. Semin Liver Dis 33(4):301–311 (NIH Public Access)

Yahagi N, Shimano H, Hasty AH, Matsuzaka T, Ide T, Yoshikawa T, Amemiya-Kudo M, Tomita S, Okazaki H, Tamura Y (2002) Absence of sterol regulatory element-binding protein-1 (SREBP-1) ameliorates fatty livers but not obesity or insulin resistance inLep ob/Lep ob mice. J Biol Chem 277(22):19353–19357

Yamaguchi K, Yang L, McCall S, Huang J, Yu XX, Pandey SK, Bhanot S, Monia BP, Li YX, Diehl AM (2007) Inhibiting triglyceride synthesis improves hepatic steatosis but exacerbates liver damage and fibrosis in obese mice with nonalcoholic steatohepatitis. Hepatology 45(6):1366–1374

Yamashita H, Takenoshita M, Sakurai M, Bruick RK, Henzel WJ, Shillinglaw W, Arnot D, Uyeda K (2001) A glucose-responsive transcription factor that regulates carbohydrate metabolism in the liver. Proc Natl Acad Sci U S A 98(16):9116–9121

Yamazaki Y, Usui I, Kanatani Y, Matsuya Y, Tsuneyama K, Fujisaka S, Bukhari A, Suzuki H, Senda S, Imanishi S (2009) Treatment with SRT1720, a SIRT1 activator, ameliorates fatty liver with reduced expression of lipogenic enzymes in MSG mice. Am J Physiol Endocrinol Metab 297(5):E1179–E1186

Yang J, Maika S, Craddock L, King JA, Liu Z-M (2008) Chronic activation of AMP-activated protein kinase-alpha1 in liver leads to decreased adiposity in mice. Biochem Biophys Res Commun 370(2):248–253

Yang J, Craddock L, Hong S, Liu ZM (2009) AMP-activated protein kinase suppresses LXR-dependent sterol regulatory element-binding protein-1c transcription in rat hepatoma McA-RH7777 cells. J Cell Biochem 106(3):414–426

Yap F, Craddock L, Yang J (2011) Mechanism of AMPK suppression of LXR-dependent Srebp-1c transcription. Int J Biol Sci 7(5):645

Yecies J (2013) SREBP: a key effector of mTORC1 signaling in metabolism and cancer. Doctoral dissertation, Harvard University

Yecies JL, Zhang HH, Menon S, Liu S, Yecies D, Lipovsky AI, Gorgun C, Kwiatkowski DJ, Hotamisligil GS, Lee C-H (2011) Akt stimulates hepatic SREBP1c and lipogenesis through parallel mTORC1-dependent and independent pathways. Cell Metab 14(1):21–32

Yellaturu CR, Deng X, Cagen LM, Wilcox HG, Mansbach CM, Siddiqi SA, Park EA, Raghow R, Elam MB (2009) Insulin enhances post-translational processing of nascent SREBP-1c by promoting its phosphorylation and association with COPII vesicles. J Biol Chem 284(12):7518–7532

Yki-Järvinen H (2005) Fat in the liver and insulin resistance. Ann Med 37(5):347–356

Yu S, Rao S, Reddy JK (2003) Peroxisome proliferator-activated receptors, fatty acid oxidation, steatohepatitis and hepatocarcinogenesis. Curr Mol Med 3(6):561–572
Zeng L, Lu M, Mori K, Luo S, Lee AS, Zhu Y, Shyy JYJ (2004) ATF6 modulates SREBP2-mediated lipogenesis. EMBO J 23(4):950–958
Zhang K, Wang S, Malhotra J, Hassler JR, Back SH, Wang G, Chang L, Xu W, Miao H, Leonardi R (2011) The unfolded protein response transducer IRE1α prevents ER stress-induced hepatic steatosis. EMBO J 30(7):1357–1375
Zhang D, Tong X, Arthurs B, Guha A, Rui L, Kamath A, Inoki K, Yin L (2014) Liver clock protein BMAL1 promotes de novo lipogenesis through insulin-mTORC2-AKT signaling. J Biol Chem 289(37):25925–25935
Zhao X, Feng D, Wang Q, Abdulla A, Xie X-J, Zhou J, Sun Y, Yang ES, Liu L-P, Vaitheesvaran B (2012) Regulation of lipogenesis by cyclin-dependent kinase 8-mediated control of SREBP-1. J Clin Invest 122(7):2417
Zhou G, Myers R, Li Y, Chen Y, Shen X, Fenyk-Melody J, Wu M, Ventre J, Doebber T, Fujii N (2001) Role of AMP-activated protein kinase in mechanism of metformin action. J Clin Invest 108(8):1167

Chapter 2
The Molecular Basis of Hepatic De Novo Lipogenesis in Insulin Resistance

Mengwei Zang

Abstract Humans with obesity and type 2 diabetes exhibit the classic triad of hyperinsulinemia, hyperglycemia, and hypertriglyceridemia. The paradox of selective insulin resistance in the liver, in which the gluconeogenic pathway becomes insensitive to insulin but the lipogenesis pathway remains sensitive to insulin, leads to an elevation in hepatic and plasma levels of fatty acids and triglyceride and makes detrimental contributions to the development of insulin resistance. However, the precise mechanism for selective insulin resistance remains largely unknown. AMP-activated protein kinase (AMPK) is an energy sensor that regulates metabolic homeostasis. Recently, elucidating the role of AMPK leads to surprising findings and helps identify novel downstream effectors of AMPK. Cellular and molecular biological approach and obese, diabetic mouse models are utilized to characterize that sterol regulatory element binding protein (SREBP), a family of the transcription regulator of lipid synthesis, functions as a conserved substrate of AMPK. AMPK specifically interacts with and phosphorylates SREBP-1c and SREBP-2. AMPK and its pharmacological activators, such as metformin and polyphenols, inhibit the cleavage processing of SREBP-1c and SREBP-2, decrease the nuclear translocation, and reduce the transcription of target genes involved in the biosynthesis of fatty acid, triglyceride, and cholesterol at least in part through AMPK-dependent inhibition of SREBP in hepatocytes. Strikingly, integrated inhibition of AMPK and stimulation of SREBP are implicated on hepatic lipogenesis and steatosis. In contrast, suppression of the de novo lipogenesis by AMPK in the liver results from an increase in SREBP-1 phosphorylation and a reduction in its cleavage processing and transcriptional activity in insulin resistance. These studies provide mechanistic insight into the development of potential therapeutic strategies to target the nutrient sensing AMPK-SREBP pathway for treating type 2 diabetes and related metabolic disorders.

M. Zang, M.D., Ph.D. (✉)
Department of Medicine, Whitaker Cardiovascular Institute, Boston University School of Medicine, 650 Albany Street, X725, Boston, MA 02118, USA
e-mail: mwzang1@bu.edu

J.M. Ntambi (ed.), *Hepatic De Novo Lipogenesis and Regulation of Metabolism*,
DOI 10.1007/978-3-319-25065-6_2

Keywords Fatty acid synthesis • Hepatic lipogenesis • Sterol regulatory element binding protein 1 (SREBP-1) • AMP-activated protein kinase (AMPK) Non-alcoholic fatty liver disease • Hepatic selective insulin resistance • Type 2 diabetes • Obesity

1 Introduction

In mammals, the liver is the principle organ responsible for the regulation of glucose and lipid metabolism. Fatty acids are substrates for cellular metabolism and are essential components of cellular membranes. In cellular metabolism, sources of fatty acids are exogenously-derived (dietary) fatty acids and endogenously-synthesized fatty acids.

The biosynthesis of endogenous fatty acids, known as de novo lipogenesis, is the metabolic pathway that converts excess carbohydrates (glucose) into fatty acids, which are ultimately esterified with glycerol 3-phosphate to form triglyceride in the liver (Fig. 2.1). During the de novo lipogenic process, glucose is first converted to pyruvate, which enters the tricarboxylic acid (TCA) cycle in the mitochondria to yield citrate. Citrate is then transported into the cytosol and broken down by ATP citrate lyase to yield acetyl-CoA and oxaloacetate. In the rate-limiting step in the lipogenesis pathway, the conversion of acetyl-CoA to malonyl-CoA is catalyzed by acetyl-CoA carboxylase (ACC). Fatty acid synthase (FAS) can synthesize long-chain fatty acids by using acetyl-CoA as a primer, malonyl-CoA as a two-carbon donor, and NADPH as a reducing equivalent. Palmitic acid (C16:0), the predominant fatty acid, is generated by FAS. Palmitic acid is desaturated by stearoyl-CoA desaturase-1 (SCD-1) to produce palmitoleic acid or it can be elongated to yield stearic acid (C18:0). SCD-1 catalyzes the conversion of stearoyl-CoA to oleoyl-CoA, which is a major substrate for triglyceride synthesis. Oleic acid (C18:1) is formed as a result of desaturation of stearic acid and it is thought to be the end product of de novo fatty acid synthesis (Dentin et al. 2005).

The activity of the lipogenic pathway is regulated upon nutritional conditions (Kim et al. 1998). A diet rich in carbohydrates stimulates the lipogenic pathway in the liver, whereas starvation decreases hepatic lipogenesis. Activities of lipogenic enzymes are controlled by post-translational and transcriptional mechanisms. It is established that the transcription of lipogenic enzymes is regulated by insulin and glucose (Hasty et al. 2000; Foretz et al. 1999a). Sterol regulatory element binding protein-1c (SREBP-1c) is implicated in the transcriptional regulation of multiple lipogenic enzymes such as acetyl-CoA carboxylase (ACC), fatty acid synthase (FAS), and stearoyl-CoA desaturase-1 (SCD-1). Aberrant activation of hepatic de novo lipogenesis is observed in obesity-induced insulin resistance in animals. In this chapter, we discuss nutrient sensing mechanisms and signaling crosstalk responsible for the regulation of de novo lipogenesis in the liver and discuss how these nutrient sensing pathways become deregulated in metabolic disease.

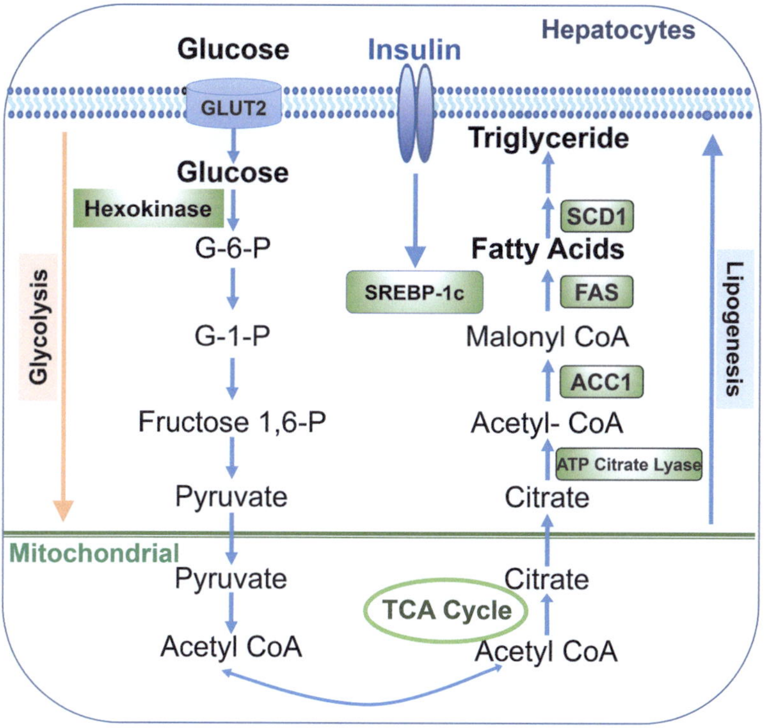

Fig. 2.1 De novo lipogenesis is regulated by insulin and glucose in hepatocytes. Metabolic flux in the liver reflects the activation of major pathways including glycolysis and lipogenesis. Acetyl-CoA is produced from glucose through a glycolytic process; it can be used as a substrate for fatty acid synthesis. The major function of glycolysis in the liver provides carbons from glucose for promoting de novo lipogenesis. The transcription of SREBP-1c and its target genes involving lipogenic enzymes is induced by high concentrations of glucose and insulin

2 Mammalian AMPK

2.1 The Regulation and Function of AMPK

AMP-activated protein kinase (AMPK) is an intracellular energy and nutrient sensor that has been implicated in the regulation of glucose and lipid homeostasis (Kahn et al. 2005; Hardie et al. 2012). AMPK serine/threonine protein kinase is a heterotrimeric protein complex that consists of α catalytic subunit and β and γ regulatory subunits (Hardie et al. 2012). Phosphorylation of a conserved Thr-172 site within the activation loop of the kinase domain of the α subunit is required for AMPK activity. The regulation of AMPK is complex: it involves allosteric activation by elevated AMP or ADP concentrations; and AMPK activity is increased by ATP depletion. AMPK activity can also be regulated by upstream kinases such as

LKB1, CaMKKβ, and TAK1 (Hardie et al. 2012; Kahn et al. 2005; Hawley et al. 2005; Momcilovic et al. 2006). The protein kinase LKB1, the Peutz-Jeghers syndrome tumor-suppressor, can directly phosphorylate and activate AMPK (Shaw et al. 2004; Sakamoto et al. 2004; Lizcano et al. 2004). Both LKB1 and CaMKKβ phosphorylate the same residue (Thr-172) on the α subunit of AMPK. AMPK is activated by LKB1 in response to increased intracellular levels of AMP or ADP and depleted intracellular levels of ATP. Binding of AMP to the γ subunit of AMPK allosterically activates the kinase while preventing dephosphorylation of the α subunit of AMPK. On the other hand, AMPK is also activated by CaMKKβ in response to elevated intracellular concentrations of Ca^{2+} (Fig. 2.1).

AMPK activity is induced by physiological stimuli such as exercise and adiponectin. AMPK is also activated by pharmacological agents such as metformin and xenobiotics. Additionally, AMPK can be activated by metabolic stresses that inhibit mitochondrial ATP production or accelerate ATP consumption. In numerous systems, AMPK serves as a protective response to energy stress. Once activated, AMPK restores cellular energy balance by switching off ATP-consuming anabolic pathways including the synthesis of fatty acid, triglyceride and cholesterol as well as by switching on ATP-generating catabolic pathways, such as fatty acid oxidation and lipolysis (Fig. 2.2). Notably, AMPK is thought to be a checkpoint that maintains energy balance in both cells and organisms. AMPK also plays a role in the regulation of whole-body energy metabolism through decreased energy production and increased energy expenditure. Specifically, AMPK can increase whole-body energy expenditure via actions on the hypothalamus. However, the dysregulation of AMPK is implicated in the pathogenesis of diabetes as well as its related metabolic disorders and some cancers.

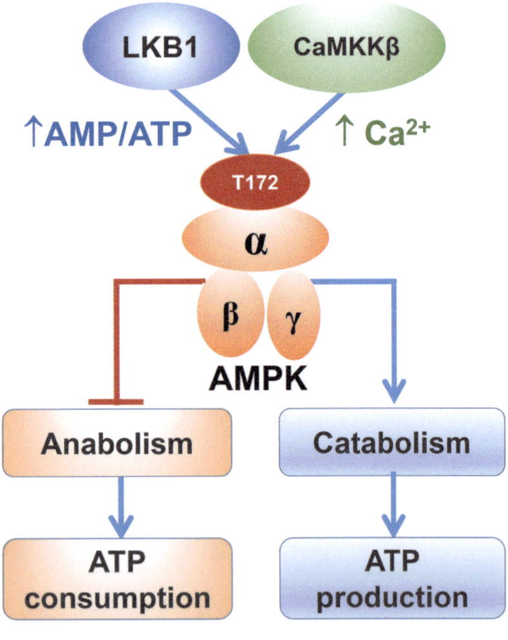

Fig. 2.2 AMPK serves as a master energy sensor that controls metabolic switch

2.2 *AMPK Activators*

Pharmacological activation of AMPK by metformin and other compounds holds promising application that may improve metabolic abnormalities associated with obesity and type 2 diabetes. Given the fact that the correction of the deregulated LKB1/AMPK pathway can reduce the Warburg effect in cancer (Faubert et al. 2013), these studies suggest that AMPK has the potential as a drug target for cancer prevention and treatment (Luo et al. 2010; Shackelford and Shaw 2009). Activation of AMPK by naturally occurring compounds also prevents the development of chronic diseases related to aging (Martin-Montalvo et al. 2013). These findings raise the possibility of metformin-based interventions that extend lifespan and promote healthy aging.

2.2.1 AMPK and Metformin

Type 2 diabetes is associated with insulin resistance, accompanied by elevated hepatic glucose production, hyperglycemia, and hyperlipidemia. Since the 1950s, one of the few classes of therapeutics effective in reducing hepatic glucose production has been the biguanides (Viollet et al. 2012). Biguanide class of antidiabetic drugs contains two linked guanidine rings that are originally derived from galegine (isoamylene guanidine), a guanidine derivative found in the French lilac Galega officinalis. The biguanide drugs include phenformin and metformin (N, N-dimethylbiguanide). Particularly, metformin is the most widely used antidiabetic agent and is currently recommended as the first line drug for patients with type 2 diabetes worldwide.

Metformin is an oral antidiabetic drug that improves systemic insulin sensitivity with reduced plasma glucose levels in patients with type 2 diabetes. Metformin may increase hepatic insulin sensitivity through the upregulation of insulin signaling by increased tyrosine phosphorylation of the insulin receptor and insulin receptor substrate 1, leading to reduced hepatic gluconeogenesis and glucose output (Viollet et al. 2012).

Beyond its effect on glucose metabolism, metformin has been reported to improve fatty liver and cardiovascular complications associated with type 2 diabetes. Metformin has also emerged as an adjunct treatment for cancer due to its role in cancer metabolism. However, the precise mechanism by which metformin lowers lipids is not fully understood. To explore the mechanism of metformin action, an in vitro cell model of high glucose-induced insulin resistance and triglyceride accumulation is initially developed (Zang et al. 2004). An in vitro insulin-resistant state, as characterized by a decrease in insulin-stimulated phosphorylation of Akt and its downstream effector, glycogen synthase kinase 3α/β in hepatocytes, is induced by high concentrations of glucose (30 mm), which mimics an in vivo hyperglycemia state. Under these conditions, metformin dose- and time-dependently increases phosphorylation of AMPK at Thr-172, the major stimulatory phosphorylation site of the AMPK α subunit, and its downstream target acetyl-CoA carboxylase 1 (ACC1)at Ser-79 and ACC2 at Ser-212, major inhibitory phosphorylation sites of ACC. A decrease in AMPK activity is correlated with an elevation in hepatocellular lipid content in an insulin-resistant state. These results indicate that the inhibition of AMPK may represent a critical mechanism for hepatocellular lipid accumulation associated with insulin resistance (Zang et al. 2004). Strikingly, the impairment of AMPK and accumulation of lipids caused by high glucose are prevented by metformin treatment. The protective effects of metformin are mimicked by overexpression of the constitutively active form of AMPKα and blocked by expressing the kinase-inactive form of AMPKα. These studies provide biochemical evidence that while AMPK activation is required for the lipid-lowering effect of metformin in insulin-resistant hepatocytes, it is up for debate whether the glucose-lowering effect of metformin is dependent on AMPK (Zhou et al. 2001; Shaw et al. 2005; Foretz et al. 2010; Fullerton et al. 2013; Shaw 2013), Given that the salutary effect of metformin on hepatocellular lipid accumulation is primarily mediated by AMPK in hepatocytes exposed to high glucose, it suggests that this cell model not only provides a useful tool to screen or develop potential therapeutic agents to target AMPK for the treatment of fatty liver and insulin resistance, but also enables us to seek new drug targets for these diseases.

Previous studies have shown that AMPK activation by either AICAR or metformin stimulates fatty acid oxidation, because AICAR reduces [^{14}C]-oleate and [^{3}H]-glycerol incorporation into triacylglycerol in rat hepatocytes (Zhou et al. 2001; Muoio et al. 1999). ACC, the major downstream target of AMPK, is a key enzyme that catalyzes the conversion of acetyl-CoA to malonyl-CoA. Malonyl-CoA serves as a potent inhibitor of carnitine palmitoyltransferase 1 α (CPT-1α), which is the rate-limiting enzyme for mitochondrial fatty acid oxidation, as evidenced by increased fatty acid oxidation and leanness in mice deficient in ACC2 (Abu-Elheiga et al. 2001). On the other hand, malonyl-CoA produced by ACC acts as the initial substrate for fatty acid biosynthesis. Interestingly, the ability of metformin to lower triglyceride levels coincides with an increase in inhibitory phosphorylation of ACC1 and ACC2 by AMPK in vitro. The lipid-lowering effect of metformin can be explained by increased fatty acid oxidation and decreased fatty acid synthesis possibly in part through phosphorylation and inactivation of ACC1 and ACC2 (Zang et al. 2004). Recent studies from the Steinberg and Kemp group confirm these early findings in a novel mouse model with alanine knock-in mutations in both ACC1 at Ser79 and ACC2 at Ser212 (ACC double knock-in, ACC-DKI) (Fullerton et al. 2013). Compared to wild-type mice, the ACC-DKI mice fed a control diet display a complete loss of phosphorylation of ACC1 and ACC2 as well as have lowered fatty acid oxidation and elevated lipogenesis. Strikingly, ACC-DKI mice challenged with high fat feeding are refractory to the lipid-lowering and insulin-sensitizing effects of metformin. Taken together, these in vitro and in vivo findings illustrate that AMPK-dependent phosphorylation and inactivation of ACC at least partially contributes to metformin-induced improvements in insulin resistance in the setting of obesity (Fullerton et al. 2013).

Glucose production in the liver is essential for providing a substrate for the glucose utilization and survival of critical tissues, such as the brain and heart, during prolonged fasting. The inability of insulin to suppress hepatic glucose output contributes to the pathological process of hyperglycemia in type 2 diabetes. Over a decade ago, metformin was demonstrated to activate AMPK by increasing phosphorylation of the AMPK catalytic α subunit at Thr-172 in primary hepatocytes (Zhou et al. 2001; Zang et al. 2004; Kahn et al. 2005). A subsequent study showed that hepatic knockout of LKB1, the upstream kinase for AMPK, abolishes metformin-mediated suppression of hepatic glucose production. These studies also indicate that LKB1 mediates hepatic AMPK activation and therapeutic effects of metformin in mice (Shaw et al. 2005). Recent studies provide a novel mechanism by which metformin antagonizes the action of glucagon and thus reduces fasting glucose levels. Metformin is also shown to reduce levels of cyclic AMP and protein kinase A (PKA) activity that abrogates phosphorylation of critical protein targets of PKA and that blocks glucagon-dependent glucose output in hepatocytes (Miller et al. 2013).

2.2.2 AMPK and Polyphenols

Many of polyphenols are widely reported to have a favorable effect on type 2 diabetes, dyslipidemia, and cardiovascular disease in animal models with obesity (Bradamante et al. 2004; Vita 2005), but their mechanism(s) remain a mystery, limiting their therapeutic potential. In addition to galegine, a number of natural polyphenolic products—many derived from plants used as herbal medicines in Asian countries—are reported to activate AMPK. These include resveratrol from red grapes, quercetin present in many fruits and vegetables, berberine from *Coptis chinensis* (used in the Chinese herbal medicine Huanglian), epigallocatechin gallate from green tea, theaflavin from black tea, ginsenoside from *Panax ginseng*, and curcumin from *Curcuma longa* (Manach et al. 2004). Resveratrol, one of natural polyphenols, has multiple beneficial activities similar to those associated with caloric restriction (CR), such as increased life span and delayed onset of diseases associated with aging (Howitz et al. 2003). Elucidating the role of AMPK in diabetes makes some surprising discoveries and identifies novel agents that directly or indirectly activate AMPK (Kahn et al. 2005; Steinberg and Kemp 2009; Zhang et al. 2009). In vitro studies demonstrate that a synthetic polyphenol, S17834, strongly and persistently stimulates AMPK phosphorylation and activity at concentrations 50–200 times lower than those of AMPK-activating compounds, 5-aminoimidazole-4-carboxamide-1-β-d-ribofuranoside (AICAR) or metformin (Zang et al. 2006). Natural polyphenols such as resveratrol, a key component in red wine, have similar but less potent effects on hepatic AMPK activity. Mechanistic studies demonstrate that the activation of AMPK by polyphenols is dependent on LKB1, but not on CaMKKβ (Hou et al. 2008). These findings suggest that LKB1 functions as an upstream regulator for AMPK signaling in response to polyphenols. As a consequence, S17834 and resveratrol suppress hepatocellular lipid accumulation caused by high glucose largely through an AMPK-dependent mechanism (Zang et al. 2006). Further animal studies show decreased phosphorylation and activity of hepatic AMPK, elevated hepatic lipids, increased hyperlipidemia, and accelerated aortic atherosclerosis in type 1 diabetic low density lipoprotein receptor (LDLR) knockout mice. In contrast, treatment with S17834 in these animals prevents impaired hepatic AMPK and elevated hepatic and serum lipids and thereby suppresses the acceleration of atherosclerosis caused by diabetes. These studies define that AMPK activation may represent a novel molecular mechanism by which polyphenols, like S17834 and resveratrol, attenuate the deleterious effect of diabetic milieu on hepatic steatosis, hyperlipidemia, and atherogenesis (Zang et al. 2006).

These early and important findings have been confirmed by several independent groups. Results from Sinclair's group indicate that resveratrol activates hepatic AMPK, improves insulin sensitivity, and modulates known longevity pathways (Baur et al. 2006). By using the mice deficient in the catalytic subunit of AMPK α1 or α2, Chung's group indicates that the effect of resveratrol on insulin sensitivity, glucose tolerance, and physical endurance are diminished in the absence of either AMPK α1 or α2. Consistent with this, expression of genes involved in mitochondrial biogenesis was not induced by resveratrol in AMPK-deficient mice (Um et al. 2010). Strikingly, a clinical trial study shows calorie restriction-like effects of resveratrol supplementation on energy metabolism and on the metabolic profile in obese humans, where healthy subjects and obese patients are given with either a placebo or 150 mg/day resveratrol in a randomized double-blind crossover study for 30 days. Mechanistically, administration of resveratrol to obese patients leads to activated AMPK, increased SIRT1 and PGC-1α protein levels, and improved mitochondrial respiration in skeletal muscle (Timmers et al. 2011). Taken together, these animal experiments and clinical studies provide strong biochemical evidence that AMPK is the central target for the metabolic effects of resveratrol.

Berberine, a natural plant product, activates AMPK and benefits metabolic disorder in obese, insulin-resistant mice (Lee et al. 2006; Turner et al. 2008). Berberine treatment results in increased AMPK activity in 3T3-L1 adipocytes and L6 myotubes. Clinical trials of berberine in newly diagnosed type 2 diabetic subjects reveal favorable effects on plasma glucose, lipids, and HbA_{1c}, although plasma berberine levels and AMPK activity are not assessed (Zhang et al. 2008; Yin et al. 2008).

3 Mammalian SREBP

3.1 The Structure of SREBP

Sterol regulatory element binding protein (SREBP) is a key regulator of intracellular lipid metabolism that belongs to the basic helix-loop-helix–leucine zipper (bHLH-Zip)family of transcription factors (Brown and Goldstein 1997, 2009; Horton et al. 2002; Goldstein and Brown 2008). SREBP differs from other bHLH-Zip proteins because it is synthesized as an inactive precursor bound to the endoplasmic reticulum (ER). The SREBP precursor with about 1150 amino acids consists of three domains: (i) an NH_2-terminal domain of about 480 amino acids that contains the transactivation domain—a region of the bHLH-Zip domain for DNA binding and dimerization; (ii) two hydrophobic transmembrane-spanning segments interrupted by a short loop of about 30 amino acids that projects into the lumen of the ER; and (iii) a COOH-terminal domain of about 590 amino acids—a regulatory region that binds to the COOH-terminal domain of an escort protein known as SREBP cleavage activating protein (SCAP). The SREBP family has three isoforms: SREBP-1a, SREBP-1c, and SREBP-2. SREBP-1c is also referred to as adipocyte determination and differentiation dependent factor 1/ADD1. SREBP-1a and SREBP-1c are derived from a single gene by using alternative transcription start sites that produce alternate forms of the first exon. SREBP-1c is expressed in most adult tissues, with especially high expression in the liver; it is considered to be more physiologically relevant. Interestingly, SREBP-1c and SREBP-2 are highly expressed in the liver and also expressed in other tissues of adult animals. SREBP-1a appears to be constitutively expressed at low levels in almost all tissues of adult animals.

3.2 *The Function of SREBP on Lipid Biosynthesis*

SREBP family controls lipid metabolism by stimulating expression of more than 30 genes encoding a range of enzymes required for the synthesis or uptake of fatty acids, triacylglycerol, cholesterol, and phospholipid (Brown and Goldstein 1997; Horton et al. 2002; Goldstein and Brown 2008). Three SREBP isoforms have isoform-specific functions on lipid biosynthetic processes. In vivo studies using transgenic and knockout mice demonstrate that SREBP-1c, the predominant isoform in the liver, preferentially promotes a de novo lipogenic process by activating genes involved in fatty acid and triglyceride synthetic pathways, whereas SREBP-2 primarily activates a cholesterol biosynthetic pathway by stimulating expression of genes governing cholesterol synthetic pathways and uptake. In contrast, the SREBP-1a isoform, which is highly expressed in cell lines and tissues with a high capacity for cell proliferations such as intestine and spleen (Shimomura et al. 1997), is likely involved in both pathways.

3.3 *The Regulation of SREBP Activity*

The regulation of SREBP activity occurs at posttranscriptional and transcriptional levels (Brown and Goldstein 1997; Horton et al. 2002). The sterol-sensitive feedback inhibition process is mainly involved in the regulation of SREBP-2. Accumulating evidence suggests that the SREBP-1c isoform appears to be mainly regulated by glucose and insulin.

3.3.1 The Posttranscriptional Regulation of SREBP Activity

Sterols such as cholesterol are fundamental components of cellular membranes and precursors of molecules such as steroid hormones. Cholesterol can be obtained from the diet as well as endogenously synthesized de novo. Hence, energetically demanding cholesterol biosynthetic pathway is only active when external supply and internal levels of sterols are low. Adequate sensing of internal cholesterol levels involves the posttranscriptional regulation of SREBP. The control occurs through SREBP cleavage processing and movement from the endoplasmic reticulum (ER) to the Golgi apparatus. Proteolytic activation of SREBP is controlled by sterols through two intracellular sterol sensors: SCAP and Insulin-induced gene (Insig) (Ye and DeBose-Boyd 2011). SCAP contains an NH_2-terminal domain with eight transmembrane helices and a cytosolic COOH-terminal domain that mediates the complex formation with the COOH-terminal regulatory domain of SREBP (Matsuda et al. 2001; Moon et al. 2012). Insig is a polytopic membrane protein with six transmembrane helices that binds to SCAP for the retention of the SCAP-SREBP complex in the ER membrane (Yang et al. 2002; Engelking et al. 2004; Flury et al. 2005; Rawson 2003b).

SREBP protein is initially synthesized as an inactive precursor that is inserted into ER membrane. The retention of SCAP-SREBP complex in the ER membrane is mediated by an increase in the binding of SCAP to the Insig protein and a decrease in the interaction of SCAP with the COPII vesicle-formation proteins Sar1, Sec23 and Sec24 (Rawson 2003a, b; Osborne and Espenshade 2009). Upon activation, the newly synthesized precursor form of SREBP (~125 kDa) migrates from the ER membrane to the Golgi apparatus where SREBP precursor undergoes a sequential two-step proteolytic processing mediated by the site 1 (S1P) and site 2 (S2P) proteases. This cleavage process can lead to the release of the transcriptionally active NH_2-terminal domain of the protein (~68 kDa) in the Golgi apparatus. Once the mature, active nuclear form of SREBP, designated nuclear SREBP, is translocated into the nucleus, the nuclear SREBP can bind to the sterol regulatory element (SRE) sequence (5′-TCACNCCAC-3′) present in promoters/enhancers of SREBP and its target genes that activates the transcription of SREBP-responsive genes and promotes the lipid synthetic process in the liver (Horton et al. 2002; Raghow et al. 2008; Amemiya-Kudo et al. 2000).

Since the discovery of SREBP in 1993, the molecular mode of SREBP-2 regulation is well characterized. The proteolytic processing of SREBP-2 is mainly controlled by cellular sterols through the feedback inhibition (Fig. 2.3). When intracellular sterols rise, cholesterol binds to its sensor SCAP in the ER that leads to the conformation change of SCAP and promotes the binding of SCAP to the ER-resident protein Insig. Under this condition, the SCAP-SREBP complex, which is not incorporated into ER transport vesicles, cannot migrate to the Golgi apparatus. Because the bHLH-Zip domain cannot be released from the membrane, the transcription of SREBP-2 target genes is suppressed. When intracellular sterols decline, the precursor of SREBP-2 can be cleaved to produce the nuclear, active form of SREBP-2, which in turn induces cholesterogenic gene expression. Consequently,

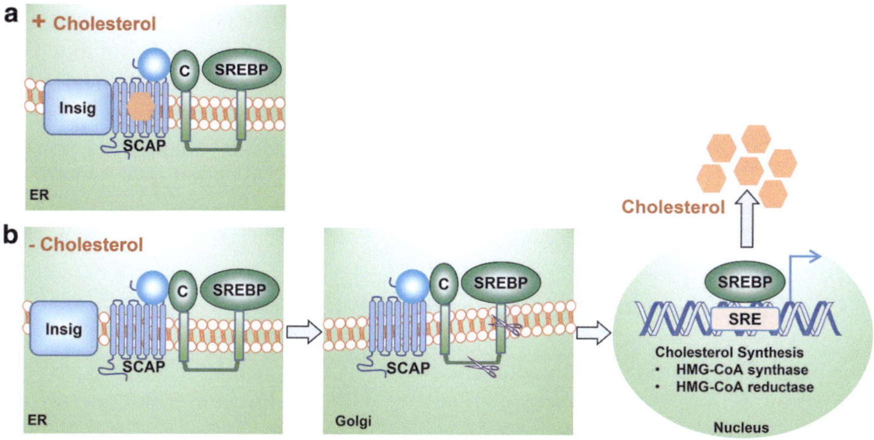

Fig. 2.3 The nutrient regulation of SREBP in the liver. (**a**) In the presence of cholesterol, cholesterol binds to the intracellular sterol sensor SCAP. The SCAP–SREBP complex binds to the Insig protein in the ER membrane and remains anchored in the ER. (**b**) In the absence of cholesterol, the SCAP–SREBP complex does not interact with the Insig protein, and the complex traffics to the Golgi where the cytoplasmic tail of SREBP is released by the proteolytic cleavage. For instance, the nuclear, active form of SREBP-2 triggers a transcriptional program in the nucleus through its binding to the SRE motif on promoters of its target genes such as HMG-CoA synthase and HMG-CoA reductase

newly synthesized cholesterol leads to the feedback inhibition of SREBP-2 activity in order to maintain cholesterol homeostasis (Nohturfft et al. 2000; Yang et al. 2002). Additionally, sterol-regulated ubiquitination and degradation of Insig-1 create a convergent mechanism for the negative feedback control of SREBP-2 activity and its function on cholesterol synthesis and uptake (Gong et al. 2006).

SREBP-1 is likely regulated by fatty acids through the negative feedback regulation. In mice, polyunsaturated fatty acids (PUFA)inhibit mRNA expression of hepatic SREBP-1c, but they are not affect SREBP-1a gene expression (Yoshikawa et al. 2002). The transcription of SREBP-1c is antagonized by PUFA through the effect of liver X receptor (LXR), because two LXR-responsive elements (LXREs) are identified on the promoter of the SREBP-1c gene, but not on that of the SREBP-1a gene (Repa et al. 2000). Strikingly, unsaturated fatty acids also inhibit proteolytic activation of SREBP-1 (Hannah et al. 2001). Further studies indicate that unsaturated fatty acids can have the greatest inhibitory effect on SREBP-1 (Hannah et al. 2001), whereas sterols have the greatest inhibitory effects on SREBP-2.

Mature nuclear SREBP is highly unstable owing to its ubiquitin-dependent degradation. Results from Ericsson's group indicate that the nuclear form of SREBP is rapidly degraded by the ubiquitin-proteasome pathway through the action of an E3 ubiquitin ligase Fbw7 (Sundqvist et al. 2005). Fbw7 negatively regulates the stability and function of nuclear SREBP-1 and SREBP-2 by promoting the ubiquitination and proteasome-mediated degradation via GSK3 dependent-phosphorylation of SREBP (Sundqvist et al. 2005). Because GSK3 activity is shown to be inhibited by insulin signaling, it suggests this non-transcriptional mechanism may be involved in insulin-induced SREBP. Moreover, the coactivator PGC-1β activity can bind to SREBP-1c, which can explain SREBP-induced lipogenesis in response to a fat-rich diet (Lin et al. 2005). Importantly, these factors permit the modulation of SREBP-1c expression independently of sterol-regulated proteolytic processing of SREBP-1.

3.3.2 The Transcriptional Regulation of SREBP Activity

The central role of SREBP in controlling lipid synthesis has recently been highlighted by the multiple inputs to regulate SREBP activity from different nutrient sensing pathways. In addition to the posttranscriptional regulation of SREBP, the transcriptional regulation of the SREBP is more complex (Eberle et al. 2004; Osborne and Espenshade 2009). SREBP-1a appears to be constitutively expressed at low levels in most tissues. SREBP-1c and SREBP-2 in the liver are transcriptionally regulated through at least three mechanisms.

In the first mechanism, a feed-forward transcriptional regulation of SREBP-1c or SREBP-2 acts through the binding of nuclear SREBP to the SRE motif (5′-TCACNCCAC-3′) present in the promoter/enhancer region of each gene (Dif et al. 2006). Elevated nuclear SREBP-1c and SREBP-2 activate the transcription of their own genes via the auto-regulatory loop (Horton et al. 2002). The feed-forward stimulation may explain increased hepatic mRNA levels of SREBP-1c and SREBP-2 observed in transgenic animals with overexpression of nuclear SREBP-1a and in obesity-induced insulin-resistant mice (Horton et al. 2003; Li et al. 2011). In contrast, when the nuclear SREBP declines, a secondary decline in mRNA expression of SREBP-1c and SREBP-2 is also seen in the liver of S1P- and SCAP-deficient animals (Yang et al. 2001; Matsuda et al. 2001).

In the second mechanism, the transcription of SREBP-1c is selectively stimulated by nutrients and hormones (Horton et al. 2002; Brown and Goldstein 2008). This concept can be illustrated by examining the effects of nutrient availability and deprivation on SREBP-1c. Prolonged fasting or food deprivation leads to a decrease in plasma glucose and insulin levels and an increase in plasma-free fatty acid concentrations. Hyperglycemia stimulates lipogenesis via three possible mechanisms: (i) glucose itself could be a substrate for the de novo lipogenesis in hepatocytes, because glucose promotes fatty acid synthesis through its conversion to acetyl-CoA via a glycolytic pathway; (ii) glucose induces the expression of lipogenic genes, the mechanisms of which are explained by activating SREBP-1c; and (iii) glucose increases lipogenesis by stimulating the release of insulin and inhibiting the release of glucagon from the pancreas. The metabolic adaption to fasting is associated with changes in plasma hormone concentrations, such as a decrease in plasma insulin levels and an increase in plasma glucagon levels. This notion is supported by the observation that the transcription of SREBP-1c is controlled positively by insulin and negatively by glucagon and cyclic AMP in primary hepatocytes, establishing a link between this transcription factor and carbohydrate availability (Foretz et al. 1999b)

In the third mechanism, the transcription of SREBP-1c is selectively regulated through the cooperation of nuclear receptors with DNA-binding properties. Nuclear receptors, such as liver X receptor (LXR) and retinoid X receptor (RXR), function as a heterodimer to selectively upregulate SREBP-1c (Liang et al. 2002; Dif et al. 2006). The induction of SREBP-1c by LXR is supported by the observation that expression of hepatic SREBP-1c and fatty acid synthetic genes is increased by synthetic LXR agonists and these effects are reduced in LXRα- and LXRβ-deficient mice (Repa et al. 2000). Similarly, critical fatty acid synthetic genes are not upregulated by LXR agonists in SREBP-1c knockout mice (Liang et al. 2002). One major function of the liver is to convert excess carbohydrates to fatty acids for storage as triglycerides. It has been shown that insulin stimulates fatty acid synthesis in response to excess carbohydrates and that the lipogenic effect of insulin in the liver is mediated by SREBP-1c. Although a complete understanding of the regulation of the SREBP-1c by insulin requires further investigation, a study shows that the stimulatory effect of insulin on SREBP-1c transcription is mediated by LXR and liver X receptor (LXR)-responsive elements (LXREs) in the promoter region of SREBP-1c, but not that of SREBP-1a (Okazaki et al. 2010; Chen et al. 2004). Insulin activates the transcription of SREBP-1c promoter primarily by increasing LXR activity, possibly through production of a ligand that activates LXR or their heterodimerizing partner, the retinoid X receptor. Interestingly, polyunsaturated fatty acids (PUFA) inhibit SREBP-1c and fatty acid synthesis by antagonizing LXR-dependent activation of SREBP-1c (Yoshikawa et al. 2002).

The unique regulation and functions of each SREBP isoform facilitate the coordinate control of lipid metabolism; however, the signaling pathways that specifically regulate each SREBP isoform are not fully understood.

The dysregulation of SREBP-1c has been implicated in the pathogenesis of hepatic steatosis and dyslipidemia in type 2 diabetes (Raghow et al. 2008). This transcription factor is identified as an attractive target for the development of new pharmaceutical interventions for metabolic disorders such as hypertriglyceridemia and obesity (Tang et al. 2011). Elucidating this mechanism for the regulation of SREBP will be fundamental to understanding the molecular basis of de novo lipogenesis and selective insulin resistance.

4 Role of AMPK-SREBP Nutrient Signaling in the Regulation of Hepatic Lipid Synthesis

Previous studies show that there is an inverse correlation between AMPK activity and SREBP-1c mRNA expression in livers of mice following re-feeding (Zhou et al. 2001; Foretz et al. 2005). However, little is known about how SREBP is downregulated by AMPK and how this regulation plays a role in the regulation of liver and systemic lipid homeostasis and slows the progression of insulin resistance and type 2 diabetes. Another important but unanswered question is how elevated hepatocellular cholesterol levels, a condition that is known to cause the negative feedback regulation of SREBP in physiological stimuli such as refeeding, lead to the aberrant activation of SREBP in pathological conditions such as insulin resistance and type 2 diabetes.

4.1 AMPK Suppresses the Transcriptional Activity of SREBP-1 and SREBP-2

Recent studies demonstrate that hepatic AMPK activity, as determined by phosphorylation of AMPK and ACC, is repressed in insulin-resistant mice fed on a high fat, high sucrose diet. This impairment is reversed by treatment with S17834, a polyphonic AMPK activator (Li et al. 2011), which is consistent with increased hepatic AMPK activity seen in S17834-treated type 1 diabetic mice (Zang et al. 2006). Importantly, this study shows that AMPK activation by S17834 leads to a decrease in proteolytic processing of SREBP-1c and in mRNA expression of SREBP-1c in the liver of insulin-resistant mice, without affecting gene expression of SREBP-1a (Li et al. 2011). Consistently, mRNA expression of SREBP-1c target genes including acetyl-CoA carboxylase (ACC1), fatty acid synthase (FAS), and stearoyl-CoA desaturase-1 (SCD1) is suppressed, suggesting decreased rates of fatty acid and triglyceride synthesis in the liver. As a consequence, hepatic and plasma triglyceride concentrations are significantly reduced by the AMPK activator. These in vivo studies suggest that increasing AMPK activity reduces hepatic fatty acid and triglyceride synthesis and ameliorates hepatic steatosis by inhibiting SREBP-1c-dependent lipogenesis (Li et al. 2011).

SREBP-2 has been implicated in the regulation of cholesterol synthesis and uptake in the liver. SREBP-2 drives a cholesterol synthetic process by stimulating gene expression of cholesterol biosynthetic enzymes including rate-limited enzymes, 3’-hydroxy-methylglutaryl coenzyme A synthase (HMG-CoA synthase) and 3’-hydroxy-methylglutaryl coenzyme A reductase (HMG-CoA reductase) (Shimano et al. 1997). To further elucidate the precise mechanism by which AMPK regulates lipid homeostasis, recent studies indicate that high levels of nuclear active form and mRNA expression of hepatic SREBP-2 are associated with AMPK inhibition in obesity-induced insulin-resistant mice. Interestingly, nuclear accumulation of SREBP-2 is reduced by the AMPK activator, which correlates with a decrease in hepatic expression of SREBP-2 and its target genes, HMG-CoA synthase and HMG-CoA reductase. The suppression of SREBP-2 causes a reduction in hepatic cholesterol synthesis and content. The decrease in hepatic cholesterol synthesis in the liver is likely the major driving force leading to the lipid-lowering effect of AMPK (Li et al. 2011). Collectively, AMPK-mediated inhibition of SREBP-1 and SREBP-2 in the liver may explain the favorable effect of pharmacological activation of AMPK on obesity-induced hepatic steatosis, insulin resistance, and hyperlipidemia (Li et al. 2011).

To support the hypothesis that AMPK is an upstream regulator to inhibit SREBP-1 activity at the posttranslational level, in vitro mechanistic experiments are performed to obtain several important observations in freshly isolated mouse hepatocytes or human HepG2 hepatoma cells (Li et al. 2011). First, hepatic triglyceride content is dramatically increased in hepatocytes exposed to high glucose or high glucose plus insulin that mimic hyperglycemia and insulin resistance in vivo (Zang et al. 2004, 2006; Hou et al. 2008; Li et al. 2011). Under this condition, cleavage processing of SREBP-1 and mRNA expression of SREBP-1c are increased, whereas gene expression of SREBP-1a is rarely affected. This suggests the feed-forward stimulation of SREBP-1c caused by high concentrations of glucose and insulin. Interestingly, polyphenols and metformin, two pharmacological activators of AMPK (Zhou et al. 2001; Zang et al. 2006; Um et al. 2010), suppress nuclear translocation and accumulation of SREBP-1 and reduce gene expression of SREBP-1c, but not SREBP-1a, leading to the inhibition of a feed-forward regulation of SREBP-1c. Second, the ability of AMPK activators such as resveratrol to repress the cleavage processing of SREBP-1 and expression of lipogenic genes are largely abrogated by overexpression of a dominant negative form of AMPK, which encodes a catalytically inactive AMPKα2 bearing a mutation altering lysine 45 to arginine (K45R) in the kinase domain of AMPKα2 subunit that exerts a dominant negative effect on AMPKα1 and AMPKα2 activities (Zang et al. 2004). These results suggest that resveratrol represses SREBP activity in an AMPK-dependent manner. Third, overexpression of the constitutively active mutant of AMPK, which contains a carboxyl-terminal truncated form of AMPKα2 lacking the auto-inhibitory domain and the region interacting with β and γ regulatory subunits of AMPK (Zang et al. 2004), increases AMPK activity in hepatocytes. The constitutively active form of AMPK also decreases mRNA expression of SREBP-1c and FAS. Fourth, unlike other bHLH-Zip transcription factors that contain a well conserved arginine residue in the bHLH-Zip domain, SREBP has a conserved tyrosine residue that enables SREBP to bind on both E-boxes, like all bHLH proteins, and SRE sequences on target genes of SREBP-1c (Kim et al. 1995). To determine whether SREBP-1c is required for AMPK-medicated inhibition of the lipogenic process, a dominant negative mutant of SREBP-1c, which disrupts the binding of SREBP-1c to the SRE motif owning to a point mutation of tyrosine 320 to arginine on a carboxyl-terminal truncated SREBP-1c (1-403) (Kim et al. 1995), is used to be transduced into hepatocytes. In vitro experiments demonstrate that AMPK suppresses the transcriptional activity of FAS promoter in a SREBP-1c-dependent manner. Lastly, to precisely elucidate the mechanism by which AMPK inhibits SREBP transcription, in vitro experiments is performed using luciferase reporters encoding different lengths of human SREBP-1c promoters (Dif et al. 2006). AMPK activators decrease the transcriptional activity of the wild-type SREBP-1c promoter (−1470/+90 and −257/+90). Conversely, disruption of the SRE motif present on the SREBP-1c promoter reduces the basal transcription of SREBP-1c and prevents the effect of AMPK activators on SREBP-1c promoter activity. These results indicate that the SRE motif is responsible for AMPK inhibiting the auto-regulation of SREBP-1c. Taken together, these biochemical studies provide strong evidence that AMPK inhibits de novo lipogenesis by downregulating the cleavage processing of SREBP-1c and suppressing its feed-forward regulation in hepatocytes and in vivo insulin-resistant states (Li et al. 2011).

4.2 AMPK Interacts with and Directly Phosphorylates SREBP

Recent studies provide evidence that SREBP serves as a conserved substrate of AMPK (Li et al. 2011). The modulation of protein-protein interaction dynamically regulates protein phosphorylation, activity, subcellular localization, and stability. To delineate the mechanism by which AMPK downregulates SREBP activity, co-immunoprecipitation experiments show that the α catalytic subunit of AMPK physically interacts with the precursor and nuclear form of SREBP-1. In addition to the repression of SREBP-2 by AMPK in mouse livers, biochemical experiments also demonstrate that the AMPKα subunit physically associates with the SREBP-2 precursor (Li et al. 2011). Because the formation of the AMPK-SREBP complex is also increased by a constitutively active form of AMPK, it postulates that this interaction makes the kinase domain of the AMPKα subunit more accessible to putative substrates, SREBP-1 and SREBP-2, for the phosphorylation modification. It has been characterized that phosphorylation at Thr-172 in the activation loop of the kinase domain of the AMPKα subunit can trigger an increase in AMPK kinase activity by converting this enzyme from an inactive conformation to a catalytically active conformation (Hardie et al. 2012). AMP or ADP can bind to the AMPK γ regulatory subunit that promotes AMPK phosphorylation by upstream kinases as well as protects the AMPK enzyme against dephosphorylation by phosphatases. AMP can also cause allosteric activation of AMPK (Hardie et al. 2012). Furthermore, biochemical studies have discoverd the crystal structure of human AMPKα1 subunit and provided structure insight into that the AMPKα1 subunit as being held in an inactive conformation that renders the AMPK enzyme inactive owning to the association of the auto-inhibitory domain with the kinase domain of the AMPKα1 subunit (Chen et al. 2009). When intracellular AMP levels rise, AMP is bound to the AMPKγ subunit. Consequently, the inhibitory domain of the AMPKα1 subunit is released from its kinase domain. This disassociation leads to an active conformation of AMPK, which allows the upstream kinase such as LKB1 to phosphorylate the Thr-172 site on the kinase domain of the AMPKα subunit (Young 2009). Such a specific active conformation of the AMPKα subunit may possibly enable the kinase domain of the AMPKα subunit to be more accessible to its substrates. This possibility is supported by the fact that the active AMPKα subunit preferentially binds to and phosphorylates SREBP-1c and SREBP-2 (Li et al. 2011).

It is known that phosphorylation sites tend to be located on protein–protein binding interfaces and may modulate protein activity. Bioinformatics tools including Scansite (http://mit.scansite.edu) are used to identify candidate substrates bearing optimal AMPK motifs, in which the target serine and its critical flanking residues are conserved broadly throughout eukaryotes (Gwinn et al. 2008). To further search a consensus phosphorylation motif of AMPK on human SREBP-1c sequence, human SREBP-1c is identified to have two putative AMPK sites, Ser-336 and Ser-372 located in the N-terminal region of SREBP-1c, which match the AMPK consensus motif. Because a high degree of conservation in consensus sequences is important for AMPK phosphorylation, only one serine site of SREBP-1c—the Ser-372 on human SREBP-1c, is conserved. Strikingly, critical residues flanking Ser-372 of human SREBP-1c are highly conserved across mouse, rat, dog, bovine as well as in zebrafish (Li et al. 2011). Sequence alignments also show the evolutionary conservation of the motif sequences surrounding Ser-372 on human SREBP-1c and surrounding Ser-792 on Raptor, the best-established substrate of AMPK (Gwinn et al. 2008). The remarkable conservation in the candidate AMPK phosphorylation site on SREBP-1 may represent an AMPK target that dictates the responsiveness of AMPK across mammals.

To test whether SREBP is directly phosphorylated by AMPK, in vitro kinase assays with purified AMPK in rat livers are performed using recombinant SREBP-1 or SREBP-2 as a substrate in the presence of AMP as described previously (Inoki et al. 2006; Greer et al. 2007; Gwinn et al. 2008). Active AMPK substantially phosphorylates SREBP-1c or SREBP-2 in the presence of [γ-^{32}P]-ATP in vitro. Mutagenic studies demonstrate that a point mutation of Serine 327 to alanine (S3721A) on human SREBP-1c, but not wild-type SREBP-1c, eliminates AMPK-triggered phosphorylation of SREBP-1c. To characterize the biological importance for the phosphorylation of SREBP-1c, a specific phospho-specific antibody against Ser-372 on human SREBP-1c is generated. Active AMPK potently stimulate Ser-372 phosphorylation on wild-type SREBP-1c, but not that on the S372A mutant, indicating that the Ser-372 on SREBP-1c is a major phosphorylation target of AMPK in vitro (Li et al. 2011).

To determine the functional consequence of the Ser-327 phosphorylation of SREBP-1c, further studies have demonstrated that AMPK is necessary for the induction of phosphorylation at Ser-372 in response to pharmacological activators of AMPK in cultured cells, since the ability of metformin and polyphenols to increase the phosphorylation at Ser-372 is largely diminished by a dominant-negative AMPK form in hepatocytes (Li et al. 2011). Moreover, like phosphorylation of AMPK and ACC, phosphorylation of Ser-327 is stimulated by AMPK activators in AMPK$^{+/+}$ MEFs. In contrast, phosphorylation of SREBP-1c is diminished in cells lacking AMPKα1 and AMPKα2 subunits. These results strongly demonstrate that Ser-372 is potently phosphorylated by AMPK in vitro and in intact cells. To define whether AMPK-stimulated phosphorylation of SREBP-1c is physiologically relevant, mechanistic studies are performed using the phosphorylation-defective S372A mutant. Luciferase reporter assays show that the S372A mutant strongly diminishes the effect of AMPK activators on SREBP-1c activity. The results presented above indicate that AMPK inhibits the transcriptional activity of SREBP-1c in a Ser-372 phosphorylation-dependent manner. Furthermore, consistent with the changes in hepatic AMPK activity, hepatic phosphorylation of

SREBP-1c at Ser-372 is reduced in insulin-resistant mice, and defective AMPK-SREBP signaling is prevented by treatment with the AMPK-activating polyphenol, revealing the in vivo physiological significance of AMPK-dependent phosphorylation of SREBP-1c (Li et al. 2011). Although AMPK can cause phosphorylation of SREBP-1c at Ser-372 in intact cells, it cannot rule out that other phosphorylation sites on SREBP-1c may be involved in the metabolic action of AMPK.

An additional mechanism for understanding the role of AMPK in the regulation of cholesterol metabolism has recently emerged (Li et al. 2011). The induction of the cleavage processing and gene expression of hepatic SREBP-2 is reduced in insulin-resistant mice treated with the AMPK activator in a similar manner to that of SREBP-1. Because both SREBP-1c and SREBP-2 exert similar cleavage processing and feed-forward regulation, phosphorylation of SREBP-2 is possibly involved in AMPK actions. Further work is required for the identification of potential phosphorylation sites present on SREBP-2 and confirm that SREBP-2 is a direct target of AMPK in intact cells. While previous studies demonstrate that AMPK directly phosphorylates acetyl-CoA carboxylase and HMG-CoA reductase, the rate-limited enzymes that regulate fatty acid and cholesterol biosynthesis, the phosphorylation regulation of SREBP by AMPK highlights the biological importance for the multilayered control of hepatic lipid biosynthesis under conditions of metabolic stresses. Therefore, the identification of the nutrient signaling crosstalk between AMPK and SREBP represents an important advance in our understanding of the molecular mechanisms linking hepatic lipogenesis to insulin resistance, hepatic steatosis, and hyperlipidemia (Fig. 2.4) (Li et al. 2011).

4.3 Role of Other Nutrient Sensors in the Regulation of Hepatic SREBP Activity

The molecular mechanisms linking obesity to insulin resistance are complex. The increase in hepatic glucose production in the liver, is possibly attributed to the inhibition of insulin-mediated suppression of gluconeogenesis. Paradoxically, insulin-induced nuclear accumulation and expression of hepatic SREBP-1c contribute substantially to hepatic lipogenesis observed in obesity-induced type 2 diabetic mice (Laplante and Sabatini 2010). However, the molecular mechanisms underlying hepatic selective insulin resistance have not yet been characterized. Brown, Goldstein and co-workers have reported that the mammalian target of rapamycin complex 1 (mTORC1)plays an essential role in mediating the ability of hyperinsulinemia to stimulate SREBP-1c and lipogenesis in type 2 diabetes (Li et al. 2010). Interestingly, inhibition of mTORC1 by rapamycin reduces the mRNA expression of SREBP-1c in primary hepatocytes, whereas hepatic gluconeogenesis is unaffected (Li et al. 2010). The role of mTORC1 in controlling insulin-regulated hepatic lipogenesis, but not in insulin-mediated suppression of gluconeogenesis, provides a molecule basis for understanding the selective nature of hepatic insulin resistance.

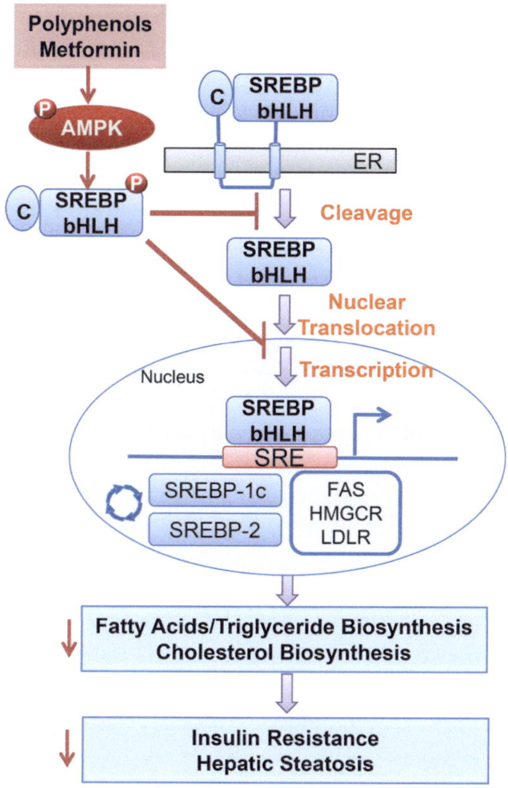

Fig. 2.4 A proposed model of the phosphorylation regulation of SREBP by AMPK in the liver: potential therapeutic implication in hepatic steatosis and insulin resistance. AMPKα subunit binds to precursors of SREBP-1c and SREBP-2 and makes them better substrates for phosphorylation. Phosphorylation of SREBP may trigger a conformation change of SREBP that inhibits its proteolytic cleavage and the release of the transcriptionally active N-terminal bHLH-Zip domain and thereby reduces its nuclear translocation in hepatocytes. Consequently, AMPK suppresses the feed-forward activation of SREBP-1c and SREBP-2 and represses the transcription of their target lipogenic genes at least in part by reduced binding of SREBP to SRE in the promoters of SREBP-responsive genes including fatty acid synthase (FAS), HMG CoA reductase (HMGCR), and LDLR. AMPK-dependent phosphorylation and repression of SREBP in the liver may represent a novel mechanism for AMPK activators, such as polyphenols and metformin, to protect against hepatic steatosis and type 2 diabetes. Figure modified from Li et al, *Cell Metabolism*, 2011, 13(4): 376–388

Recently, Sabatin's group has reported that mTORC1-induced nuclear translocation of Lipin 1, a phosphatidic acid phosphatase, controls the transcriptional activity of SREBP-1 in hepatocytes and that activation of mTORC1-Lipin 1 signaling contributes to diet-induced hepatic steatosis and hypercholesterolemia in mice (Peterson et al. 2011).

Selective insulin resistance is a very challenging question in anti-diabetic therapies, because treatment of type 2 diabetes patients with large doses of insulin can overwhelm the insulin resistance and control the blood glucose levels, but insulin further enhances hepatic lipid synthesis and secretion while increasing lipotoxicity. To better understand the mechanism of selective insulin resistance, recent studies suggest that dysregulation of AMPK signaling may contribute to the induction of hepatic SREBP-1c and lipogenesis, since the transcription of SREBP-1c and FAS is upregulated in cells lacking AMPKα1 and α2 (Li et al. 2011). Given the demonstration that AMPK is a key kinase that negatively regulates mTORC1 (Inoki et al. 2006; Gwinn et al. 2008), it would be important to evaluate whether SREBP-1c processing is affected by AMPK through the inhibition of mTORC1 in hepatocytes. Alternatively, the possibility that mTORC1 could interfere with the action of LXR on SREBP-1c gene is considered because activation of AMPK suppresses SREBP-1c activity in a LXR-dependent fashion (Yap et al. 2011).

Kemper's group have reported that SREBP-1c is an in vivo target of SIRT1 (Ponugoti et al. 2010). SIRT1 interaction with SREBP-1c is increased by fasting. Consistently, decreased SREBP-1c acetylation levels in mouse livers are associated with decreased lipogenic gene expression during fasting. In vivo knockdown of hepatic SIRT1 increases the acetylation of SREBP-1c, accompanied with elevated lipogenic gene expression. Tandem mass spectrometry and mutagenesis studies further reveal that SREBP-1c is acetylated by p300 at Lys-289 and Lys-309. Mechanistic studies using acetylation-defective mutants indicate that SIRT1 deacetylates and inhibits SREBP-1c transactivation by decreasing its stability and its occupancy at the lipogenic genes. Näär's group has also reported a conserved role of SIRT1 orthologs in fasting-dependent inhibition of SREBP. These results demonstrate that deacetylation and degradation of nuclear SREBP-1c and SREBP-2 by SIRT1 occur in response to fasting, which in turn inhibits hepatic lipogenic and cholesterol synthetic processes (Walker et al. 2010). It has been shown that SIRT1 inhibits fatty acid synthesis and lipid accumulation caused by high glucose through activation of AMPK in cultured hepatocytes (Hou et al. 2008). According to the results presented above, AMPK inhibits SREBP-1c activity in a phosphorylation-dependent fashion. Further work is of interest to determine whether the effect of SIRT1 on SREBP-1c is possibly mediated by AMPK-dependent phosphorylation of SREBP-1c.

AMPK phosphorylates SREBP-1c at Ser-372, which is different from the residue regulated by other protein kinases. For instance, protein kinase A has been previously identified to attenuate the binding of nuclear SREBP-1c to SRE-containing promoters of SREBP-1c-responsive genes through phosphorylation of nuclear SREBP-1c at Ser-314, without altered its cleavage processing (Lu and Shyy 2006). GSK3 directly phosphorylates nuclear SREBP-1a at Thr-426/Thr-420 and SREBP-2 at Ser-433 that mediate Fbw7-induced ubiquitination and degradation of their nuclear forms (Sundqvist et al. 2005). Recent study provides additional insight that unlike the effect of these kinases on the nuclear form of SREBP-1c, AMPK binding to SREBP-1c precursor triggers the Ser-327 phosphorylation. The phosphorylation resistant S372A mutant abrogates the ability of AMPK to inhibit SREBP-1c cleavage and nuclear translocation (Li et al. 2011).

Recent studies indicate that hepatic AMPK regulates triglyceride and cholesterol metabolism at least partially through the downregulation of proteolytic processing of SREBP-1c and SREBP-2 in obese, insulin-resistant mice (Zang et al. 2006).

These animals share some features with SCAP knockout mice (Matsuda et al. 2001) and with transgenic mice overexpressing Insig-1 in the liver (Engelking et al. 2004) where nuclear amounts of both SREBP-1 and SREBP-2 decline, due to the interruption of proteolytic cleavage of their precursors. Since SCAP and Insig play a key role in sterol-mediated negative feedback regulation of SREBP, it would be of interest to further determine whether SCAP and Insig are involved in the ability of AMPK to regulate SREBP proteolytic processing.

5 Conclusion

Over a decade ago, AMPK is identified as a fundamental regulator of cellular metabolism and coordinates several metabolic responses in different cells. Subsequent work has shown that AMPK plays important roles in different biological processes and influences various metabolic disorders associated with diabetes. Though our understanding of nutrient-sensing mechanisms is far from complete, significant progress has been achieved. Recent studies provide compelling biochemical evidence that AMPK-dependent phosphorylation and inactivation of SREBP lead to inhibited cleavage processing and nuclear translocation of SREBP and suppressed its auto-regulatory loop and lipogenic gene's transcription. The underlying mechanism plays an important role in controlling hepatocyte and whole-body lipid homeostasis, as well as in mediating the therapeutic effects of AMPK activators on insulin resistance. While some of metabolic consequences of the nutrient-sensing AMPK-SREBP, such as those that occur in obese states, have been defined, we have yet determined whether exercise modulates this nutrient sensing pathway or how ageing affects nutrient-sensing abilities. Additionally, nutrient excess not only affects the onset of diabetes but also influences cancer development and the ageing process. Nutrient sensing and metabolism in cancer cells have received a new attention, partly due to advances in metabolomics and next-generation sequencing. Therefore, understanding nutrient-sensing mechanisms is a prerequisite for designing better interventions against human diseases such as diabetes, non-alcoholic fatty liver disease, and cancer.

Acknowledgements Studies described that were carried out in the authors' laboratory were supported by the National Institutes of Health Grants (DK076942, R01DK100603, and R21 AA021181), American Diabetes Association Basic Science Award (1-15-BS-216), and Wing Tat Lee Award (1UL1TR001430).

References

Abu-Elheiga L, Matzuk MM, Abo-Hashema KA, Wakil SJ (2001) Continuous fatty acid oxidation and reduced fat storage in mice lacking acetyl-CoA carboxylase 2. Science 291(5513):2613–2616

Amemiya-Kudo M, Shimano H, Yoshikawa T, Yahagi N, Hasty AH, Okazaki H, Tamura Y, Shionoiri F, Iizuka Y, Ohashi K, Osuga J, Harada K, Gotoda T, Sato R, Kimura S, Ishibashi S, Yamada N (2000) Promoter analysis of the mouse sterol regulatory element-binding protein-1c gene. J Biol Chem 275(40):31078–31085. doi:10.1074/jbc.M005353200

Baur JA, Pearson KJ, Price NL, Jamieson HA, Lerin C, Kalra A, Prabhu VV, Allard JS, Lopez-Lluch G, Lewis K, Pistell PJ, Poosala S, Becker KG, Boss O, Gwinn D, Wang MY, Ramaswamy S, Fishbein KW, Spencer RG, Lakatta EG, Le Couteur D, Shaw RJ, Navas P, Puigserver P, Ingram DK, de Cabo R, Sinclair DA (2006) Resveratrol improves health and survival of mice on a high-calorie diet. Nature 444(7117):337–342

Bradamante S, Barenghi L, Villa A (2004) Cardiovascular protective effects of resveratrol. Cardiovasc Drug Rev 22(3):169–188

Brown MS, Goldstein JL (1997) The SREBP pathway: regulation of cholesterol metabolism by proteolysis of a membrane-bound transcription factor. Cell 89(3):331–340. doi:10.1016/s0092-8674(00)80213-5

Brown MS, Goldstein JL (2008) Selective versus total insulin resistance: a pathogenic paradox. Cell Metab 7(2):95–96. doi:10.1016/j.cmet.2007.12.009

Brown MS, Goldstein JL (2009) Cholesterol feedback: from Schoenheimer's bottle to Scap's MELADL. J Lipid Res 50(Suppl):S15–S27. doi:10.1194/jlr.R800054-JLR200

Chen G, Liang G, Ou J, Goldstein JL, Brown MS (2004) Central role for liver X receptor in insulin-mediated activation of Srebp-1c transcription and stimulation of fatty acid synthesis in liver. Proc Natl Acad Sci U S A 101(31):11245–11250. doi:10.1073/pnas.0404297101

Chen L, Jiao ZH, Zheng LS, Zhang YY, Xie ST, Wang ZX, Wu JW (2009) Structural insight into the autoinhibition mechanism of AMP-activated protein kinase. Nature 459(7250):1146–1149. doi:10.1038/nature08075

Dentin R, Girard J, Postic C (2005) Carbohydrate responsive element binding protein (ChREBP) and sterol regulatory element binding protein-1c (SREBP-1c): two key regulators of glucose metabolism and lipid synthesis in liver. Biochimie 87(1):81–86

Dif N, Euthine V, Gonnet E, Laville M, Vidal H, Lefai E (2006) Insulin activates human sterol-regulatory-element-binding protein-1c (SREBP-1c) promoter through SRE motifs. Biochem J 400:179–188

Eberle D, Hegarty B, Bossard P, Ferre P, Foufelle F (2004) SREBP transcription factors: master regulators of lipid homeostasis. Biochimie 86(11):839–848

Engelking LJ, Kuriyama H, Hammer RE, Horton JD, Brown MS, Goldstein JL, Liang G (2004) Overexpression of Insig-1 in the livers of transgenic mice inhibits SREBP processing and reduces insulin-stimulated lipogenesis. J Clin Invest 113(8):1168–1175

Faubert B, Boily G, Izreig S, Griss T, Samborska B, Dong ZF, Dupuy F, Chambers C, Fuerth BJ, Viollet B, Mamer OA, Avizonis D, DeBerardinis RJ, Siegel PM, Jones RG (2013) AMPK is a negative regulator of the Warburg effect and suppresses tumor growth in vivo. Cell Metab 17(1):113–124. doi:10.1016/j.cmet.2012.12.001

Flury I, Garza R, Shearer A, Rosen J, Cronin S, Hampton RY (2005) INSIG: a broadly conserved transmembrane chaperone for sterol-sensing domain proteins. EMBO J 24(22):3917–3926

Foretz M, Guichard C, Ferre P, Foufelle F (1999a) Sterol regulatory element binding protein-1c is a major mediator of insulin action on the hepatic expression of glucokinase and lipogenesis-related genes. Proc Natl Acad Sci U S A 96(22):12737–12742

Foretz M, Pacot C, Dugail I, Lemarchand P, Guichard C, Le Liepvre X, Berthelier-Lubrano C, Spiegelman B, Kim JB, Ferre P, Foufelle F (1999b) ADD1/SREBP-1c is required in the activation of hepatic lipogenic gene expression by glucose. Mol Cell Biol 19(5):3760–3768

Foretz M, Ancellin N, Amdreelli F, Saintillan Y, Grondin P, Kahn A, Thorens B, Vaulont S, Viollet B (2005) Short-term overexpression of a constitutively active form of AMP-activated protein kinase in the liver leads to mild hypoglycemia and fatty liver. Diabetes 54(5):1331–1339

Foretz M, Hebrard S, Leclerc J, Zarrinpashneh E, Soty M, Mithieux G, Sakamoto K, Andreelli F, Viollet B (2010) Metformin inhibits hepatic gluconeogenesis in mice independently of the LKB1/AMPK pathway via a decrease in hepatic energy state. J Clin Invest 120(7):2355–2369. doi:10.1172/jci40671

Fullerton MD, Galic S, Marcinko K, Sikkema S, Pulinilkunnil T, Chen ZP, O'Neill HM, Ford RJ, Palanivel R, O'Brien M, Hardie DG, Macaulay SL, Schertzer JD, Dyck JR, van Denderen BJ, Kemp BE, Steinberg GR (2013) Single phosphorylation sites in Acc1 and Acc2 regulate lipid homeostasis and the insulin-sensitizing effects of metformin. Nat Med 19(12):1649–1654. doi:10.1038/nm.3372

Goldstein JL, Brown MS (2008) From fatty streak to fatty liver: 33 years of joint publications in the JCI. J Clin Invest 118(4):1220–1222. doi:10.1172/jci34973

Gong Y, Lee JN, Lee PC, Goldstein JL, Brown MS, Ye J (2006) Sterol-regulated ubiquitination and degradation of Insig-1 creates a convergent mechanism for feedback control of cholesterol synthesis and uptake. Cell Metab 3(1):15–24. doi:10.1016/j.cmet.2005.11.014

Greer EL, Oskoui PR, Banko MR, Maniar JM, Gygi MP, Gygi SP, Brunet A (2007) The energy sensor AMP-activated protein kinase directly regulates the mammalian FOXO3 transcription factor. J Biol Chem 282(41):30107–30119

Gwinn DM, Shackelford DB, Egan DF, Mihaylova MM, Mery A, Vasquez DS, Turk BE, Shaw RJ (2008) AMPK phosphorylation of raptor mediates a metabolic checkpoint. Mol Cell 30(2):214–226

Hannah VC, Ou J, Luong A, Goldstein JL, Brown MS (2001) Unsaturated fatty acids down-regulate srebp isoforms 1a and 1c by two mechanisms in HEK-293 cells. J Biol Chem 276(6):4365–4372. doi:10.1074/jbc.M007273200

Hardie DG, Ross FA, Hawley SA (2012) AMPK: a nutrient and energy sensor that maintains energy homeostasis. Nat Rev Mol Cell Biol 13(4):251–262. doi:10.1038/nrm3311

Hasty AH, Shimano H, Yahagi N, Amemiya-Kudo M, Perrey S, Yoshikawa T, Osuga J, Okazaki H, Tamura Y, Iizuka Y, Shionoiri F, Ohashi K, Harada K, Gotoda T, Nagai R, Ishibashi S, Yamada N (2000) Sterol regulatory element-binding protein-1 is regulated by glucose at the transcriptional level. J Biol Chem 275(40):31069–31077. doi:10.1074/jbc.M003335200

Hawley SA, Pan DA, Mustard KJ, Ross L, Bain J, Edelman AM, Frenguelli BG, Hardie DG (2005) Calmodulin-dependent protein kinase kinase-beta is an alternative upstream kinase for AMP-activated protein kinase. Cell Metab 2(1):9–19. doi:10.1016/j.cmet.2005.05.009

Horton JD, Goldstein JL, Brown MS (2002) SREBPs: activators of the complete program of cholesterol and fatty acid synthesis in the liver. J Clin Invest 109(9):1125–1131

Horton JD, Shimomura I, Ikemoto S, Bashmakov Y, Hammer RE (2003) Overexpression of sterol regulatory element-binding protein-1a in mouse adipose tissue produces adipocyte hypertrophy, increased fatty acid secretion, and fatty liver. J Biol Chem 278(38):36652–36660

Hou X, Xu S, Maitland-Toolan KA, Sato K, Jiang B, Ido Y, Lan F, Walsh K, Wierzbicki M, Verbeuren TJ, Cohen RA, Zang M (2008) SIRT1 regulates hepatocyte lipid metabolism through activating AMP-activated protein kinase. J Biol Chem 283(29):20015–20026

Howitz KT, Bitterman KJ, Cohen HY, Lamming DW, Lavu S, Wood JG, Zipkin RE, Chung P, Kisielewski A, Zhang LL, Scherer B, Sinclair DA (2003) Small molecule activators of sirtuins extend Saccharomyces cerevisiae lifespan. Nature 425(6954):191–196

Inoki K, Ouyang H, Zhu T, Lindvall C, Wang Y, Zhang X, Yang Q, Bennett C, Harada Y, Stankunas K, Wang CY, He X, MacDougald OA, You M, Williams BO, Guan KL (2006) TSC2 integrates Wnt and energy signals via a coordinated phosphorylation by AMPK and GSK3 to regulate cell growth. Cell 126(5):955–968

Kahn BB, Alquier T, Carling D, Hardie DG (2005) AMP-activated protein kinase: ancient energy gauge provides clues to modern understanding of metabolism. Cell Metab 1(1):15–25

Kim JB, Spotts GD, Halvorsen YD, Shih HM, Ellenberger T, Towle HC, Spiegelman BM (1995) Dual DNA-binding specificity of ADD1/SREBP1 controlled by a single amino-acid in the basic helix-loop-helix domain. Mol Cell Biol 15(5):2582–2588

Kim JB, Sarraf P, Wright M, Yao KM, Mueller E, Solanes G, Lowell BB, Spiegelman BM (1998) Nutritional and insulin regulation of fatty acid synthetase and leptin gene expression through ADD1/SREBP1. J Clin Invest 101(1):1–9

Laplante M, Sabatini DM (2010) mTORC1 activates SREBP-1c and uncouples lipogenesis from gluconeogenesis. Proc Natl Acad Sci U S A 107(8):3281–3282. doi:10.1073/pnas.1000323107

Lee YS, Kim WS, Kim KH, Yoon MJ, Cho HJ, Shen Y, Ye JM, Lee CH, Oh WK, Kim CT, Hohnen-Behrens C, Gosby A, Kraegen EW, James DE, Kim JB (2006) Berberine, a natural plant product, activates AMP-activated protein kinase with beneficial metabolic effects in diabetic and insulin-resistant states. Diabetes 55(8):2256–2264. doi:10.2337/db06-0006

Li S, Brown MS, Goldstein JL (2010) Bifurcation of insulin signaling pathway in rat liver: mTORC1 required for stimulation of lipogenesis, but not inhibition of gluconeogenesis. Proc Natl Acad Sci U S A 107(8):3441–3446. doi:10.1073/pnas.0914798107

Li Y, Xu SQ, Mihaylova MM, Zheng B, Hou XY, Jiang BB, Park O, Luo ZJ, Lefai E, Shyy JYJ, Gao B, Wierzbicki M, Verbeuren TJ, Shaw RJ, Cohen RA, Zang MW (2011) AMPK phosphorylates and inhibits SREBP activity to attenuate hepatic steatosis and atherosclerosis in diet-induced insulin-resistant mice. Cell Metab 13(4):376–388

Liang GS, Yang J, Horton JD, Hammer RE, Goldstein JL, Brown MS (2002) Diminished hepatic response to fasting/refeeding and liver X receptor agonists in mice with selective deficiency of sterol regulatory element-binding protein-1c. J Biol Chem 277(11):9520–9528

Lin JD, Yang RJ, Tarr PT, Wu PH, Handschin C, Li SM, Yang WL, Pei LM, Uldry M, Tontonoz P, Newgard CB, Spiegelman BM (2005) Hyperlipidemic effects of dietary saturated fats mediated through PGC-1 beta coactivation of SREBP. Cell 120(2):261–273

Lizcano JM, Goransson O, Toth R, Deak M, Morrice NA, Boudeau J, Hawley SA, Udd L, Makela TP, Hardie DG, Alessi DR (2004) LKB1 is a master kinase that activates 13 kinases of the AMPK subfamily, including MARK/PAR-1. EMBO J 23(4):833–843

Lu M, Shyy JYJ (2006) Sterol regulatory element-binding protein 1 is negatively modulated by PKA phosphorylation. Am J Physiol Cell Physiol 290(6):C1477–C1486

Luo ZJ, Zang MW, Guo W (2010) AMPK as a metabolic tumor suppressor: control of metabolism and cell growth. Future Oncol 6(3):457–470

Manach C, Scalbert A, Morand C, Remesy C, Jimenez L (2004) Polyphenols: food sources and bioavailability. Am J Clin Nutr 79(5):727–747

Martin-Montalvo A, Mercken EM, Mitchell SJ, Palacios HH, Mote PL, Scheibye-Knudsen M, Gomes AP, Ward TM, Minor RK, Blouin MJ, Schwab M, Pollak M, Zhang YQ, Yu YB, Becker KG, Bohr VA, Ingram DK, Sinclair DA, Wolf NS, Spindler SR, Bernier M, de Cabo R (2013) Metformin improves healthspan and lifespan in mice. Nat Commun 4:2192. doi:10.1038/ncomms3192

Matsuda M, Korn BS, Hammer RE, Moon YA, Komuro R, Horton JD, Goldstein JL, Brown MS, Shimomura I (2001) SREBP cleavage-activating protein (SCAP) is required for increased lipid synthesis in liver induced by cholesterol deprivation and insulin elevation. Genes Dev 15(10):1206–1216

Miller RA, Chu Q, Xie J, Foretz M, Viollet B, Birnbaum MJ (2013) Biguanides suppress hepatic glucagon signalling by decreasing production of cyclic AMP. Nature 494(7436):256–260. doi:10.1038/nature11808

Momcilovic M, Hong SP, Carlson M (2006) Mammalian TAK1 activates Snf1 protein kinase in yeast and phosphorylates AMP-activated protein kinase in vitro. J Biol Chem 281(35):25336–25343

Moon YA, Liang GS, Xie XF, Frank-Kamenetsky M, Fitzgerald K, Koteliansky V, Brown MS, Goldstein JL, Horton JD (2012) The Scap/SREBP pathway is essential for developing diabetic fatty liver and carbohydrate-induced hypertriglyceridemia in animals. Cell Metab 15(2):240–246

Muoio DM, Seefeld K, Witters LA, Coleman RA (1999) AMP-activated kinase reciprocally regulates triacylglycerol synthesis and fatty acid oxidation in liver and muscle: evidence that sn-glycerol-3-phosphate acyltransferase is a novel target. Biochem J 338(Pt 3):783–791

Nohturfft A, Yabe D, Goldstein JL, Brown MS, Espenshade PJ (2000) Regulated step in cholesterol feedback localized to budding of SCAP from ER membranes. Cell 102(3):315–323

Okazaki H, Goldstein JL, Brown MS, Liang G (2010) LXR-SREBP-1c-phospholipid transfer protein axis controls very low density lipoprotein (VLDL) particle size. J Biol Chem 285(9):6801–6810. doi:10.1074/jbc.M109.079459

Osborne TF, Espenshade PJ (2009) Evolutionary conservation and adaptation in the mechanism that regulates SREBP action: what a long, strange tRIP it's been. Genes Dev 23(22):2578–2591. doi:10.1101/gad.1854309

Peterson TR, Sengupta SS, Harris TE, Carmack AE, Kang SA, Balderas E, Guertin DA, Madden KL, Carpenter AE, Finck BN, Sabatini DM (2011) mTOR complex 1 regulates lipin 1 localization to control the SREBP pathway. Cell 146(3):408–420

Ponugoti B, Kim DH, Xiao Z, Smith Z, Miao J, Zang MW, Wu SY, Chiang CM, Veenstra TD, Kemper JK (2010) SIRT1 deacetylates and inhibits SREBP-1C activity in regulation of hepatic lipid metabolism. J Biol Chem 285(44):33959–33970

Raghow R, Yellaturu C, Deng X, Park EA, Elam MB (2008) SREBPs: the crossroads of physiological and pathological lipid homeostasis. Trends Endocrinol Metab 19(2):65–73

Rawson RB (2003a) Control of lipid metabolism by regulated intramembrane proteolysis of sterol regulatory element binding proteins (SREBPs). Biochem Soc Symp 70:221–231

Rawson RB (2003b) The SREBP pathway—insights from Insigs and insects. Nat Rev Mol Cell Biol 4(8):631–640

Repa JJ, Liang G, Ou J, Bashmakov Y, Lobaccaro JM, Shimomura I, Shan B, Brown MS, Goldstein JL, Mangelsdorf DJ (2000) Regulation of mouse sterol regulatory element-binding protein-1c gene (SREBP-1c) by oxysterol receptors, LXRalpha and LXRbeta. Genes Dev 14(22):2819–2830

Sakamoto K, Goransson O, Hardie DG, Alessi DR (2004) Activity of LKB1 and AMPK-related kinases in skeletal muscle: effects of contraction, phenformin, and AICAR. Am J Physiol Endocrinol Metab 287(2):E310–E317

Shackelford DB, Shaw RJ (2009) The LKB1-AMPK pathway: metabolism and growth control in tumour suppression. Nat Rev Cancer 9(8):563–575. doi:10.1038/nrc2676

Shaw RJ (2013) Metformin trims fats to restore insulin sensitivity. Nat Med 19(12):1570–1572. doi:10.1038/nm.3414

Shaw RJ, Kosmatka M, Bardeesy N, Hurley RL, Witters LA, DePinho RA, Cantley LC (2004) The tumor suppressor LKB1 kinase directly activates AMP-activated kinase and regulates apoptosis in response to energy stress. Proc Natl Acad Sci U S A 101(10):3329–3335

Shaw RJ, Lamia KA, Vasquez D, Koo SH, Bardeesy N, DePinho RA, Montminy M, Cantley LC (2005) The kinase LKB1 mediates glucose homeostasis in liver and therapeutic effects of metformin. Science 310(5754):1642–6

Shimano H, Shimomura I, Hammer RE, Herz J, Goldstein JL, Brown MS, Horton JD (1997) Elevated levels of SREBP-2 and cholesterol synthesis in livers of mice homozygous for a targeted disruption of the SREBP-1 gene. J Clin Invest 100(8):2115–2124

Shimomura I, Shimano H, Horton JD, Goldstein JL, Brown MS (1997) Differential expression of exons 1a and 1c in mRNAs for sterol regulatory element binding protein-1 in human and mouse organs and cultured cells. J Clin Invest 99(5):838–845

Steinberg GR, Kemp BE (2009) AMPK in health and disease. Physiol Rev 89(3):1025–1078

Sundqvist A, Bengoechea-Alonso MT, Ye X, Lukiyanchuk V, Jin JP, Harper JW, Ericsson J (2005) Control of lipid metabolism by phosphorylation-dependent degradation of the SREBP family of transcription factors by SCFFbw7. Cell Metab 1(6):379–391

Tang JJ, Li JG, Qi W, Qiu WW, Li PS, Li BL, Song BL (2011) Inhibition of SREBP by a small molecule, betulin, improves hyperlipidemia and insulin resistance and reduces atherosclerotic plaques. Cell Metab 13(1):44–56

Timmers S, Konings E, Bilet L, Houtkooper RH, van de Weijer T, Goossens GH, Hoeks J, van der Krieken S, Ryu D, Kersten S, Moonen-Kornips E, Hesselink MK, Kunz I, Schrauwen-Hinderling VB, Blaak EE, Auwerx J, Schrauwen P (2011) Calorie restriction-like effects of 30 days of resveratrol supplementation on energy metabolism and metabolic profile in obese humans. Cell Metab 14(5):612–622. doi:10.1016/j.cmet.2011.10.002

Turner N, Li JY, Gosby A, To SWC, Cheng Z, Miyoshi H, Taketo MM, Cooney GJ, Kraegen EW, James DE, Hu LH, Li J, Ye JM (2008) Berberine and its more biologically available derivative, dihydroberberine, inhibit mitochondrial respiratory complex I: a mechanism for the action of berberine to activate AMP-activated protein kinase and improve insulin action. Diabetes 57(5):1414–1418. doi:10.2337/db07-1552

Um JH, Park SJ, Kang H, Yang ST, Foretz M, McBurney MW, Kim MK, Viollet B, Chung JH (2010) AMP-activated protein kinase-deficient mice are resistant to the metabolic effects of resveratrol. Diabetes 59(3):554–563

Viollet B, Guigas B, Sanz Garcia N, Leclerc J, Foretz M, Andreelli F (2012) Cellular and molecular mechanisms of metformin: an overview. Clin Sci (Lond) 122(6):253–270. doi:10.1042/cs20110386

Vita JA (2005) Polyphenols and cardiovascular disease: effects on endothelial and platelet function. Am J Clin Nutr 81(1):292S–297S
Walker AK, Yang FJ, Jiang KR, Ji JY, Watts JL, Purushotham A, Boss O, Hirsch ML, Ribich S, Smith JJ, Israelian K, Westphal CH, Rodgers JT, Shioda T, Elson SL, Mulligan P, Najafi-Shoushtari H, Black JC, Thakur JK, Kadyk LC, Whetstine JR, Mostoslavsky R, Puigserver P, Li XL, Dyson NJ, Hart AC, Naar AM (2010) Conserved role of SIRT1 orthologs in fasting-dependent inhibition of the lipid/cholesterol regulator SREBP. Genes Dev 24(13):1403–1417
Yang J, Goldstein JL, Hammer RE, Moon YA, Brown MS, Horton JD (2001) Decreased lipid synthesis in livers of mice with disrupted Site-1 protease gene. Proc Natl Acad Sci U S A 98(24):13607–13612
Yang T, Espenshade PJ, Wright ME, Yabe D, Gong Y, Aebersold R, Goldstein JL, Brown MS (2002) Crucial step in cholesterol homeostasis: Sterols promote binding of SCAP to INSIG-1, a membrane protein that facilitates retention of SREBPs in ER. Cell 110(4):489–500
Yap F, Craddock L, Yang J (2011) Mechanism of AMPK suppression of LXR-dependent Srebp-1c transcription. Int J Biol Sci 7(5):645–650
Ye J, DeBose-Boyd RA (2011) Regulation of cholesterol and fatty acid synthesis. Cold Spring Harb Perspect Biol 3(7). doi:10.1101/cshperspect.a004754
Yin J, Xing H, Ye J (2008) Efficacy of berberine in patients with type 2 diabetes mellitus. Metabolism 57(5):712–717. doi:10.1016/j.metabol.2008.01.013
Yoshikawa T, Shimano H, Yahagi N, Ide T, Amemiya-Kudo M, Matsuzaka T, Nakakuki M, Tomita S, Okazaki H, Tamura Y, Iizuka Y, Ohashi K, Takahashi A, Sone H, Osuga Ji J, Gotoda T, Ishibashi S, Yamada N (2002) Polyunsaturated fatty acids suppress sterol regulatory element-binding protein 1c promoter activity by inhibition of liver X receptor (LXR) binding to LXR response elements. J Biol Chem 277(3):1705–1711. doi:10.1074/jbc.M105711200
Young LH (2009) A crystallized view of AMPK activation. Cell Metab 10(1):5–6
Zang MW, Zuccollo A, Hou XY, Nagata D, Walsh K, Herscovitz H, Brecher P, Ruderman NB, Cohen RA (2004) AMP-activated protein kinase is required for the lipid-lowering effect of metformin in insulin-resistant human HepG2 cells. J Biol Chem 279(46):47898–47905
Zang MW, Xu SQ, Maitland-Toolan KA, Zuccollo A, Hou XY, Jiang BB, Wierzbicki M, Verbeuren TJ, Cohen RA (2006) Polyphenols stimulate AMP-activated protein kinase, lower lipids, and inhibit accelerated atherosclerosis in diabetic LDL receptor-deficient mice. Diabetes 55(8):2180–2191
Zhang Y, Li X, Zou D, Liu W, Yang J, Zhu N, Huo L, Wang M, Hong J, Wu P, Ren G, Ning G (2008) Treatment of type 2 diabetes and dyslipidemia with the natural plant alkaloid berberine. J Clin Endocrinol Metab 93(7):2559–2565. doi:10.1210/jc.2007-2404
Zhang BB, Zhou GC, Li C (2009) AMPK: an emerging drug target for diabetes and the metabolic syndrome. Cell Metab 9(5):407–416
Zhou GC, Myers R, Li Y, Chen YL, Shen XL, Fenyk-Melody J, Wu M, Ventre J, Doebber T, Fujii N, Musi N, Hirshman MF, Goodyear LJ, Moller DE (2001) Role of AMP-activated protein kinase in mechanism of metformin action. J Clin Invest 108(8):1167–1174

Chapter 3
Regulation of Hepatic Metabolism by the Mechanistic Target of Rapamycin

Dudley W. Lamming

Abstract Lipids are essential for many cellular and organismal processes, yet an excess of lipids can cause toxicity. The liver is a key organ for the maintenance of lipid homeostasis, performing lipogenesis as well as mediating the exchange of lipoproteins with peripheral tissues. In this chapter, we focus primarily on the regulation of lipogenesis in the liver by the protein kinase mTOR (mechanistic Target of Rapamycin), a central sensor of environmental cues that coordinates growth, protein synthesis and metabolism with nutrient availability. The mTOR protein kinase is found in two distinct complexes, each of which plays a role in the regulation of lipogenesis. We discuss the regulation of lipogenesis both directly by hepatic mTOR, indirectly by mTOR in other organs, and the regulation of hepatic lipogenesis by hormones and growth factors that regulate mTOR. This regulation by mTOR is an extremely complex process that we are only now beginning to fully understand.

Keywords mTOR • Rapamycin • Liver • Lipogenesis • mTORC1 • mTORC2 • Akt • Hormones

1 Introduction

1.1 Two Distinct mTOR Complexes

The mechanistic Target of Rapamycin (mTOR) is a serine/threonine protein kinase that is widely conserved in eukaryotes, and is found in yeast, *C. elegans*, *D. melanogaster*, mice and humans as well as in plants (Chantranupong et al. 2015). mTOR

D.W. Lamming (✉)
Division of Endocrinology, Department of Medicine, University of Wisconsin-Madison and William S. Middleton Memorial Veterans Hospital, 2500 Overlook Terrace, WSM VAH Room C3127 Research 151, Madison, WI 53705, USA
e-mail: dlamming@medicine.wisc.edu

J.M. Ntambi (ed.), *Hepatic De Novo Lipogenesis and Regulation of Metabolism*,
DOI 10.1007/978-3-319-25065-6_3

itself was discovered as a result of research into understanding the anti-proliferative effects of rapamycin, an FDA-approved immunosuppressant and anti-cancer agent originally isolated from bacteria discovered in the soil of Easter Island (Vezina et al. 1975). Subsequent research has determined that the mTOR protein kinase is found in two discrete complexes, mTOR complex 1 (mTORC1) and mTORC2, with distinct protein components as well as different subcellular localizations and substrates (Laplante and Sabatini 2012). mTORC1 is defined by the interaction of mTOR with Raptor and mLST8/GβL. Additional proteins that interact with the core of mTORC1 include PRAS40 and DEPTOR, which also regulate mTORC1 activity (Sancak et al. 2007; Peterson et al. 2009). Structural studies have determined that mTORC1 is an obligate dimer, with the interface defined by mTOR/Raptor interactions (Yip et al. 2010).

mTORC1 regulates numerous processes involved in cell growth and metabolism via the control of ribosomal biogenesis and translation, mediated by substrates that include S6K1 and 4E-BP1 (Laplante and Sabatini 2012). More recently, phosphoproteomic approaches have identified new mTORC1 substrates with key roles in the regulation of insulin signaling, translation, and autophagy, including Grb10, Larp1, Patl1, and Ulk1 (Egan et al. 2011; Hsu et al. 2011; Kang et al. 2013). Much of the research on the localization of mTORC1 has concentrated on its recruitment to the lysosome in response to amino acids, which is mediated by the Rag/Ragulator family of proteins and the GATOR1/2 complex (Bar-Peled and Sabatini 2014). However, new visualization technology has identified mTORC1 activity throughout the cytoplasm and nucleus (Zhou et al. 2015).

mTORC2 is defined by the interaction of mTOR with Rictor and mLST8/GβL (Sarbassov et al. 2004), but much like mTORC1, additional regulatory interacting proteins have been identified. These include DEPTOR, mSin1 and Protor 1/2 (Frias et al. 2006; Pearce et al. 2007; Peterson et al. 2009). mTORC2 has primarily been characterized as a downstream effector of the insulin/IGF-1 signaling pathway, and several of its substrates have been characterized. This includes three distinct sites on AKT that are important for its activation—AKT T450, the turn-motif, AKT S473, the hydrophobic motif, and AKT S477/479 near the C-terminal end (Sarbassov et al. 2005; Ikenoue et al. 2008; Liu et al. 2014a). mTORC2 also phosphorylates the hydrophobic motif of SGK1 (serum- and glucocorticoid-induced protein kinase 1) S422 (Garcia-Martinez and Alessi 2008). Finally, mTORC2 phosphorylates several members of the protein kinase C (PKC) family, including PKCα (Sarbassov et al. 2004), PKCδ (Gan et al. 2012), and PKCε (Ikenoue et al. 2008). While the localization of mTORC2 is still a subject of research, at least some mTORC2 is localized to the mitochondria-associated endoplasmic reticulum membrane (Betz et al. 2013), and the activity of mTORC2 may depend upon its association with ribosomes (Zinzalla et al. 2011).

1.2 Molecular Regulation of mTOR Signaling

The mTOR pathway serves as a central integrator of environmental cues that indicate if conditions are favorable for growth. Cells in a multicellular organism must integrate signals regarding both the local ability of nutrients and resources that can serve as building blocks for protein translation and cell division, and signals from the organism as a whole. mTOR signaling integrates these two types of signals by placing mTORC2, which is sensitive to growth factors such as insulin and IGF-1, upstream of mTORC1, which is sensitive to stimuli which includes amino acids, glucose, cellular energy, and oxygen (Fig. 3.1).

mTORC2 activity is stimulated by growth factors, most importantly by the insulin/IGF-1 receptor. The mechanism through which PI3K actually modulates mTORC2 is unknown. However, it has been suggested that mTORC2 may be

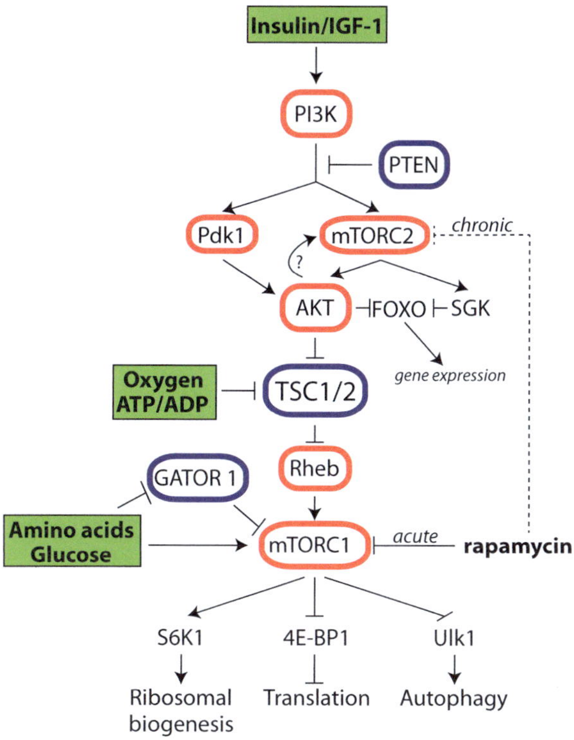

Fig. 3.1 The insulin/IGF-1/Akt/mTORC1 signaling pathway. A schematic outline of the insulin/IGF-1/Akt/mTORC1 signaling pathway. Environmental stimuli which positively regulate mTOR signaling, including insulin, IGF-1, oxygen, a high ATP/ADP ratio, amino acids and glucose, are boxed in *green*. Amino acids and the insulin/IGF-1 pathway promote mTORC1 activity via distinct mechanisms. Upstream signaling pathway members that promote mTORC1 activity are highlighted in *red*, while inhibitors of the pathway are highlighted in *blue*. Rapamycin is an acute inhibitor of mTORC1, and when given chronically also inhibits mTORC2

regulated by association with ribosomal subunits (Zinzalla et al. 2011), and the GTPase Rac1 has also been linked to mTORC2 activation (Saci et al. 2011). A more recent finding suggests that Akt itself—which as mentioned above is an mTORC2 substrate—may also positively regulate mTORC2 through the phosphorylation of mSin1 (Humphrey et al. 2013). While insulin/IGF-1 are major regulators of mTORC2 signaling, we have recently discussed the possibility that in a physiological context, leptin may act to regulate mTORC2 activity directly via Jak2 mediated phosphorylation of insulin receptor substrate (IRS) and indirectly via activation of IGF-1 signaling [reviewed in (Lamming 2014)].

mTORC2 activity promotes mTORC1 activity via the activation of Akt. Akt activates mTORC1 signaling through two different mechanisms. First, Akt phosphorylates PRAS40, a protein which, when unphosphorylated, binds to and inhibits mTORC1 (Sancak et al. 2007). Secondly, Akt phosphorylates TSC2, a member of the tuberous sclerosis complex which is composed of TSC1 (hamartin) and TSC2 (tuberin) (Tee et al. 2003). The TSC complex acts as a GTPase-activating protein (GAP) upon Rheb, a GTPase that stimulates mTORC1 activity (Inoki et al. 2003). However, the TSC complex is also responsive to many other stimuli and functions as a mini-regulatory hub upstream of mTORC1. Regulators of TSC include AMPK, ERK and GSK3 which phosphorylate TSC2 (Rolfe et al. 2005; Inoki et al. 2006), and IKKβ which phosphorylates TSC1 (Lee et al. 2007).

The regulation of Rheb by TSC is controlled spatially, with TSC and Rheb co-localizing to the lysosome in the absence of insulin. Insulin stimulation induces TSC to leave the lysosome, thereby derepressing Rheb and allowing it to activate mTORC1 (Menon et al. 2014). However, in order for Rheb to activate mTORC1, mTORC1 must also be present at the lysosome. An extensive body of work has been compiled over the last several years on this topic (Bar-Peled and Sabatini 2014). Briefly, the presence of amino acids, particularly the branched chain amino acid leucine, promotes the localization of mTORC1 to the lysosome. This process is mediated by the Rag family of GTPases and the Ragulator complex, which recruits mTORC1 to the lysosome and promote its interaction with the vacuolar ATPase (Sancak et al. 2008, 2010; Zoncu et al. 2011; Bar-Peled et al. 2012).

The pathways that regulate the Rags and Ragulator are an area of active investigation by numerous labs. Several protein complexes that act as GAPs for the Rags have been identified, including the GATOR1/2 complex, as well as a complex composed of folliculin and FNIP1/2 (Bar-Peled et al. 2013; Tsun et al. 2013). The amino acid transporter SLC38A9 was recently identified as a Ragulator-binding protein that acts as an arginine sensor for the mTORC1 pathway (Rebsamen et al. 2015; Wang et al. 2015). Separately, it was recently determined that glutamine promotes mTORC1 localization to the lysosome independently of Rags and Ragulator, suggesting that different amino acids may stimulate mTORC1 activity via separate signaling mechanisms (Jewell et al. 2015).

1.3 Regulation of Metabolism and Physiology by mTOR

Our understanding of the physiological role of mTOR signaling has benefited in recent years from the steadily decreasing price of rapamycin—currently available for less than $500 per gram—and the availability of mouse models in which mTOR signaling can be manipulated genetically. Early experiments with conventional knockout mouse models demonstrated that *mTOR*, *Raptor*, *mLST8* and *Rictor* are all essential for embryonic development (Guertin et al. 2006). Subsequent experiments have revealed essential roles as well for at least some members of the Rag family of GTPases (Efeyan et al. 2014). Constitutive activation of the mTORC1 signaling pathway, by deletion of either *TSC1* or *TSC2*, likewise leads to embryonic lethality (Onda et al. 1999; Kobayashi et al. 2001). However, the use of conditional knockout mouse models has recently enabled us to understand the tissue specific roles of each mTOR complex.

Mice were first given rapamycin in 1975, establishing that acute toxicity of rapamycin was extremely low, with an intraperitoneal LD_{50} of approximately 600 mg/kg (Baker et al. 1978). Since that time, the effects of rapamycin on the metabolism of mice, rats, and humans has been extensively researched. Metabolic effects of rapamycin that have been noticed in rodents and humans include elevated cholesterol and free fatty acids, hyperlipidemia and hyperglycemia (Morrisett et al. 2002; Houde et al. 2010; Yu et al. 2015). Other effects of rapamycin that may or may not be based in metabolism include immunosuppression, hypertension, and thrombocytopenia (Hong and Kahan 2000; Reis et al. 2009; Powell et al. 2012). Mouth ulcers are another frequently observed and painful complication of rapamycin therapy (Fricain et al. 2008). There is also a significant accumulating evidence that rapamycin impairs beta cell proliferation and function (Barlow et al. 2013).

Originally, all of the side effects of rapamycin were attributed to mTORC1, which was until recently the only identified mTOR-containing complex. Even after the discovery of mTORC2, it was believed that mTORC2 was rapamycin insensitive and that rapamycin acted solely upon mTORC1 (Sarbassov et al. 2004). However, it was soon realized that mTORC2 could be disrupted by prolonged treatment with rapamycin in certain cell culture lines (Sarbassov et al. 2006), likely by sequestering newly translated mTOR and blocking the formation of new mTORC2 (Sarbassov et al. 2006). This suggested the possibility that rapamycin treatment for a prolonged period of time might also inhibit mTORC2 in vivo, but surprisingly this possibility was not pursued vigorously for several years.

1.4 Mechanistic Exploration of Rapamycin Side Effects Identify a Role of mTORC2 in Metabolism

One of the most dramatic effects of rapamycin in rodents is glucose intolerance and fasting hyperglycemia following rapamycin administration for 2 weeks or more, with a statistically significant effect after only 2 days (Cunningham et al. 2007;

Houde et al. 2010; Lamming et al. 2012). In an effort to understand the mechanistic basis of the glucose intolerance, we performed a hyperinsulinemic-euglycemic clamp on rapamycin treated mice (Lamming et al. 2012), which is the gold standard for assessing gluconeogenesis and insulin sensitivity in vivo. We determined that rapamycin causes glucose intolerance primarily through the induction of hepatic insulin resistance, resulting in increased fasting gluconeogenesis that is not properly suppressed by insulin. Surprisingly though, we found that mice lacking hepatic Raptor—and thus hepatic mTORC1 activity (Sengupta et al. 2010)—were glucose tolerant (Lamming et al. 2012).

We were therefore forced to consider the possibility that prolonged rapamycin treatment was impairing hepatic insulin sensitivity through inhibition of mTORC2. We determined that hepatic deletion of Rictor, which blocked hepatic mTORC2 activity, resulted in significant glucose intolerance, a finding subsequently confirmed by two independent groups (Hagiwara et al. 2012; Lamming et al. 2012; Yuan et al. 2012). We also generated a mouse in which we could conditionally delete Rictor in the whole body of an adult mouse through treatment with tamoxifen. This mouse was also glucose intolerant, and importantly did not experience further glucose intolerance following chronic treatment with rapamycin. Hyperinsulinemic-euglycemic clamp analysis determined that, as with wild-type rapamycin treated mice, the principle defect was hepatic insulin resistance (Lamming et al. 2012).

If this one metabolic effect of rapamycin was due to mTORC2, which other side effects of rapamycin may be attributable to inhibition of mTORC2? It is important to note that chronic rapamycin inhibits mTORC2 signaling in many tissues, not only the liver. Chronic rapamycin at 2 mg/kg/day inhibits mTORC2 signaling in liver, skeletal muscle, and white adipose tissue (Lamming et al. 2012), with greater inhibitory effects in the context of a high-fat diet (Liu et al. 2014b). Chronic rapamycin treatment with 8 mg/kg on alternate days inhibits mTORC2 signaling in liver, skeletal muscle and white adipose tissue as well as heart, pancreas, spleen, lung (Schreiber et al. 2015).

Detailed studies of the immune system have revealed that both mTOR complexes likely contribute to the immunosuppressive effects of rapamycin. Both mTOR complexes are involved in $CD4^+$ T cell differentiation, with mTORC1 activity promoting the T helper type 1 (Th1) lineage, mTORC2 promoting Th2 type cells, and both mTORC1 and mTORC2 repressing regulatory T cells (Tregs) [reviewed in (Powell et al. 2012)]. Both mTOR complexes are also required to properly regulate macrophage polarization in response to proinflammatory stimuli (Byles et al. 2013; Festuccia et al. 2014). Deletion of Rictor, inhibiting mTORC2, also impairs B cell maturation and in vivo cell numbers, but rapamycin has additive effects with Rictor deletion, suggesting that both mTOR complexes are also involved in B cell function [reviewed in (Fantus and Thomson 2015)].

1.5 mTOR Regulates Ketogenesis and Lipogenesis in the Liver

As mentioned above, a major side effect of rapamycin treatment in humans is hyperlipidemia as well as elevated cholesterol and free fatty acids. Some of the first studies in this area were conducted in cell culture using primary hepatocytes exposed to rapamycin. Rapamycin treatment was shown to reduce the expression and activation of sterol regulatory element-binding proteins (SREBPs) in response to insulin (Ma et al. 2007; Li et al. 2010). SREBPs are a class of transcription factors that are conserved from yeast to humans that play an extremely important role in the regulation of lipid metabolism (Osborne and Espenshade 2009). While there are three SREBPs in mammals, SREBP-1c is the predominant isoform in the liver as well as in adipose tissue (Jeon and Osborne 2012).

SREBPs are translated in an inactive and immature form that localizes to the endoplasmic reticulum (Yecies and Manning 2011). In response to insulin and other stimuli, SREBPs mature via a highly elaborate process that results in their proteolytic cleavage at two distinct sites (Jeon and Osborne 2012). The mature SREBPs can then localize to the nucleus, activating lipogenic genes by binding to SREBP binding sites in the promoters of those genes. Work by Porstmann and colleagues demonstrated that this process was sensitive to rapamycin, demonstrating that mTORC1 had an important role in lipogenesis, and they found that mTORC1 regulated the nuclear accumulation of SREBPs (Porstmann et al. 2008). In agreement with this finding, activation of mTORC1 signaling by deletion of *TSC1* or *TSC2* increases SREBP activity (Duvel et al. 2010).

The exact mechanism by which mTORC1 regulates the SREBPs has proven elusive. Using a chemical inhibitor of S6K, it was determined that the insulin-induced posttranscriptional processing of SREBP1c is dependent upon S6K, but that inhibition of S6K did not regulate SREBP1c transcription (Owen et al. 2012). In agreement with this finding, depletion of S6K1/2 in mouse liver using RNAi techniques protects from high fat diet-induced hepatic steatosis (Bae et al. 2012). Bae and colleagues also observed decreased expression of SREBP1c as well as decreased mature SREBP-1c protein. mTORC1 also likely regulates SREBPs through the phosphorylation of Lipin1 (Peterson et al. 2011). Lipin1 is a phosphatidic acid phosphatase that promotes triglyceride synthesis and is downregulated in obese humans, while mutation of Lipin1 leads to lipodystrophy (Laplante and Sabatini 2009a). Lipin1 is also a transcriptional coactivator for peroxisome proliferator-activated receptor γ (PPARγ), and thereby regulates adipogenesis (Laplante and Sabatini 2009a). Fascinatingly, it was recently discovered that Lipin1 is an mTORC1 substrate, which is sequestered in the cytoplasm by mTORC1 activity. Dephosphorylated, nuclear Lipin1 acts to repress nuclear SREBP levels and thus repress SREBP-dependent gene transcription (Peterson et al. 2011). As shown in Fig. 3.2, mTORC1 thereby regulates hepatic lipogenesis through the phosphorylation of two independent substrates which both converge to regulate the level and activity of Srebp1.

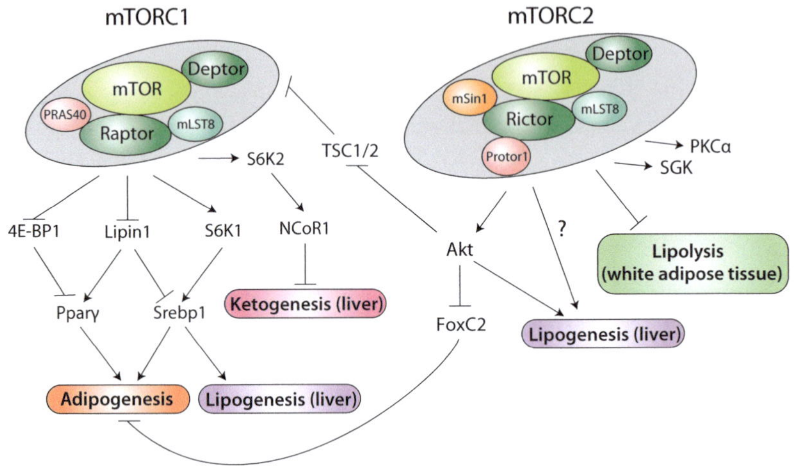

Fig. 3.2 mTOR in lipid homeostasis. A model of the actions of mTORC1 and mTORC2 in lipid metabolism, including adipogenesis, lipogenesis and lipolysis. mTORC1 also controls ketogenesis, which may have a reciprocal relationship with lipogenesis, via NCoR1, but the exact mechanism by which it regulates NCoR1 is unknown. The mechanism by which mTORC2 regulates lipogenesis and lipolysis are likewise not currently known, although Akt may be involved in the regulation of lipogenesis. This figure was originally published in Lamming and Sabatini (2013), *Cell Metabolism*

The lipogenesis process has significant feedback mechanisms to enable fine-tuning of this critical cellular process. First, Lipin1 gene transcription is regulated by the SREBPs (Jeon and Osborne 2012). In addition to the SREBPs, Lipin1 may also regulate mTOR at the protein level, potentially acting to coordinate mTOR activity and SREBP-dependent lipogenesis (Zhang et al. 2014). Still further regulation of this process is provided by AMPK, which directly phosphorylates the SREBPs as well as inhibiting mTORC1 signaling through phosphorylation of Raptor and TSC2 (Laplante and Sabatini 2009b; Jeon and Osborne 2012).

While mTORC1 activity is essential for SREBP-dependent lipogenic gene expression, it is not sufficient, as demonstrated by mice lacking hepatic *TSC1*. Due to feedback inhibition of insulin signaling and Akt activity, mice lacking hepatic TSC1 have decreased SREBP1c levels and lipogenesis despite massively increased mTORC1 activity (Yecies et al. 2011). Deletion of Akt2 in the liver is sufficient to impair lipogenesis, likely through pathways that involve not only mTORC1 and SREBP1c, but also through the FOXO family of transcription factors (Wan et al. 2011). The necessity of Akt activity for lipogenesis clearly suggests that mTORC2 is also essential for hepatic lipogenesis. Over the past several years, several studies have been done on mice lacking hepatic Rictor, an essential subunit of mTORC2. Deletion of hepatic *Rictor* protects mice against high fat diet induced fatty liver (Hagiwara et al. 2012; Yuan et al. 2012; Lamming and Sabatini 2013). Deletion of *Rictor* in the liver also leads to decreased expression of numerous proteins involved in lipid synthesis, including acetyl-CoA carboxylase, fatty acid synthase, PPARγ, and SREBP1c.

However, it is unlikely that defective lipid synthesis is the only pathway responsible for the decreased lipid accumulation observed when placed on a high fat diet, as fatty acid uptake gene expression is reduced while genes related to fatty acid oxidation such as PPARα are upregulated. The combined effect of these alterations in lipogenesis, lipid uptake and lipid oxidation is decreased level of lipid and protection from high-fat diet induced fatty liver. It is unclear if all of these effects are mediated by Akt, as Yuan et al. found that constitutively active Akt did not rescue the lipid defect of mice lacking hepatic Rictor (Yuan et al. 2012), while Hagiwara et al. came to the opposite conclusion (Hagiwara et al. 2012). While one potential explanation for these divergent reports is that mTORC1 activity differed between the two studies (Lamming and Sabatini 2013), we recently observed that mTORC1 activity is essentially normal in mice lacking hepatic Rictor (Lamming et al. 2014b). A recent report found that hepatic insulin sensitivity is not required for feeding induced expression of SREBP1c, but that it is required for obesity induced expression (Haas et al. 2012), highlighting complexities within this pathway.

We recently reported that mice lacking hepatic Rictor had profound alterations in gluconeogenesis, showing significant increased glucose output in vivo and increased expression of gluconeogenic genes including *G6p* (Lamming et al. 2014a). It is likely that this shift of fuel into gluconeogenesis directly accounts for the decrease in lipid synthesis, as lipogenic substrates including glucose and pyruvate are exported from the liver into the blood. Ingenuity Pathway Analysis (IPA) performed on gene expression data from the livers of mice lacking hepatic *Rictor* suggested that many transcription factors were involved in the response. Interestingly, this included not only the SREBPs, but also FOXO1 and FOXA2, transcription factors recently highlighted as mTORC1-independent regulators of lipogenesis (Wan et al. 2011). IPA analysis also suggested altered regulation of the transcription factor PPARγ (Lamming et al. 2014a). While mTORC1 has been known to regulate PPARγ and thus control adipogenesis for some time, mice lacking hepatic Rictor have essentially normal mTORC1 activity (Lamming et al. 2014b). This suggests that mTORC2 may mediate PPARγ activation and/or SREBP activity, perhaps via Akt (Kim et al. 2010). While no direct link between mTORC2 and SREBP has yet been identified, mTORC2 localizes to the endoplasmic reticulum (Betz et al. 2013), which is the location where SREBP processing and maturation occurs.

Interestingly, when we performed a phosphoproteomic analysis of Rictor depleted livers we found altered phosphorylation of glycogen synthase (Gys2) (Lamming et al. 2014a), a key enzyme in the conversion of glucose to glycogen. Gys2 is phosphorylated by many key factors in central energy sensing pathways, including AMPK and GSK3 (Bultot et al. 2012; von Wilamowitz-Moellendorff et al. 2013). While the phosphorylation of some sites of Gys2 showed decreased abundance (S645, SS652) in mice lacking hepatic Rictor, other sites showed an increase in phosphorylation (S6S11, S8S11). The exact mechanism by which mTORC2 regulates Gys2 has yet to be determined.

1.6 Indirect Regulation of Hepatic Lipid Metabolism by mTOR

Mammals possess integrated physiological systems in which the metabolism and function of one organ, such as the liver, interacts with and can be regulated by other organs. With respect to hepatic lipogenesis, mediators of this effect include not only lipids and other metabolites, but also circulating hormones. Thus far, we have primarily focused on the role of mTOR within the liver on hepatic lipogenesis; however, mTOR also regulates many other aspects of lipid metabolism in mammals (Caron et al. 2015). Here, we will first discuss the role of mTOR in adipose tissue, and then discuss the regulation of hepatic function by two secreted liver hormones, insulin-like growth factor 1 (IGF-1) and fibroblast growth factor 21 (FGF21). We will also discuss the indirect regulation of hepatic lipogenesis by adipose tissue and the brain, mediated by Growth Hormone (GH) and IGF-1 signaling, as well as by adiponectin and leptin. We recently reviewed how these factors regulate the PI3K/Akt/mTOR signaling pathway with respect to aging and longevity (Lamming 2014); but these factors also regulate hepatic lipogenesis.

mTOR signaling is critical to the formation and metabolism of adipose tissue. mTORC1 has been known to play a role in adipogenesis for over 10 years, with experiments in 2001 demonstrating that genetic or pharmacological inhibition of mTORC1 impairs adipogenesis in tissue culture systems, and activation of mTORC1 via deletion of *TSC2* promotes adipogenesis (Zhang et al. 2009; Lamming and Sabatini 2013). Oddly though, the effects of rapamycin on adipogenesis in vivo are significantly less apparent, with most investigators concluding that rapamycin has at most a transient effect on body composition (Liu et al. 2014b). One possible explanation for the divergence is that in vivo rapamycin treatment is begun much later in life, typically at 3–4 months of age, while experiments in vitro mimic conditions found during pre and post-natal development.

Mechanistically, the effects of mTORC1 on adipogenesis are mediated by 4E-BP1, Lipin1, and S6K1 (Fig. 3.2), which in turn regulates the activity of PPARγ and SREBPs (Fig. 3.2). The effect of mTORC2 on adipogenesis is more subtle, and deletion of *Rictor* in adipose tissue was initially reported to have essentially no effect on adipogenesis in mice (Cybulski et al. 2009; Kumar et al. 2010). However, a new study in which *Rictor* deletion was performed in very early adipogenesis was recently shown to substantially block adipogenesis (Yao et al. 2013). This early effect is mediated by Akt through the transcriptional repression of the transcription factor FoxC2 (Fig. 3.2).

One effect of *Rictor* deletion in adipose tissue is an increase in whole body size and the size of many individual organs (Cybulski et al. 2009). The driver of this effect is increased hepatic expression of insulin-like growth factor 1 (IGF-1), as well as IGF binding protein 3 (IGFBP3), although the mechanism through which adipose tissue regulates the hepatic expression of these proteins in unknown. Adipose tissue signals to the liver directly, through the production of the circulating hormones adiponectin and leptin. Adipose tissue also promotes the expression of growth hormone via leptin, which in turn regulates the hepatic expression of IGF-1.

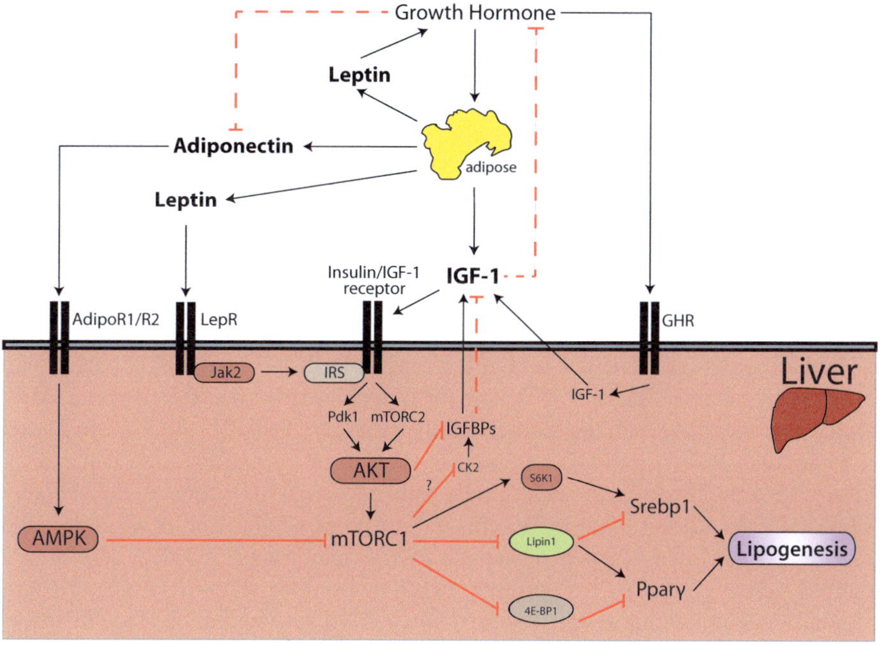

Fig. 3.3 Regulation of hepatic lipogenesis by the growth hormone/IGF-1 axis. Growth hormone signaling promotes the expression of IGF-1 by the liver and adipose tissue, which signals through the liver insulin/IGF-1 receptor to promote PI3K/Akt/mTORC1 signaling and hepatic lipogenesis. Growth hormone also normally represses adiponectin, a hormone from white adipose tissue that inhibits mTORC1 activity by activating AMPK. Leptin promotes PI3K/Akt/mTOR signaling via the Jak2-mediated phosphorylation of insulin receptor substrate. Leptin also promotes IGF-1 signaling by stimulating the GH/IGF-1 axis. Finally, mTORC1 and Akt regulate the expression and phosphorylation of IGFBPs, which regulate the activity and bioavailability of IGF-1. This figure is adapted from a figure originally published in Lamming (2014), *Springerplus*

Figure 3.3 illustrates the connections between these hormones and mTOR-mediated hepatic lipogenesis, and these hormones are discussed in detail below.

1.7 Insulin-Like Growth Factor 1 (IGF-1)

Insulin is one of the most potent activators of lipogenesis (Laplante and Sabatini 2009a). Signaling through the insulin receptor is intimately connected with signaling by the closely related hormone insulin-like growth factor 1 (IGF-1). Insulin and IGF-1 can both signal through the insulin and IGF-1 receptor as well as through hybrid insulin-IGF-1 receptor complexes (van Heemst 2010). The effect of IGF-1 on Akt phosphorylation is similar to insulin in many tissues, including the liver (Park et al. 1999; Hvid et al. 2011).

The effect of IGF-1 on lipogenesis in hepatocytes has not been investigated, but there is evidence that IGF-1 regulates lipogenesis in other tissues. In 2008, it was shown that IGF-1 regulates SREBP-1 expression and lipogenesis in sebocytes—cells derived from the sebaceous gland—through a PI3K dependent pathway (Smith et al. 2008). It is logical that since the liver is also responsive to IGF-1, lipogenesis would likewise be activated. Insulin and IGF-1 can both stimulate glycogen synthesis in hepatocytes (Park et al. 1999). While glycogenesis and lipogenesis are distinct pathways, recent work has shown that they are coordinately regulated processes, and it was recently demonstrated that glycogen stimulates the mTORC1/SREBP1 pathway to promote lipogenesis (Lu et al. 2014).

IGF-1 does not circulate freely in the blood, and is usually found bound to an IGF binding protein (IGFBP), and in the case of IGFBP-3 or IGFBP-5, also to an acid labile subunit (ALS). IGF-1 activity is regulated by the IGFBPs and ALS, which act to regulate both the bioavailability and serum half-life of IGF-1 [reviewed in (Lamming 2014)]. IGFBP-1 is repressed by insulin signaling, and it has been known for a number of years that mTOR pathway activity is critical to the suppression of *Igfbp1* transcription by insulin in hepatocytes (Mounier et al. 2006). We recently showed that deletion of hepatic *Rictor*, inhibiting mTORC2, results in a tenfold increase in *Igfbp1* transcription in the mouse liver (Lamming et al. 2014a). In addition to regulating *Igfbp1* transcription, mTOR activity inhibits the phosphorylation of IGFBP-1 by repressing the activity of CK2 (Abu Shehab et al. 2014). Phosphorylated IGFPB-1 binds IGF-1 with higher affinity, repressing IGF-1 action (Jones et al. 1993; Yu et al. 1998); therefore mTOR activity promotes IGF-1 activity by inhibiting this phosphorylation. In Fig. 3.3, we have highlighted the phosphorylation of hepatic IGFBP-1 by mTOR, but IGFBP-1 produced by other tissues could be regulated similarly by mTOR in those tissues.

Hepatic mTORC1 activity is regulated directly and indirectly by the adipose-derived hormones leptin and adiponectin. Indirectly, leptin promotes mTORC1 signaling through activation of the growth hormone/IGF-1 axis (Watanobe and Habu 2002), which then acts as discussed above to activate PI3K/Akt/mTOR/SREBP lipogenesis. Directly, leptin acts through the leptin receptor, which is expressed by hepatocytes, to activate mTORC1 via Jak2 mediated phosphorylation of insulin receptor substrate (Park and Ahima 2014). In contrast to leptin, which promotes mTORC1 activity, adiponectin acts to inhibit mTORC1 signaling via the activation of AMPK. AMPK acts to inhibit mTORC1 signaling via the activating phosphorylation of TSC1 and the inhibitory phosphorylation of Raptor, a protein component of mTORC1 (Laplante and Sabatini 2012).

We have illustrated a model of how these hormones may impact mTOR signaling and hepatic lipogenesis in Fig. 3.3. However, the complex physiological environment in which these hormones act in vivo means that this model is limited. For instance, while this model seems to predict that decreased hepatic growth hormone signaling would lead to decreased lipogenesis, growth hormone also stimulates the production of IGF-1 in the liver which suppresses the production of growth hormone (Fig. 3.3). Thus, deletion of the liver growth hormone receptor blocks this feedback loop and increases levels of growth hormone systemically, resulting in steatosis and hepatocellular carcinoma (Mueller et al. 2011). Feedback loops may also act to limit the magnitude of observed effects. Notably, mTORC1 negatively regulates insulin/IGF-1 action via Grb10 and the S6K-mediated phosphorylation of insulin receptor substrate (Hsu et al. 2011; Yu et al. 2011).

1.8 Other Hormones May also Regulate Hepatic Lipogenesis via mTOR

We have previously discussed the regulation of mTOR pathway signaling by hormones including cholecystokinin, fibroblast growth factor 21 (FGF21), glucagon-like peptide 1 (GLP-1), ghrelin, humanin, and resistin [reviewed in (Lamming 2014)]. Many of these hormones have been demonstrated to regulate mTOR activity in specific target tissues, including the pancreas [e.g., GLP-1 (Kelly et al. 2010)] and the brain. While it is not clear that these tissues directly regulate hepatic mTOR signaling, it is likely that at least some of these hormones regulate lipogenesis indirectly through the regulation of metabolism.

Fibroblast growth factor 21 (FGF21) is a particularly interesting hormone that does act on the liver—indeed, FGF21 is a hormone primarily produced by the liver, although FGF21 production from other tissues has been reported (Izumiya et al. 2008). FGF21 is produced in response to the stimuli such as fatty acids which stimulate peroxisome proliferator-activated receptor alpha (PPARα) dependent gene transcription [reviewed in (Emanuelli et al. 2014)]. FGF21 agonists have been pursued for the treatment of type 2 diabetes on the basis of exciting data from rodent, primate, and even human studies of FGF21 agonists (Kharitonenkov and Adams 2014).

FGF21 is an insulin sensitizing hormone (Xu et al. 2009), and thus naively one might expect that FGF21 increases hepatic lipogenesis. However, FGF21 is also induced by fasting, a condition in which lipogenesis is unlikely to be beneficial. In vivo and in vitro data suggest that FGF21 actually promotes hepatic lipid oxidation and triglyceride clearance (Badman et al. 2007). Studies in cell culture lines have determined that FGF21 represses the expression of SREBP1c at the transcriptional level, and also inhibits SREBP1c processing and maturation (Zhang et al. 2011).

2 Conclusion

Hepatic lipogenesis is an extremely important process with consequences not only for the liver, but the physiology and health of the entire organism. Lipogenesis has therefore been placed under the control of the mTOR protein kinase, linking it firmly to environmental cues and nutrient availability. In addition to direct control by hepatic mTOR, lipogenesis in the liver is responsive to hormones including insulin, IGF-1, adiponectin and leptin that are produced by the pancreas, liver, and adipose tissue and modulated by other peripheral tissues and the brain. Other hormones indirectly regulate mTOR-mediated hepatic lipogenesis by acting upon these peripheral tissues or by modulating the metabolic rate. Future research will further clarify how diverse metabolic inputs coordinate the lipogenic response to dietary intake and environmental stimuli.

Acknowledgements We would like to thank Emma Baar, Joseph Baur and Mathieu Laplante for their comments, all the members of the Lamming laboratory for their assistance and insight, and Dena Cohen for advice and consultation. We have previously explored some of the concepts and ideas in this chapter (Lamming and Sabatini 2013; Lamming 2014), and Figs. 3.2 and 3.3 are adapted from our recent publications on lipid metabolism and the regulation of the mTOR signaling pathway by the GH/IGF-1 axis (Lamming and Sabatini 2013; Lamming 2014). The Lamming lab is supported by a K99/R00 Pathway to Independence Award to D.W.L. from the National Institute of Health/National Institute on Aging (AG041765), as well as startup funds from the UW-Madison School of Medicine and Public Health and the UW-Madison Department of Medicine. This work was supported using facilities and resources at the William S. Middleton Memorial Veterans Hospital. This work does not represent the views of the Department of Veterans Affairs or the United States Government.

References

Abu Shehab M, Damerill I, Shen T, Rosario FJ, Nijland M, Nathanielsz PW, Kamat A, Jansson T, Gupta MB (2014) Liver mTOR controls IGF-I bioavailability by regulation of protein kinase CK2 and IGFBP-1 phosphorylation in fetal growth restriction. Endocrinology 155:1327–1339

Badman MK, Pissios P, Kennedy AR, Koukos G, Flier JS, Maratos-Flier E (2007) Hepatic fibroblast growth factor 21 is regulated by PPARalpha and is a key mediator of hepatic lipid metabolism in ketotic states. Cell Metab 5:426–437

Bae EJ, Xu J, Oh DY, Bandyopadhyay G, Lagakos WS, Keshwani M, Olefsky JM (2012) Liver-specific p70 S6 kinase depletion protects against hepatic steatosis and systemic insulin resistance. J Biol Chem 287:18769–18780

Baker H, Sidorowicz A, Sehgal SN, Vezina C (1978) Rapamycin (AY-22,989), a new antifungal antibiotic. III. In vitro and in vivo evaluation. J Antibiot (Tokyo) 31:539–545

Barlow AD, Nicholson ML, Herbert TP (2013) Evidence for rapamycin toxicity in pancreatic beta-cells and a review of the underlying molecular mechanisms. Diabetes 62:2674–2682

Bar-Peled L, Sabatini DM (2014) Regulation of mTORC1 by amino acids. Trends Cell Biol 24:400–406

Bar-Peled L, Schweitzer LD, Zoncu R, Sabatini DM (2012) Ragulator is a GEF for the rag GTPases that signal amino acid levels to mTORC1. Cell 150:1196–1208

Bar-Peled L, Chantranupong L, Cherniack AD, Chen WW, Ottina KA, Grabiner BC, Spear ED, Carter SL, Meyerson M, Sabatini DM (2013) A tumor suppressor complex with GAP activity for the Rag GTPases that signal amino acid sufficiency to mTORC1. Science 340:1100–1106

Betz C, Stracka D, Prescianotto-Baschong C, Frieden M, Demaurex N, Hall MN (2013) Feature article: mTOR complex 2-Akt signaling at mitochondria-associated endoplasmic reticulum membranes (MAM) regulates mitochondrial physiology. Proc Natl Acad Sci U S A 110:12526–12534

Bultot L, Guigas B, Von Wilamowitz-Moellendorff A, Maisin L, Vertommen D, Hussain N, Beullens M, Guinovart JJ, Foretz M, Viollet B, Sakamoto K, Hue L, Rider MH (2012) AMP-

activated protein kinase phosphorylates and inactivates liver glycogen synthase. Biochem J 443:193–203
Byles V, Covarrubias AJ, Ben-Sahra I, Lamming DW, Sabatini DM, Manning BD, Horng T (2013) The TSC-mTOR pathway regulates macrophage polarization. Nat Commun 4:2834
Caron A, Richard D, Laplante M (2015) The Roles of mTOR Complexes in Lipid Metabolism. Annu Rev Nutr 35:321–348
Chantranupong L, Wolfson RL, Sabatini DM (2015) Nutrient-sensing mechanisms across evolution. Cell 161:67–83
Cunningham JT, Rodgers JT, Arlow DH, Vazquez F, Mootha VK, Puigserver P (2007) mTOR controls mitochondrial oxidative function through a YY1-PGC-1alpha transcriptional complex. Nature 450:736–740
Cybulski N, Polak P, Auwerx J, Ruegg MA, Hall MN (2009) mTOR complex 2 in adipose tissue negatively controls whole-body growth. Proc Natl Acad Sci U S A 106:9902–9907
Duvel K, Yecies JL, Menon S, Raman P, Lipovsky AI, Souza AL, Triantafellow E, Ma Q, Gorski R, Cleaver S, Vander Heiden MG, MacKeigan JP, Finan PM, Clish CB, Murphy LO, Manning BD (2010) Activation of a metabolic gene regulatory network downstream of mTOR complex 1. Mol Cell 39:171–183
Efeyan A, Schweitzer LD, Bilate AM, Chang S, Kirak O, Lamming DW, Sabatini DM (2014) RagA, but not RagB, is essential for embryonic development and adult mice. Dev Cell 29(3):321–329
Egan D, Kim J, Shaw RJ, Guan KL (2011) The autophagy initiating kinase ULK1 is regulated via opposing phosphorylation by AMPK and mTOR. Autophagy 7:643–644
Emanuelli B, Vienberg SG, Smyth G, Cheng C, Stanford KI, Arumugam M, Michael MD, Adams AC, Kharitonenkov A, Kahn CR (2014) Interplay between FGF21 and insulin action in the liver regulates metabolism. J Clin Invest 124:515–527
Fantus D, Thomson AW (2015) Evolving perspectives of mTOR complexes in immunity and transplantation. Am J Transplant 15:891–902
Festuccia WT, Pouliot P, Bakan I, Sabatini DM, Laplante M (2014) Myeloid-specific Rictor deletion induces M1 macrophage polarization and potentiates in vivo pro-inflammatory response to lipopolysaccharide. PLoS One 9:e95432
Frias MA, Thoreen CC, Jaffe JD, Schroder W, Sculley T, Carr SA, Sabatini DM (2006) mSin1 is necessary for Akt/PKB phosphorylation, and its isoforms define three distinct mTORC2s. Curr Biol 16:1865–1870
Fricain JC, Cellerie K, Sibaud V, Catros S, Taieb A, Merville P (2008) Oral ulcers in kidney allograft recipients treated with sirolimus. Ann Dermatol Venereol 135:737–741
Gan X, Wang J, Wang C, Sommer E, Kozasa T, Srinivasula S, Alessi D, Offermanns S, Simon MI, Wu D (2012) PRR5L degradation promotes mTORC2-mediated PKC-delta phosphorylation and cell migration downstream of Galpha12. Nat Cell Biol 14:686–696
Garcia-Martinez JM, Alessi DR (2008) mTOR complex 2 (mTORC2) controls hydrophobic motif phosphorylation and activation of serum- and glucocorticoid-induced protein kinase 1 (SGK1). Biochem J 416:375–385
Guertin DA, Stevens DM, Thoreen CC, Burds AA, Kalaany NY, Moffat J, Brown M, Fitzgerald KJ, Sabatini DM (2006) Ablation in mice of the mTORC components raptor, rictor, or mLST8 reveals that mTORC2 is required for signaling to Akt-FOXO and PKCalpha, but not S6K1. Dev Cell 11:859–871
Haas JT, Miao J, Chanda D, Wang Y, Zhao E, Haas ME, Hirschey M, Vaitheesvaran B, Farese RV Jr, Kurland IJ, Graham M, Crooke R, Foufelle F, Biddinger SB (2012) Hepatic insulin signaling is required for obesity-dependent expression of SREBP-1c mRNA but not for feeding-dependent expression. Cell Metab 15:873–884
Hagiwara A, Cornu M, Cybulski N, Polak P, Betz C, Trapani F, Terracciano L, Heim MH, Ruegg MA, Hall MN (2012) Hepatic mTORC2 activates glycolysis and lipogenesis through Akt, glucokinase, and SREBP1c. Cell Metab 15:725–738
Hong JC, Kahan BD (2000) Sirolimus-induced thrombocytopenia and leukopenia in renal transplant recipients: risk factors, incidence, progression, and management. Transplantation 69:2085–2090

Houde VP, Brule S, Festuccia WT, Blanchard PG, Bellmann K, Deshaies Y, Marette A (2010) Chronic rapamycin treatment causes glucose intolerance and hyperlipidemia by upregulating hepatic gluconeogenesis and impairing lipid deposition in adipose tissue. Diabetes 59:1338–1348

Hsu PP, Kang SA, Rameseder J, Zhang Y, Ottina KA, Lim D, Peterson TR, Choi Y, Gray NS, Yaffe MB, Marto JA, Sabatini DM (2011) The mTOR-regulated phosphoproteome reveals a mechanism of mTORC1-mediated inhibition of growth factor signaling. Science 332:1317–1322

Humphrey SJ, Yang G, Yang P, Fazakerley DJ, Stockli J, Yang JY, James DE (2013) Dynamic adipocyte phosphoproteome reveals that Akt directly regulates mTORC2. Cell Metab 17:1009–1020

Hvid H, Fels JJ, Kirk RK, Thorup I, Jensen HE, Hansen BF, Oleksiewicz MB (2011) In situ phosphorylation of Akt and ERK1/2 in rat mammary gland, colon, and liver following treatment with human insulin and IGF-1. Toxicol Pathol 39:623–640

Ikenoue T, Inoki K, Yang Q, Zhou X, Guan KL (2008) Essential function of TORC2 in PKC and Akt turn motif phosphorylation, maturation and signalling. EMBO J 27:1919–1931

Inoki K, Li Y, Xu T, Guan KL (2003) Rheb GTPase is a direct target of TSC2 GAP activity and regulates mTOR signaling. Genes Dev 17:1829–1834

Inoki K, Ouyang H, Zhu T, Lindvall C, Wang Y, Zhang X, Yang Q, Bennett C, Harada Y, Stankunas K, Wang CY, He X, MacDougald OA, You M, Williams BO, Guan KL (2006) TSC2 integrates Wnt and energy signals via a coordinated phosphorylation by AMPK and GSK3 to regulate cell growth. Cell 126:955–968

Izumiya Y, Bina HA, Ouchi N, Akasaki Y, Kharitonenkov A, Walsh K (2008) FGF21 is an Akt-regulated myokine. FEBS Lett 582:3805–3810

Jeon TI, Osborne TF (2012) SREBPs: metabolic integrators in physiology and metabolism. Trends Endocrinol Metab 23:65–72

Jewell JL, Kim YC, Russell RC, Yu FX, Park HW, Plouffe SW, Tagliabracci VS, Guan KL (2015) Metabolism. Differential regulation of mTORC1 by leucine and glutamine. Science 347:194–198

Jones JI, Busby WH Jr, Wright G, Clemmons DR (1993) Human IGFBP-1 is phosphorylated on 3 serine residues: effects of site-directed mutagenesis of the major phosphoserine. Growth Regul 3:37–40

Kang SA, Pacold ME, Cervantes CL, Lim D, Lou HJ, Ottina K, Gray NS, Turk BE, Yaffe MB, Sabatini DM (2013) mTORC1 phosphorylation sites encode their sensitivity to starvation and rapamycin. Science 341:1236566

Kelly P, Bailey CL, Fueger PT, Newgard CB, Casey PJ, Kimple ME (2010) Rap1 promotes multiple pancreatic islet cell functions and signals through mammalian target of rapamycin complex 1 to enhance proliferation. J Biol Chem 285:15777–15785

Kharitonenkov A, Adams AC (2014) Inventing new medicines: the FGF21 story. Mol Metab 3:221–229

Kim SP, Ha JM, Yun SJ, Kim EK, Chung SW, Hong KW, Kim CD, Bae SS (2010) Transcriptional activation of peroxisome proliferator-activated receptor-gamma requires activation of both protein kinase A and Akt during adipocyte differentiation. Biochem Biophys Res Commun 399:55–59

Kobayashi T, Minowa O, Sugitani Y, Takai S, Mitani H, Kobayashi E, Noda T, Hino O (2001) A germ-line Tsc1 mutation causes tumor development and embryonic lethality that are similar, but not identical to, those caused by Tsc2 mutation in mice. Proc Natl Acad Sci U S A 98:8762–8767

Kumar A, Lawrence JC Jr, Jung DY, Ko HJ, Keller SR, Kim JK, Magnuson MA, Harris TE (2010) Fat cell-specific ablation of rictor in mice impairs insulin-regulated fat cell and whole-body glucose and lipid metabolism. Diabetes 59:1397–1406

Lamming DW (2014) Diminished mTOR signaling: a common mode of action for endocrine longevity factors. Springerplus 3:735

Lamming DW, Sabatini DM (2013) A central role for mTOR in lipid homeostasis. Cell Metab 18:465–469

Lamming DW, Ye L, Katajisto P, Goncalves MD, Saitoh M, Stevens DM, Davis JG, Salmon AB, Richardson A, Ahima RS, Guertin DA, Sabatini DM, Baur JA (2012) Rapamycin-induced insulin resistance is mediated by mTORC2 loss and uncoupled from longevity. Science 335:1638–1643

Lamming DW, Demirkan G, Boylan JM, Mihaylova MM, Peng T, Ferreira J, Neretti N, Salomon A, Sabatini DM, Gruppuso PA (2014a) Hepatic signaling by the mechanistic target of rapamycin complex 2 (mTORC2). FASEB J 28:300–315

Lamming DW, Mihaylova MM, Katajisto P, Baar EL, Yilmaz OH, Hutchins A, Gultekin Y, Gaither R, Sabatini DM (2014b) Depletion of Rictor, an essential protein component of mTORC2, decreases male lifespan. Aging Cell 13:911–917

Laplante M, Sabatini DM (2009a) An emerging role of mTOR in lipid biosynthesis. Curr Biol 19:R1046–R1052

Laplante M, Sabatini DM (2009b) mTOR signaling at a glance. J Cell Sci 122:3589–3594

Laplante M, Sabatini DM (2012) mTOR signaling in growth control and disease. Cell 149:274–293

Lee DF, Kuo HP, Chen CT, Hsu JM, Chou CK, Wei Y, Sun HL, Li LY, Ping B, Huang WC, He X, Hung JY, Lai CC, Ding Q, Su JL, Yang JY, Sahin AA, Hortobagyi GN, Tsai FJ, Tsai CH, Hung MC (2007) IKK beta suppression of TSC1 links inflammation and tumor angiogenesis via the mTOR pathway. Cell 130:440–455

Li S, Brown MS, Goldstein JL (2010) Bifurcation of insulin signaling pathway in rat liver: mTORC1 required for stimulation of lipogenesis, but not inhibition of gluconeogenesis. Proc Natl Acad Sci U S A 107:3441–3446

Liu P, Begley M, Michowski W, Inuzuka H, Ginzberg M, Gao D, Tsou P, Gan W, Papa A, Kim BM, Wan L, Singh A, Zhai B, Yuan M, Wang Z, Gygi SP, Lee TH, Lu KP, Toker A, Pandolfi PP, Asara JM, Kirschner MW, Sicinski P, Cantley L, Wei W (2014a) Cell-cycle-regulated activation of Akt kinase by phosphorylation at its carboxyl terminus. Nature 508:541–545

Liu Y, Diaz V, Fernandez E, Strong R, Ye L, Baur JA, Lamming DW, Richardson A, Salmon AB (2014b) Rapamycin-induced metabolic defects are reversible in both lean and obese mice. Aging (Albany NY) 6:742–754

Lu B, Bridges D, Yang Y, Fisher K, Cheng A, Chang L, Meng ZX, Lin JD, Downes M, Yu RT, Liddle C, Evans RM, Saltiel AR (2014) Metabolic crosstalk: molecular links between glycogen and lipid metabolism in obesity. Diabetes 63:2935–2948

Ma KL, Ruan XZ, Powis SH, Chen Y, Moorhead JF, Varghese Z (2007) Sirolimus modifies cholesterol homeostasis in hepatic cells: a potential molecular mechanism for sirolimus-associated dyslipidemia. Transplantation 84:1029–1036

Menon S, Dibble CC, Talbott G, Hoxhaj G, Valvezan AJ, Takahashi H, Cantley LC, Manning BD (2014) Spatial control of the TSC complex integrates insulin and nutrient regulation of mTORC1 at the lysosome. Cell 156:771–785

Morrisett JD, Abdel-Fattah G, Hoogeveen R, Mitchell E, Ballantyne CM, Pownall HJ, Opekun AR, Jaffe JS, Oppermann S, Kahan BD (2002) Effects of sirolimus on plasma lipids, lipoprotein levels, and fatty acid metabolism in renal transplant patients. J Lipid Res 43:1170–1180

Mounier C, Dumas V, Posner BI (2006) Regulation of hepatic insulin-like growth factor-binding protein-1 gene expression by insulin: central role for mammalian target of rapamycin independent of forkhead box O proteins. Endocrinology 147:2383–2391

Mueller KM, Kornfeld JW, Friedbichler K, Blaas L, Egger G, Esterbauer H, Hasselblatt P, Schlederer M, Haindl S, Wagner KU, Engblom D, Haemmerle G, Kratky D, Sexl V, Kenner L, Kozlov AV, Terracciano L, Zechner R, Schuetz G, Casanova E, Pospisilik JA, Heim MH, Moriggl R (2011) Impairment of hepatic growth hormone and glucocorticoid receptor signaling causes steatosis and hepatocellular carcinoma in mice. Hepatology 54:1398–1409

Onda H, Lueck A, Marks PW, Warren HB, Kwiatkowski DJ (1999) Tsc2(+/−) mice develop tumors in multiple sites that express gelsolin and are influenced by genetic background. J Clin Invest 104:687–695

Osborne TF, Espenshade PJ (2009) Evolutionary conservation and adaptation in the mechanism that regulates SREBP action: what a long, strange tRIP it's been. Genes Dev 23:2578–2591

Owen JL, Zhang Y, Bae SH, Farooqi MS, Liang G, Hammer RE, Goldstein JL, Brown MS (2012) Insulin stimulation of SREBP-1c processing in transgenic rat hepatocytes requires p70 S6-kinase. Proc Natl Acad Sci U S A 109:16184–16189

Park HK, Ahima RS (2014) Leptin signaling. F1000Prime Rep 6:73

Park BC, Kido Y, Accili D (1999) Differential signaling of insulin and IGF-1 receptors to glycogen synthesis in murine hepatocytes. Biochemistry 38:7517–7523

Pearce LR, Huang X, Boudeau J, Pawlowski R, Wullschleger S, Deak M, Ibrahim AF, Gourlay R, Magnuson MA, Alessi DR (2007) Identification of Protor as a novel Rictor-binding component of mTOR complex-2. Biochem J 405:513–522

Peterson TR, Laplante M, Thoreen CC, Sancak Y, Kang SA, Kuehl WM, Gray NS, Sabatini DM (2009) DEPTOR is an mTOR inhibitor frequently overexpressed in multiple myeloma cells and required for their survival. Cell 137:873–886

Peterson TR, Sengupta SS, Harris TE, Carmack AE, Kang SA, Balderas E, Guertin DA, Madden KL, Carpenter AE, Finck BN, Sabatini DM (2011) mTOR complex 1 regulates lipin 1 localization to control the SREBP pathway. Cell 146:408–420

Porstmann T, Santos CR, Griffiths B, Cully M, Wu M, Leevers S, Griffiths JR, Chung YL, Schulze A (2008) SREBP activity is regulated by mTORC1 and contributes to Akt-dependent cell growth. Cell Metab 8:224–236

Powell JD, Pollizzi KN, Heikamp EB, Horton MR (2012) Regulation of immune responses by mTOR. Annu Rev Immunol 30:39–68

Rebsamen M, Pochini L, Stasyk T, de Araujo ME, Galluccio M, Kandasamy RK, Snijder B, Fauster A, Rudashevskaya EL, Bruckner M, Scorzoni S, Filipek PA, Huber KV, Bigenzahn JW, Heinz LX, Kraft C, Bennett KL, Indiveri C, Huber LA, Superti-Furga G (2015) SLC38A9 is a component of the lysosomal amino acid sensing machinery that controls mTORC1. Nature 519(7544):477–481

Reis F, Parada B, Teixeira de Lemos E, Garrido P, Dias A, Piloto N, Baptista S, Sereno J, Eufrasio P, Costa E, Rocha-Pereira P, Santos-Silva A, Figueiredo A, Mota A, Teixeira F (2009) Hypertension induced by immunosuppressive drugs: a comparative analysis between sirolimus and cyclosporine. Transplant Proc 41:868–873

Rolfe M, McLeod LE, Pratt PF, Proud CG (2005) Activation of protein synthesis in cardiomyocytes by the hypertrophic agent phenylephrine requires the activation of ERK and involves phosphorylation of tuberous sclerosis complex 2 (TSC2). Biochem J 388:973–984

Saci A, Cantley LC, Carpenter CL (2011) Rac1 regulates the activity of mTORC1 and mTORC2 and controls cellular size. Mol Cell 42:50–61

Sancak Y, Thoreen CC, Peterson TR, Lindquist RA, Kang SA, Spooner E, Carr SA, Sabatini DM (2007) PRAS40 is an insulin-regulated inhibitor of the mTORC1 protein kinase. Mol Cell 25:903–915

Sancak Y, Peterson TR, Shaul YD, Lindquist RA, Thoreen CC, Bar-Peled L, Sabatini DM (2008) The Rag GTPases bind raptor and mediate amino acid signaling to mTORC1. Science 320:1496–1501

Sancak Y, Bar-Peled L, Zoncu R, Markhard AL, Nada S, Sabatini DM (2010) Ragulator-Rag complex targets mTORC1 to the lysosomal surface and is necessary for its activation by amino acids. Cell 141:290–303

Sarbassov DD, Ali SM, Kim DH, Guertin DA, Latek RR, Erdjument-Bromage H, Tempst P, Sabatini DM (2004) Rictor, a novel binding partner of mTOR, defines a rapamycin-insensitive and raptor-independent pathway that regulates the cytoskeleton. Curr Biol 14:1296–1302

Sarbassov DD, Guertin DA, Ali SM, Sabatini DM (2005) Phosphorylation and regulation of Akt/PKB by the rictor-mTOR complex. Science 307:1098–1101

Sarbassov DD, Ali SM, Sengupta S, Sheen JH, Hsu PP, Bagley AF, Markhard AL, Sabatini DM (2006) Prolonged rapamycin treatment inhibits mTORC2 assembly and Akt/PKB. Mol Cell 22:159–168

Schreiber KH, Ortiz D, Academia EC, Anies AC, Liao CY, Kennedy BK (2015) Rapamycin-mediated mTORC2 inhibition is determined by the relative expression of FK506-binding proteins. Aging Cell 14(2):265–273

Sengupta S, Peterson TR, Laplante M, Oh S, Sabatini DM (2010) mTORC1 controls fasting-induced ketogenesis and its modulation by ageing. Nature 468:1100–1104
Smith TM, Gilliland K, Clawson GA, Thiboutot D (2008) IGF-1 induces SREBP-1 expression and lipogenesis in SEB-1 sebocytes via activation of the phosphoinositide 3-kinase/Akt pathway. J Invest Dermatol 128:1286–1293
Tee AR, Anjum R, Blenis J (2003) Inactivation of the tuberous sclerosis complex-1 and -2 gene products occurs by phosphoinositide 3-kinase/Akt-dependent and -independent phosphorylation of tuberin. J Biol Chem 278:37288–37296
Tsun ZY, Bar-Peled L, Chantranupong L, Zoncu R, Wang T, Kim C, Spooner E, Sabatini DM (2013) The folliculin tumor suppressor is a GAP for the RagC/D GTPases that signal amino acid levels to mTORC1. Mol Cell 52:495–505
van Heemst D (2010) Insulin, IGF-1 and longevity. Aging Dis 1:147–157
Vezina C, Kudelski A, Sehgal SN (1975) Rapamycin (AY-22,989), a new antifungal antibiotic. I. Taxonomy of the producing streptomycete and isolation of the active principle. J Antibiot (Tokyo) 28:721–726
von Wilamowitz-Moellendorff A, Hunter RW, Garcia-Rocha M, Kang L, Lopez-Soldado I, Lantier L, Patel K, Peggie MW, Martinez-Pons C, Voss M, Calbo J, Cohen PT, Wasserman DH, Guinovart JJ, Sakamoto K (2013) Glucose-6-phosphate-mediated activation of liver glycogen synthase plays a key role in hepatic glycogen synthesis. Diabetes 62:4070–4082
Wan M, Leavens KF, Saleh D, Easton RM, Guertin DA, Peterson TR, Kaestner KH, Sabatini DM, Birnbaum MJ (2011) Postprandial hepatic lipid metabolism requires signaling through Akt2 independent of the transcription factors FoxA2, FoxO1, and SREBP1c. Cell Metab 14:516–527
Wang S, Tsun ZY, Wolfson RL, Shen K, Wyant GA, Plovanich ME, Yuan ED, Jones TD, Chantranupong L, Comb W, Wang T, Bar-Peled L, Zoncu R, Straub C, Kim C, Park J, Sabatini BL, Sabatini DM (2015) Metabolism. Lysosomal amino acid transporter SLC38A9 signals arginine sufficiency to mTORC1. Science 347:188–194
Watanobe H, Habu S (2002) Leptin regulates growth hormone-releasing factor, somatostatin, and alpha-melanocyte-stimulating hormone but not neuropeptide Y release in rat hypothalamus in vivo: relation with growth hormone secretion. J Neurosci 22:6265–6271
Xu J, Lloyd DJ, Hale C, Stanislaus S, Chen M, Sivits G, Vonderfecht S, Hecht R, Li YS, Lindberg RA, Chen JL, Jung DY, Zhang Z, Ko HJ, Kim JK, Veniant MM (2009) Fibroblast growth factor 21 reverses hepatic steatosis, increases energy expenditure, and improves insulin sensitivity in diet-induced obese mice. Diabetes 58:250–259
Yao Y, Suraokar M, Darnay BG, Hollier BG, Shaiken TE, Asano T, Chen CH, Chang BH, Lu Y, Mills GB, Sarbassov D, Mani SA, Abbruzzese JL, Reddy SA (2013) BSTA promotes mTORC2-mediated phosphorylation of Akt1 to suppress expression of FoxC2 and stimulate adipocyte differentiation. Sci Signal 6:ra2
Yecies JL, Manning BD (2011) Transcriptional control of cellular metabolism by mTOR signaling. Cancer Res 71:2815–2820
Yecies JL, Zhang HH, Menon S, Liu S, Yecies D, Lipovsky AI, Gorgun C, Kwiatkowski DJ, Hotamisligil GS, Lee CH, Manning BD (2011) Akt stimulates hepatic SREBP1c and lipogenesis through parallel mTORC1-dependent and independent pathways. Cell Metab 14:21–32
Yip CK, Murata K, Walz T, Sabatini DM, Kang SA (2010) Structure of the human mTOR complex I and its implications for rapamycin inhibition. Mol Cell 38:768–774
Yu J, Iwashita M, Kudo Y, Takeda Y (1998) Phosphorylated insulin-like growth factor (IGF)-binding protein-1 (IGFBP-1) inhibits while non-phosphorylated IGFBP-1 stimulates IGF-I-induced amino acid uptake by cultured trophoblast cells. Growth Horm IGF Res 8:65–70
Yu Y, Yoon SO, Poulogiannis G, Yang Q, Ma XM, Villen J, Kubica N, Hoffman GR, Cantley LC, Gygi SP, Blenis J (2011) Phosphoproteomic analysis identifies Grb10 as an mTORC1 substrate that negatively regulates insulin signaling. Science 332:1322–1326
Yu Z, Wang R, Fok WC, Coles A, Salmon AB, Perez VI (2015) Rapamycin and dietary restriction induce metabolically distinctive changes in mouse liver. J Gerontol A Biol Sci Med Sci 70:410–420

Yuan M, Pino E, Wu L, Kacergis M, Soukas AA (2012) Identification of Akt-independent regulation of hepatic lipogenesis by mammalian target of rapamycin (mTOR) complex 2. J Biol Chem 287:29579–29588
Zhang HH, Huang J, Duvel K, Boback B, Wu S, Squillace RM, Wu CL, Manning BD (2009) Insulin stimulates adipogenesis through the Akt-TSC2-mTORC1 pathway. PLoS One 4:e6189
Zhang Y, Lei T, Huang JF, Wang SB, Zhou LL, Yang ZQ, Chen XD (2011) The link between fibroblast growth factor 21 and sterol regulatory element binding protein 1c during lipogenesis in hepatocytes. Mol Cell Endocrinol 342:41–47
Zhang P, Verity MA, Reue K (2014) Lipin-1 regulates autophagy clearance and intersects with statin drug effects in skeletal muscle. Cell Metab 20:267–279
Zhou X, Clister TL, Lowry PR, Seldin MM, Wong GW, Zhang J (2015) Dynamic visualization of mTORC1 activity in living cells. Cell Rep. doi: 10.1016/j.celrep.2015.02.031
Zinzalla V, Stracka D, Oppliger W, Hall MN (2011) Activation of mTORC2 by association with the ribosome. Cell 144:757–768
Zoncu R, Bar-Peled L, Efeyan A, Wang S, Sancak Y, Sabatini DM (2011) mTORC1 senses lysosomal amino acids through an inside-out mechanism that requires the vacuolar H(+)-ATPase. Science 334:678–683

Chapter 4
FoxO1: A Conductor of Insulin Signaling to Glucose and Lipid Metabolism

Ting Zhang and H. Henry Dong

Abstract Excessive hepatic glucose production and very low-density lipoprotein (VLDL) secretion contributes to the pathogenesis of hyperglycemia and hypertriglyceridemia—two pathological traits that are intertwined in insulin resistant subjects with obesity and type 2 diabetes. To date, the molecular basis that links insulin resistance to hepatic overproduction of glucose and VLDL is poorly understood. Preclinical and clinical investigation leads to the characterization of the forkhead box O1 (FoxO1) as a significant transcription factor that integrates insulin signaling to hepatic glucose and lipid metabolism. FoxO1 is abundantly expressed in the liver and its transcription activity is tightly regulated by insulin. Insulin inhibits FoxO1 activity via a distinct mechanism by altering its subcellular redistribution. Insulin signaling bifurcates at FoxO1 in the liver to govern two metabolic pathways, namely gluconeogenesis and VLDL assembly. This effect helps synchronize hepatic insulin signaling to simultaneously adjust the rates of hepatic glucose production and VLDL secretion in response to nutrient availability. Such FoxO1-dependent mechanism seems pivotal for the liver to rapidly adapt to metabolic shift between fasting to feeding states for maintaining normal glucose and lipid homeostasis. In this article, we center our review on the role of FoxO1 in the liver. We provide mechanistic insights into how FoxO1 orchestrates insulin action on hepatic glucose and lipid metabolism in healthy individuals, and how FoxO1 dysregulation, resulting from insulin resistance, contributes to the dual pathogenesis of hyperglycemia and hyperlipidemia in obesity and type 2 diabetes. Finally, we discuss whether FoxO1 is a potential therapeutic target for improving blood glucose and lipid profiles in insulin resistant subjects with obesity and type 2 diabetes.

Keywords FoxO1 • Insulin • Lipogenesis • Gluconeogenesis • Hyperglycemia • Hypertriglyceridemia • Steatosis • Liver

T. Zhang • H.H. Dong, Ph.D. (✉)
Division of Pediatric Endocrinology, Department of Pediatrics, Childern's Hospital of Pittsburgh of UPMC, University of Pittsburgh School of Medicine, 4401 Penn Ave, Pittsburgh, PA 15224, USA

J.M. Ntambi (ed.), *Hepatic De Novo Lipogenesis and Regulation of Metabolism*,
DOI 10.1007/978-3-319-25065-6_4

1 The Forkhead Transcription Factor FoxO1

FoxO1 belongs to the forkhead box O subfamily that is characterized by a highly conserved DNA binding motif, known as forkhead box or winged helix domain, at the amino domain (Accili and Arden 2004). This subfamily of transcription factors comprises FoxO1, FoxO3, FoxO4 and FoxO6 in mammals (Accili and Arden 2004; Barthel et al. 2005) (Fig. 4.1). Among the four members of the FoxO subfamily, FoxO1 is the most characterized one. FoxO1 is first reported by Galili et al. (1993) as a gene fusion product of the t(2;13) (q35;q14) translocation that is associated with alveolar rhabdomyosarcoma, a soft tissue cancer that is presented mainly in children. Subsequent investigation characterizes FoxO1 as an important forkhead transcription factor that mediates the inhibitory effect of insulin or insulin-like growth factor 1 (IGF-1) on the expression of genes, whose functions are instrumental for cell metabolism, cell growth, cell oxidative stress and survival, cell differentiation in mammals (Accili and Arden 2004). This FoxO1-dependent signaling pathway is evolutionally conserved. The FoxO1 ortholog daf16 in *C. elegans* acts downstream of the insulin-like receptor daf-2 to protect against the oxidative stress (Mueller et al. 2014; Essers et al. 2005). Increased daf16 expression, resulting from attenuated insulin-like signaling, contributed to the life-span extension in *C. elegans* (Lee et al. 2003). Likewise, the FoxO1 ortholog dFoxO signaling through insulin-like receptor substrate *chico* contributes to the regulation of longevity in *Drosophila melanogaster* (Hwangbo et al. 2004).

In this article, we provide a comprehensive review on the role of FoxO1 in the liver, with focus on the physiological significance of FoxO1 in governing hepatic glucose and lipid metabolism. We will discuss the importance of hepatic regulation of FoxO1 activity in response to insulin and nutritional signaling for maintaining

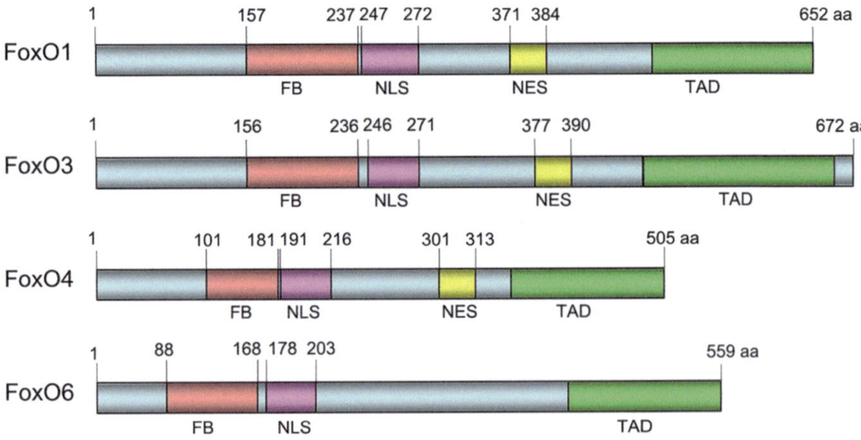

Fig. 4.1 Structural comparison of FoxO homologues. Schematic depiction of the four FoxO members in the mouse FoxO subfamily. *NLS* nuclear localization sequence, *NES* nuclear export sequence, *FB* forkhead box, *TAD trans*-activation domain

glucose and lipid homeostasis in normal individuals. In addition, we review evidence that FoxO1 dysregulation, resulting from impaired insulin signaling, contributes to aberrant glucose and lipid metabolism in the liver, contributing to the pathogenesis of hyperglycemia and hyperlipidemia in obesity and type 2 diabetes. We also discuss whether FoxO1 is a potential therapeutic target for improving blood glucose and lipid profiles in insulin resistant subjects with obesity and type 2 diabetes.

2 FoxO1 Mediates the Inhibitory Action of Insulin on Gene Expression

FoxO1 is a substrate of protein kinase B (PKB, also known as Akt) and serum/glucocorticoid-induced kinase (SGK) (Fig. 4.2). Both PKB and SGK are regulated by insulin or IGF-1 in response to nutritional cues. Insulin or IGF-1 exerts its inhibitory effect on gene expression via a highly conserved sequence (TG/ATTTT/G), termed insulin response element (IRE) in target promoters (Accili and Arden 2004). FoxO1 binds to the IRE DNA as a monomer with residues Asn^{211}, Ser^{212}, Arg^{214}, His^{215}, Asn^{216}, and Ser^{218} in the major groove of DNA helix (Brent et al. 2008). This effect is tightly regulated by insulin. In the absence of insulin, FoxO1 resides in the nucleus and binds as a trans-activator to the IRE DNA motif, enhancing promoter activity. In response to insulin, FoxO1 is phosphorylated at three highly conserved phosphorylation sites (Thr24, Ser256, and Ser319) via the PI3K-dependent pathway, resulting in its nuclear exclusion and inhibition of target gene expression (Accili and Arden 2004) (Fig. 4.3). Three additional phosphorylation sites (S322, S325, and S329) are identified and their phosphorylation by CK1 (casein kinase 1) and DYRK1A (dual-specificity tyrosine-phosphorylated and regulated kinase 1A) seems to accelerate FoxO1 nuclear exclusion (Accili and Arden 2004). FoxO1, when phosphorylated in response to insulin or IGF-1, is targeted for ubiquitination, followed by proteosome-mediated degradation (Matsuzaki et al. 2003; Aoki et al. 2004; Huang et al. 2005).

Consistent with its nuclear-cytoplasmic shuttling, FoxO1 contains a characteristic nuclear localization signal (NLS) at the amino domain and nuclear export signal (NES) at the carboxyl domain. NES is capable of associating with the chromosomal maintenance 1 (CRM-1), known as exportin-1 that is responsible for binding to the NES motif of a cargo protein and transporting the cargo protein from the nucleus to cytoplasm (Monecke et al. 2009; Dong et al. 2009). Indeed, FoxO1 is able to complex with CRM-1 in liver cells (Kim et al. 2011). In response to insulin, FoxO1 in complex with CRM-1 is translocated from the nucleus to cytoplasm. This effect is abrogated by leptomycin B, an agent that binds specifically to CRM-1 and inhibits its cargo-trafficking activity (Kim et al. 2011). These results underscore the significance of CRM-1 in facilitating insulin-dependent FoxO1 subcellular redistribution. Failure in phosphorylation of FoxO1 results in its permanent nuclear localization and constitutive trans-activation of gene expression (Accili and Arden 2004).

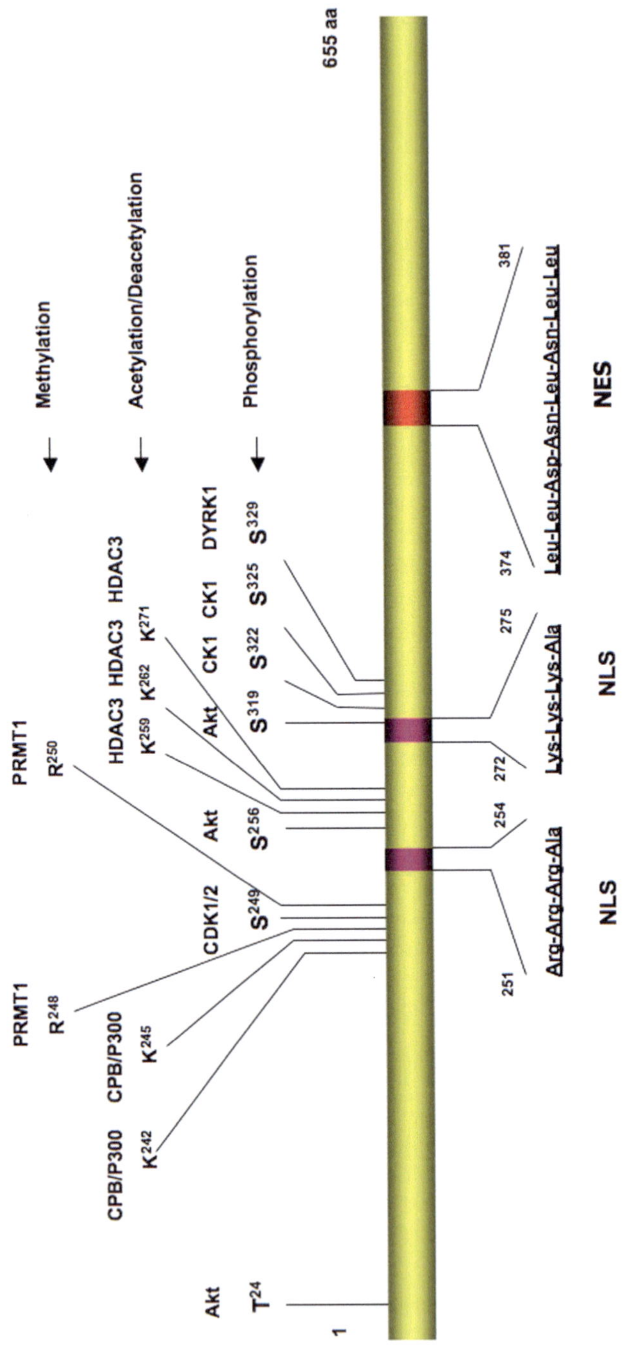

Fig. 4.2 Schematic depiction of the human FoxO1 protein. *NLS* nuclear localization signals, *NES* nuclear export signal. Highlighted are amino acid residues that undergo post-translational modifications including phosphorylation, acetylation, de-acetylation, and methylation. *Akt/PKB* protein kinase B, *HDAC* class IIa histone deacetylase, *Ck1* casein kinase 1, *DYRK1* dual specificity tyrosine-phosphorylated- and regulated-kinase 1, *CBP* cAMP response element-binding (CREB)-binding protein, *PRMT1* protein arginine N-methyltransferase 1, *Cdk* cyclin-dependent kinase

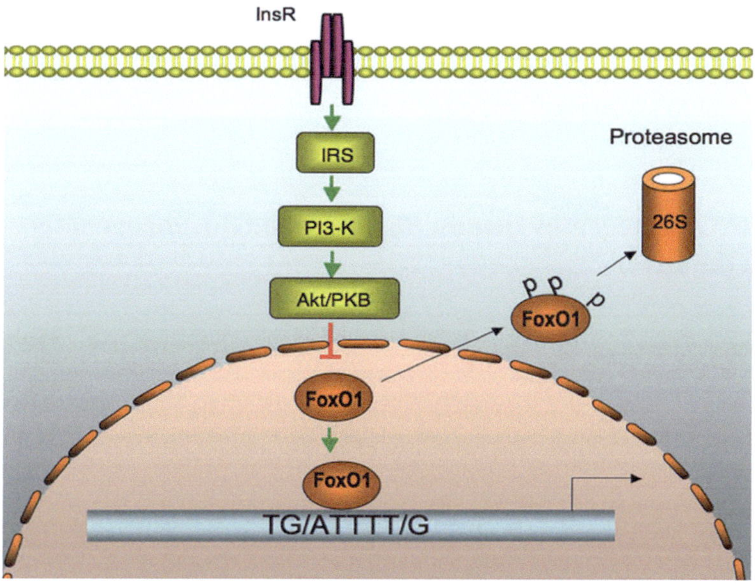

Fig. 4.3 Insulin-dependent inhibition of FoxO1 activity. In the absence of insulin, FoxO1 binds to the insulin responsive element (IRE, TG/ATTTT/G) in the target promoter and this effect stimulates target gene expression. In response to insulin, FoxO1 undergoes insulin-dependent phosphorylation and trafficking from the nucleus to the cytoplasm. Cytosolic FoxO1 proteins are destined for proteasome-mediated degradation. *InsR* insulin receptor, *IRS* insulin receptor substrate, *PI3-K* phosphoinositide 3-kinase, *Akt/PKB* protein kinase B

Although the nucleocytoplasmic shuttling constitutes a compartmental mechanism by which insulin inhibits FoxO activity in cells (Accili and Arden 2004; Kamagate and Dong 2008), insulin (or IGF-1) inhibition of FoxO1 activity can occur without necessarily altering FoxO1 subcellular redistribution, but this inhibition depends on the ability of FoxO1 to undergo insulin-dependent phosphorylation (Tsai et al. 2003; Zhang et al. 2002). Phosphorylation seems to distort FoxO1 DNA binding domain, preventing FoxO1 binding to target promoters. Consistent with this notion is the presence of both Thr24 and Ser256 phosphorylation sites in its DNA binding domain (Accili and Arden 2004). A similar mechanism is evolved in FoxO6. Unlike other members of the FoxO subfamily, FoxO6 lacks the characteristic NES motif and does not undergo insulin-dependent nuclear-cytoplasmic trafficking. Insulin inhibits FoxO6 activity by promoting its phosphorylation at Thr26 and Ser184 within the FoxO6 DNA binding domain, and this effect disables its DNA binding activity in the nucleus (Kim et al. 2011).

Interestingly, phosphorylation of *C. elegans* ortholog daf-16 does not affect its DNA binding activity (Accili and Arden 2004). Instead, daf-16 phosphorylation promotes its association with the multifunctional factor 14-3-3, perturbing daf-16 binding to target DNA. Likewise, FoxO1 is capable of interacting with 14-3-3 via two consensus sites for 14-3-3 binding (Thr24 and Ser253) (Accili and Arden

2004). A prevailing notion is that 14-3-3 in complex with FoxO1 masks its DNA binding motif and disables its binding to target promoters. In the cytoplasm, 14-3-3 serves as an anchor to sequester FoxO1, precluding its re-entry into the nucleus (Accili and Arden 2004).

3 FoxO1 Integrates Insulin Signaling to Gluconeogenesis in the Liver

Gluconeogenesis is a pivotal metabolic pathway for converting non-carbohydrate metabolites (lactate, glycerol, and amino acids) to glucose. A life-sustaining process, gluconeogenesis provides the sole fuel source for the brain, testes and erythrocytes during starvation or exercise. Gluconeogenesis takes place mainly in liver and to a much lesser extent in kidney (Edgerton et al. 2009a; Wahren and Ekberg 2007), accounting for up to 80 % of total glucose production in healthy individuals during a prolonged fast (Ekberg et al. 1999). Gluconeogenesis is tightly regulated by hormonal and nutritional cues (Wahren and Ekberg 2007; Tonelli et al. 2005; Rossetti et al. 1997; Massillon et al. 1995; Pilkis and Granner 1992; Yoon et al. 2001). In response to postprandial insulin secretion, gluconeogenic activity is suppressed to limit glucose production (Puigserver et al. 2003; Fisher and Kahn 2003; Edgerton et al. 2009b). This effect serves two purposes: (**i**) to prevent prolonged postprandial glucose excursion and (**ii**) to replenish glycogen content in liver. In response to fasting, gluconeogenesis is stimulated, resulting in increased glucose output (Koo et al. 2005; Liu et al. 2008; Jiang and Zhang 2003; Zhang et al. 2005). Such a reciprocal mechanism, orchestrated by two opposing hormones (insulin and glucagon), is crucial for rapid adaptation by the liver to metabolic shift between fed and fasting states for maintaining blood sugar levels within a normal range (Wahren and Ekberg 2007).

FoxO1 plays a pivotal role in mediating insulin action on gluconeogenesis, a metabolic pathway that takes place mainly in liver for the generation of glucose from non-carbohydrate substrates. Gluconeogenesis is controlled by phosphoenolpyruvate carboxykinase (PEPCK) and glucose-6-phosphatase (G6Pase). PEPCK catalyzes the conversion of oxaloacetate into phosphoenolpyruvate, the rate-limiting step of gluconeogenesis. G6Pase carries out the final step of gluconeogenesis by converting glucose-6-phosphate to glucose. FoxO1 stimulates PEPCK and G6Pase expression via selective binding to the insulin response element (IRE), a conserved DNA motif (TG/ATTTT/G) in the PEPCK and G6Pase promoters, promoting gluconeogenesis. Interestingly, the PEPCK promoter contains only one IRE motif whereas G6Pase promoter has three tandem repeats of IRE for FoxO1 binding (Fig. 4.4). As a result, FoxO1 gain-of-function is associated with augmented hepatic gluconeogenesis, contributing to fasting hyperglycemia in mice (Matsumoto et al. 2007; Qu et al. 2006). Furthermore, FoxO1-mediated induction of hepatic gluconeogenesis is amplified by PGC-1α that acts as a co-activator to bind and augment FoxO1 activity (Puigserver et al. 2003). Conversely, FoxO1 loss-of-function is associated with diminished hepatic gluconeogenesis, resulting in reduced fasting

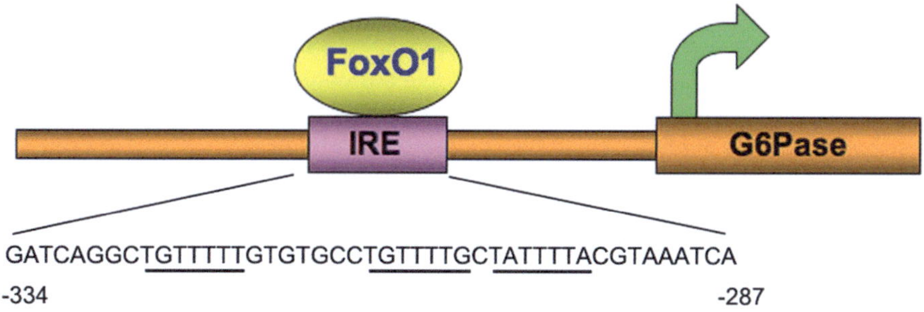

Fig. 4.4 Schematic depiction of the G6Pase promoter. *IRE* insulin responsive element. The nucleotide sequences corresponding to the consensus IRE motif within the proximal region of the G6Pase promoter are *underlined*

blood glucose levels in mice (Altomonte et al. 2003). FoxO1-dependent regulation of gluconeogenesis provides an acute mechanism for the liver to adjust the rate of glucose production under different physiological states.

Although FoxO1 is shown to play a key role in integrating insulin signaling to hepatic gluconeogenesis, FoxO1 depletion in liver does not result in abolition of hormonal regulation of gluconeogenesis in mice (Matsumoto et al. 2007). In keeping with this observation, we show that mice with FoxO1 loss-of-function in liver are associated with diminished gluconeogenic activities (by ~50 %) and impaired abilities to maintain fasting blood glucose levels within the normal range (Kamagate et al. 2008). Furthermore, FoxO1 loss-of-function reduces, but does not abrogate the responsiveness of liver to insulin or glucagon (via cAMP) signaling, suggesting that there are additional factors or compensatory mechanisms in mediating hormonal regulation of hepatic gluconeogenesis (Matsumoto et al. 2007; Kamagate et al. 2008). Consistent with this idea is that FoxO3 and FoxO4 are shown to play redundant roles in regulating hepatic glucose production in the liver (Haeusler et al. 2010). In addition, FoxO6 seems to compensate for the loss of FoxO1 function in mediating insulin action on hepatic gluconeogenesis (Kim et al. 2011, 2013). Given the crucial role of gluconeogenesis in whole-body glucose metabolism, it is conceivable that liver has evolved to possess redundant pathways for governing insulin-dependent inhibition of glucose production for rapid adaptation to metabolic shift from fasting to refeeding states.

4 FoxO1-Independent Inhibition of Gluconeogenesis in the Liver

Apart from its direct inhibitory effect on gluconeogenesis, insulin acts to suppress hepatic gluconeogenesis by indirect mechanisms. Consistent with this notion, Olefsky and colleagues show that in obese subjects, suppression of hepatic glucose production can occur in response to insulin infusion without an apparent increase in portal insulin concentrations (Prager et al. 1987). Insulin is known to exert a potent inhibitory effect on hormone-sensitive lipase, which in turn curbs adipose tissue lipolysis and reduces the availability of free fatty acid (FFA) and glycerol (Holm et al. 2000). In parallel, insulin inhibits muscle proteolysis and restrains the release of amino acids (Liu et al. 2008). FFA is used as an energy source for gluconeogenesis, and glycerol and amino acids are precursors of gluconeogenesis. A significant reduction in FFA influx to liver along with dwindling supplies of ambient gluconeogenic substrates imparts a potent indirect inhibitory effect of insulin on hepatic gluconeogenesis (Ader and Bergman 1990). In addition, insulin acts on pancreatic α-cells to suppress glucagon release, limiting hepatic glucose output (Jiang and Zhang 2003; Unger et al. 1978; Ishihara et al. 2003). It follows that insulin modulates hepatic gluconeogenesis via both direct and indirect mechanisms (Ader and Bergman 1990; Cherrington 2005).

5 FoxO1 in the Pathogenesis of Fasting Hyperglycemia in Type 2 Diabetes

Unchecked gluconeogenesis in the liver contributes to excessive glucose production in diabetes. Excessive endogenous glucose production exerts its deleterious effect on whole-body metabolism in diabetes in three fundamental ways: (1) It causes prolonged postprandial blood glucose excursion, (2) It contributes to the pathogenesis of fasting hyperglycemia, and (3) It accounts in part for diminished glycogen storage in liver. To date, the molecular basis that links impaired insulin action to unrestrained gluconeogenesis remains incompletely understood.

There are preclinical data showing that hepatic expression of FoxO1 is markedly elevated along with its increased nuclear localization in both insulin-deficient and insulin-resistant liver, coinciding with the development of fasting hyperglycemia in type 1 and type 2 diabetes (Altomonte et al. 2003, 2004; Imae et al. 2003). Similar observations are made in high fat-induced obese mice (Qu et al. 2006) and high fructose-fed hamsters (Qu et al. 2007a). Furthermore, there is clinical evidence that hepatic FoxO1 expression is abnormally upregulated in the liver, correlating with increased hepatic gluconeogenesis and insulin resistance in humans with clinical manifestation of non-alcoholic fatty liver disease (Valenti et al. 2008). Thus, type 1 and type 2 diabetes share a common pathogenic feature of FoxO1 dysregulation in the liver. Increased FoxO1 activity, resulting from insulin deficiency or insulin resistance, stimulates hepatic gluconeogenesis and contributes to the pathogenesis of fasting hyperglycemia in diabetes.

6 FoxO1 Mediates Insulin Action on TG Metabolism

In addition to its role in glucose metabolism, FoxO1 plays an important role in lipid metabolism. The role of FoxO1 in lipid metabolism culminates in its ability to mediate insulin action on hepatic expression of ApoC-III, an apolipoprotein that is synthesized in the liver and intestine (Haddad et al. 1986). ApoC-III is one of the most abundant lipoproteins in plasma, where it is present as an exchangeable moiety between high-density lipoprotein (HDL) and TG-rich particles including VLDL and chylomicrons (Shachter 2001). ApoC-III functions as inhibitor of lipoprotein lipase (LPL), a key enzyme for the hydrolysis of TG in VLDL and chylomicrons (Wang et al. 1985). Therefore, ApoC-III is a crucial determinant of plasma triglyceride levels (Shachter 2001). Elevated ApoC-III levels are associated with impaired TG hydrolysis and delayed clearance of TG-rich particles due to inhibition of lipoprotein lipase activity (Shachter 2001). At higher concentrations, ApoC-III also inhibits hepatic lipase activity (Shachter 2001). Furthermore, it affects the ApoE-mediated hepatic uptake of TG-rich remnants, resulting in the accumulation of VLDL-TG and chylomicrons in plasma and development of hypertriglyceridemia (Shachter 2001). Transgenic mice with ApoC-III overproduction develop hepatic steatosis and hypertriglyceridemia (Aalto-Setala et al. 1992; Lee et al. 2011; Qu et al. 2007b). In contrast, ApoC-III knockout mice are associated with hypotriglyceridemia (Maeda et al. 1994). Due to its crucial importance in modulating plasma TG hydrolysis and clearance, ApoC-III has been recognized as a marker of plasma TG-rich lipoprotein catabolism and proposed as an indicator of prognosis for coronary risk in healthy subjects (Onat et al. 2003). APOC-III gene variants are also linked to non-alcoholic fatty liver disease in humans (Petersen et al. 2010). In contrast, Loss-of-Function Mutations in the APOC-III gene is associated with lower plasma TG levels and reduced risk of coronary disease in humans.

Hepatic ApoC-III production is subject to insulin inhibition (Altomonte et al. 2004). To understand the underlying mechanism, Altomonte et al. (2004) report that hepatic ApoC-III expression is regulated in a FoxO1-dependent mechanism. FoxO1 physically binds to the ApoC-III promoter and functionally stimulates ApoC-III promoter activity. This effect contributes to significant upregulation of hepatic ApoC-III production. Likewise, FoxO1 mediates insulin action on intestinal ApoC-III production and TG secretion. As a result, transgenic mice expressing a FoxO1 constitutively active allele exhibit elevated plasma ApoC-III and VLDL-TG levels, and retarded postprandial TG clearance, characteristic of hypertriglyceridemia (Altomonte et al. 2004). FoxO1 activity is increased in the liver and this effect contributes to the induction of hepatic ApoC-III production and the development of hypertriglyceridemia in diabetic *db/db* mice. Thus, FoxO1 signaling through ApoC-III plays an important role in regulating TG-rich lipoprotein catabolism.

7 FoxO1 Mediates Insulin Action on VLDL-TG Production

In addition to its effect on ApoC-III production and TG catabolism, FoxO1 impacts hepatic VLDL-TG production via the microsomal triglyceride transfer protein (MTP), a molecular chaperone that resides in the endoplasmic reticulum (ER). MTP (MW, 88 kDa) functions in complex with protein disulphide isomerase (PDI, 58 kDa) transports lipid droplets to nascent apolipoprotein B (ApoB), the rate-limiting step in hepatic VLDL-TG assembly and secretion (Tietge et al. 1999; Jamil et al. 1998; Hussain et al. 2003). Human subjects with genetic MTP deficiency develop abetalipoproteinemia or Bassen-Kornzweig syndrome, a rare autosomal recessive disorder (Berriot-Varoqueaux et al. 2000; Liao et al. 2003; Bjorkegren et al. 2002; Leung et al. 2000; Raabe et al. 1999). Patients with abetalipoproteinemia manifest severe lipid disorders and multiple vitamin deficiencies, due to abnormal absorption of dietary fats, lipid-associated substances including cholesterols and fat-soluble vitamins (Berriot-Varoqueaux et al. 2000; Liao et al. 2003; Bjorkegren et al. 2002; Leung et al. 2000; Raabe et al. 1999). Similarly, homozygous $MTP^{-/-}$ mice are associated with impaired VLDL production and embryonic death (Raabe et al. 1998, 1999; Chang et al. 1999). MTP haploinsufficiency is associated with increased hepatic fat deposition due to reduced VLDL secretion in $MTP^{+/-}$ heterozygous mice (Leung et al. 2000). The human MTP promoter contains a polymorphic site at -493G/T that has been linked with altered lipoprotein metabolism and increased risk of coronary heart disease (Ledmyr et al. 2002, 2004), the phenotype of MTP-493G/T polymorphism in inflicted individuals is strongly influenced by visceral obesity, highlighting the significance of insulin action in hepatic MTP regulation (Juo et al. 2000; Garcia-Garcia et al. 2005). In addition, both hepatic and intestinal MTP mRNA levels are significantly elevated in insulin resistant non-diabetic obese Zucker rats with impaired TG metabolism (Phillips et al. 2002), consistent with the notion that MTP is inhibited by insulin (Hagan et al. 1994; Lin et al. 1995).

Kamagate et al. (2008) report that FoxO1 targets the MTP gene for *trans-activation* and this effect promotes hepatic MTP expression and VLDL-TG production. In response to insulin, FoxO1-mediated induction of MTP expression is attenuated, correlating with the ability of FoxO1 to undergo insulin-dependent phosphorylation and translocation from nucleus to the cytoplasm. FoxO1 gain-of-function is associated with enhanced MTP expression, correlating with elevated hepatic VLDL production and plasma TG levels in FoxO1 transgenic mice. Hepatic FoxO1 abundance and MTP production are increased in diabetic *db/db* mice with abnormal TG metabolism. FoxO1 loss-of-function, resulting from RNAi-mediated FoxO1 knockdown in the liver, suppresses hepatic MTP expression and curbs VLDL production in both diabetic *db/db* and hypertriglyceridemic FoxO1 transgenic mice (Kamagate et al. 2008). These data provide important insights into the mechanism by which insulin regulates hepatic VLDL-TG production in the liver. Hepatic VLDL production is induced in response to attenuated insulin action, resulting in increased VLDL secretion during prolonged fasting. In contrast, hepatic VLDL production is suppressed to limit plasma triglyceride excursion in response to postprandial insulin secretion (Malmstrom et al. 1997, 1998; Lewis and Steiner 1996; Sparks and Sparks 1994). This effect is critical for rapid adaptation by the liver to metabolic shift from fasting to refeeding in order to maintain normal TG homeostasis.

8 FoxO1 in the Pathogenesis of Hypertriglyceridemia

Hypertriglyceridemia is the most common lipid disorder in obesity and type 2 diabetes. Hypertriglyceridemia is characterized by elevated plasma VLDL-TG and chylomicrons, and decreased plasma HDL levels (Betteridge 2000; Marcoux et al. 2001). Hypertriglyceridemia along with its metabolic sequelae of the accumulation of TG-rich lipoproteins and chylomicron remnants is considered pro-atherogenic. There are clinical and preclinical data showing that hypertriglyceridemia constitutes an independent risk factor for the progression of coronary artery disease in obesity and type 2 diabetes (Stewart et al. 1994; Defronzo 1997; Krauss 1998; Brewer 1999; Bard et al. 2001; Oliveieri et al. 2002). To date, the precise cause and mechanism of hypertriglyceridemia is incompletely understood. It is clear that hypertriglyceridemia is closely associated with increased adiposity and type 2 diabetes, underscoring the importance of insulin resistance in the pathogenesis of hypertriglyceridemia (Bard et al. 2001; Lewis et al. 2002; Chan et al. 2002; Nieves et al. 2003). An important consequence of impaired insulin action in adipose tissue is unrestrained fat mobilization, resulting in elevated plasma FFA levels and increased FFA flux into non-adipose tissues such as the liver and muscle. As a result, an increased FFA uptake by the liver stimulates hepatic VLDL-TG production via MTP, exacerbating hepatic insulin resistance (Lewis et al. 2002; Kissebah et al. 1974; Taghibiglou et al. 2000). In addition, insulin resistance is associated with ApoC-III overproduction, which inhibits lipoprotein lipase and hepatic lipase activities, resulting in retarded hydrolysis/clearance of VLDL-TG and chylomicrons (Lewis et al. 2002; Adeli et al. 2001). These data are suggestive of a close relationship between insulin resistance and hypertriglyceridemia.

The revelation of FoxO1 as an integrator of insulin signaling to MTP and ApoC-III production shed light on the pathophysiology of hypertriglyceridemia. In normal individuals, insulin signaling through FoxO1 regulates hepatic MTP and ApoC-III production, adjusting the rates of hepatic VLDL-TG production vs. clearance (Fig. 4.5). FoxO1 is subject to insulin inhibition via PI3K-dependent phosphorylation and nuclear exclusion (Accili and Arden 2004). In obesity and type 2 diabetes, an impaired ability of insulin to keep FoxO1 activity in check in insulin resistant liver promotes hepatic MTP and ApoC-III overproduction. This effect in coalition with increased fatty acid influx into liver contributes to augmented VLDL-TG production and diminished clearance of TG-rich lipoprotein particles, contributing to the development of hypertriglyceridemia in obesity and diabetes (Kamagate et al. 2008; Altomonte et al. 2004).

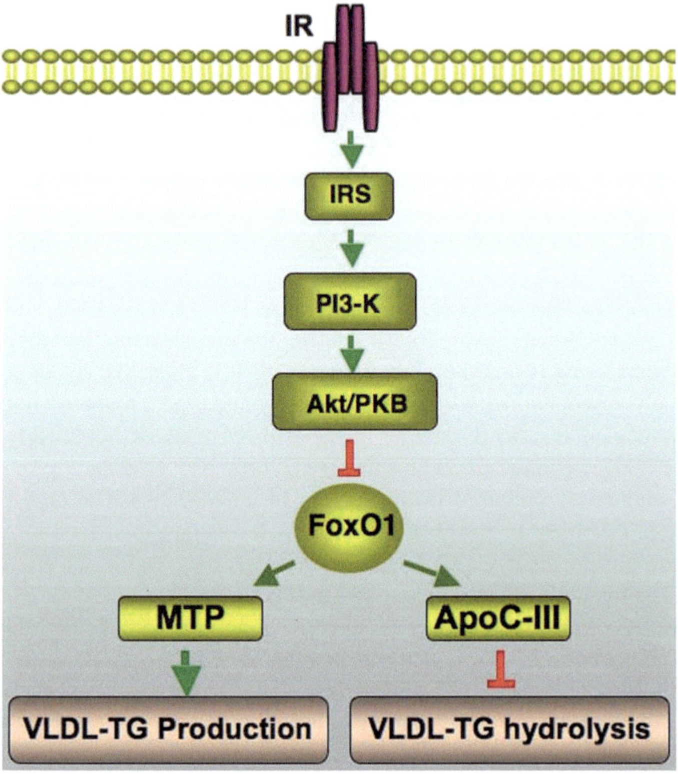

Fig. 4.5 FoxO1 integrates insulin signaling to MTP and ApoC-III Production. FoxO1 mediates insulin action on hepatic MTP and ApoC-III expression. This effect is critical for regulating hepatic VLDL-TG production and VLDL-TG hydrolysis in normal individuals. Increased FoxO1 activity, resulting from insulin resistance, promotes hepatic VLDL-TG overproduction and retards VLDL-TG clearance, contributing to hypertriglyceridemia in obesity and type 2 diabetes. *IR* insulin receptor, *IRS* insulin receptor substrate, *Akt/PKB* protein kinase B, *PI3-K* phosphoinositide 3-kinase

9 FoxO1 in Lipogenesis and Non-alcoholic Fatty Liver Disease

There is both clinical and preclinical datashowing that elevated FoxO1 production in liver is associated with increased intra-hepatic fat infiltration, characteristic of non-alcoholic fatty liver disease (NAFLD) (Altomonte et al. 2005). Patients with hepatosteatosis are associated with enhanced FoxO1 activity in the liver (Valenti et al. 2008). This effect also correlated with increased hepatic expression of FoxO1 target genes, namely PEPCK, G6Pase and ApoC-III, three key functions in the glucose and lipid metabolism (Valenti et al. 2008). Likewise, Matsumoto et al. (2006) show that hepatic lipogenic genes including sterol regulatory element binding protein (SREBP-1c), fatty acid synthase (FAS), and acetyl-CoA carboxylase (ACC) are upregulated in transgenic mice expressing a constitutively active FoxO1 allele. These findings are reproduced by two independent studies showing that adenovirus-mediated FoxO1 production results in augmented lipogenesis with concomitant fat accumulation in the liver (Qu et al. 2006; Matsumoto et al. 2006). It appears that unchecked FoxO1 activity contributes to the up-regulation of lipogenic gene expression, accounting in part for the development of NAFLD.

These observations are counterintuitive, as FoxO1 does not directly stimulate SREBP-1c expression. To address the underlying mechanism, Qu et al. show that increased FoxO1 activity stimulates the expression of peroxisome proliferators-activated receptor-γ co-activator (PGC)-1β in hepatocytes (Qu et al. 2006). PGC-1β is characterized as a co-activator (Lin et al. 2005), that is responsible for binding and activating SREBP-1c, a master nuclear transcription factor for hepatic lipogenesis (Foretz et al. 1999a, b). There is evidence that PGC-1β plays a role in NAFLD in high fructose-fed rats (Nagai et al. 2009).

However this observation seems at variance with (Zhang et al. 2006; Prager et al. 1987), who report that hepatic SREBP-1c mRNA expression is reduced by 50 %, accounting for decreased hepatic lipogenic activity in transgenic mice expressing a constitutively active FoxO1 allele. It is noted that Zhang et al. studied the expression of SREBP-1c 6 h after feeding, when lipogenesis is relatively higher and FoxO1 activity is lower under fasting conditions.

One potential mechanism for FoxO1-mediated induction of steatosis is that FoxO1 overproduction alters insulin action in the liver. Indeed, FoxO1 gain-of-function, resulting from adenovirus-mediated FoxO1 production or transgenic overexpression of its constitutively active allele, is associated with impaired insulin action in liver, accompanied by inappropriately increased hepatic glucose production (Altomonte et al. 2005; Nakae et al. 2002). FoxO1 loss-of-function, caused by hepatic expression of its dominant-negative allele or anti-sense oligonucleotide-mediated FoxO1 knockdown in liver, is associated with improved insulin action and blood glucose profiles in insulin resistant obese mice (Altomonte et al. 2003; Samuel et al. 2006). FoxO1 haploinsufficiency protects mice from developing high fat diet-induced insulin resistance (Nakae et al. 2003) and rescues the diabetic phenotype in insulin receptor substrate 2 (IRS2)-deficient diabetic mice (Okamoto et al. 2004). Furthermore, liver-specific FoxO1 knockout results in near normalization of the metabolic disorders in insulin resistant mice with genetic ablation of both IRS1 and IRS2 function in liver (Dong et al. 2008). It is likely that FoxO1-elicited induction of NAFLD is via both direct and indirect mechanisms. Further investigation is warranted to understand the mechanism by which FoxO1 dysregulation in the liver is linked to NAFLD.

10 Pharmacological FoxO1 Inhibition for Improving Hypertriglyceridemia

Hypertriglyceridemia is a chief risk factor for cardiovascular disease. Patients with hypertriglyceridemia are recommended with exercise or dietary supplements of omega-3 fatty acid-enriched food or ingestion of fish oil. When these non-pharmacological interventions fail to correct atherogenic lipid profiles, hypertriglyceridemia is commonly treated with fibrates, TG-lowering agents that act as agonists of peroxisome proliferator-activated receptor alpha (PPAR-α) to enhance fatty acid oxidation in peripheral tissues including liver and muscle (Qu et al. 2007a; Minnich et al. 2001). Fibrates can cause stomach upset and myopathy (Abourbih et al. 2009; Jun et al. 2010). Fibrates also decrease bile acid secretion and increase the incidence of cholesterol gallstones in patients on long-term therapy (Stahlberg et al. 1995; Grundy et al. 1972; Vanhanen and Miettinen 1995; Caroli-Bosc et al. 2001; Post et al. 2001). Better pharmacological agents are warranted for clinical management of hypertriglyceridemia.

Characterization of FoxO1 in the pathogenesis of hypertriglyceridemia has spurred the idea that targeted FoxO1 inhibition in the liver would translate a beneficial effect plasma TG metabolism in obesity and type 2 diabetes. In support of this idea, Samuel et al. (2006) report that targeted inhibition of FoxO1 by anti-sense oligonucleotide-mediated gene silencing approach enhances peripheral insulin sensitivity and improves glucose and lipid metabolism in high fat-induced obese mice. Altomonte et al. (2004) show that functional inhibition of FoxO1 by adenovirus-mediated production of FoxO1 dominant-negative mutant in the liver improves whole-body insulin sensitivity and reduces hyperinsulinemia, contributing to improved blood glucose profiles in adult diabetic *db/db* mice (Altomonte et al. 2003). Using mass spectrometric affinity screening, Nagashima et al. (2010) has identified an orally active small-molecule compound named AS1842856 that has the chemical property of binding and inhibiting FoxO1 activity in cells. When orally administrated, AS1842856 is effective in suppressing excessive gluconeogenesis in insulin resistant liver and improving blood glucose profiles in diabetic *db/db* mice. These results support the idea that targeted inhibition of FoxO1 activity in insulin resistant liver is sufficient to improve glucose and lipid profiles in animal models.

Although these studies provide the proof-of-principle that FoxO1 activity is amenable to pharmacological or genetic modulation, caution should be taken in developing FoxO1-targeted therapeutics. Apart from its role in carbohydrate metabolism and insulin action, FoxO1 is involved in cell growth, proliferation and differentiation. There is a lack of thorough investigation into the impact of chronic FoxO1 inhibition on central and peripheral tissues in vivo. In addition, FoxO1 is identified in a gene fusion product, resulting from chromosomal translocations in alveolar rhabdomyosarcoma tumors (Barr et al. 1996). Therefore, it is imperative to closely monitor the potential complication of tumorigenesis associated with long-term therapeutic intervention of FoxO1 activity in preclinical models.

11 Conclusion

FoxO1 has emerged as a key player in insulin action and carbohydrate metabolism. FoxO1 targets PEPCK and G6Pase genes for *trans*-activation in the liver for regulating hepatic glucose production. Likewise, FoxO1 integrates insulin signaling to MTP in modulating VLDL-TG assembly and secretion. In fasting states, FoxO1 becomes active in promoting hepatic glucose and triglyceride production. In fed states, FoxO1 activity is attenuated in response to postprandial insulin secretion, and this effect contributes to the inhibition of hepatic gluconeogenesis and VLDL-TG production. FoxO1 appears to be a checkpoint for synchronizing hepatic insulin signaling to adjust the rates of both glucose and triglyceride output for maintaining blood glucose and lipid levels within the physiological range. However, FoxO1 becomes deregulated in obesity and diabetes, accounting for its enhanced transcriptional activity for promoting unrestrained hepatic glucose and triglyceride production. This effect contributes to the dual pathogenesis of hyperglycemia and hypertriglyceridemia in obesity and type 2 diabetes. Although available preclinical studies support the notion of targeting FoxO1 for treating both hyperglycemia and hypertriglyceridemia, this concept should be treated with caution. It is known that FoxO1 is ubiquitously expressed and FoxO1 plays diverse functions including cell metabolism, survival, oxidative stress, proliferation, and differentiation. Due to space limitation, we could not provide a comprehensive survey of FoxO1 function in central and other peripheral tissues. Instead, we focus on reviewing recent research progress made in the liver toward delineating the molecular basis that link insulin resistance to the development of metabolic abnormalities in obesity and type 2 diabetes. Further in-depth characterization of FoxO1 in central and other peripheral tissues is needed for developing therapeutic avenues for better clinical management of metabolic disorders associated with obesity and diabetes.

Acknowledgement This work is supported by National Health Institute grant DK087764.

References

Aalto-Setala K, Fisher EA, Chen X, Chajek-Shaul T, Hayek T, Zechner R, Walsh A, Ramakrishnan R, Ginsberg HN, Breslow JL (1992) Mechanism of hypertriglyceridemia in human apolipoprotein (apo) CIII transgenic mice. Diminished very low density lipoprotein fractional catabolic rate associated with increased apo CIII and reduced apo E on the particles. J Clin Invest 90:1889–1900

Abourbih S, Filion KB, Joseph L, Schiffrin EL, Rinfret S, Poirier P, Pilote L, Genest J, Eisenberg MJ (2009) Effect of fibrates on lipid profiles and cardiovascular outcomes: a systematic review. Am J Med 122(962):e961–e968

Accili D, Arden KC (2004) FoxOs at the crossroads of cellular metabolism, differentiation, and transformation. Cell 117:421–426

Adeli K, Taghibiglou C, Van Iderstine SC, Lewis GF (2001) Mechanisms of hepatic very low-density lipoprotein overproduction in insulin resistance. Trends Cardiovasc Med 11:170–176

Ader M, Bergman RN (1990) Peripheral effects of insulin dominate suppression of fasting hepatic glucose production. Am J Physiol 258:E1020–E1032
Altomonte J, Richter A, Harbaran S, Suriawinata J, Nakae J, Thung SN, Meseck M, Accili D, Dong H (2003) Inhibition of Foxo1 function is associated with improved fasting glycemia in diabetic mice. Am J Physiol 285:E718–E728
Altomonte J, Cong L, Harbaran S, Richter A, Xu J, Meseck M, Dong HH (2004) Foxo1 mediates insulin action on apoC-III and triglyceride metabolism. J Clin Invest 114:1493–1503
Altomonte J, Qu S, Meseck M, Dong H (2005) Aberrant Foxo1 expression is associated with intrahepatic fat deposition and impaired insulin action in liver. Diabetes 54(Suppl 1):A334
Aoki M, Jiang H, Vogt PK (2004) Proteasomal degradation of the FoxO1 transcriptional regulator in cells transformed by the P3k and Akt oncoproteins. Proc Natl Acad Sci U S A 101:13613–13617
Bard JM, Charles MA, Juhan-Vague I, Vague P, Andre P, Safar M, Fruchart JC, Eschwege E (2001) Accumulation of triglyceride-rich lipoprotein in subjects with abdominal obesity. Arterioscler Thromb Vasc Biol 21:407–414
Barr FG, Nauta LE, Davis RJ, Schafer BW, Nycum LM, Biegel JA (1996) In vivo amplification of the PAX3-FKHR and PAX7-FKHR fusion genes in alveolar rhabdomyosarcoma. Hum Mol Genet 5:15–21
Barthel A, Schmoll D, Unterman TG (2005) FoxO proteins in insulin action and metabolism. Trends Endocrinol Metab 16:183–189
Berriot-Varoqueaux N, Aggerbeck LP, Samson-Bouma M, Wetterau JR (2000) The role of the microsomal triglygeride transfer protein in abetalipoproteinemia. Annu Rev Nutr 20:663–697
Betteridge DJ (2000) Diabetic dyslipidaemia. Diabetes Obes Metab 2(Suppl 1):S31–S36
Bjorkegren J, Beigneux A, Bergo MO, Maher JJ, Young SG (2002) Blocking the secretion of hepatic very low density lipoproteins renders the liver more susceptible to toxin-induced injury. J Biol Chem 277:5476–5483
Brent MM, Anand R, Marmorstein R (2008) Structural basis for DNA recognition by FoxO1 and its regulation by posttranslational modification. Structure 16:1407–1416
Brewer HBJ (1999) Hypertriglyceridemia: changes in the plasma lipoproteins associated with an increased risk of cardiovascular disease. Am J Cardiol 83(9B):3F–12F
Caroli-Bosc FX, Le Gall P, Pugliese P, Delabre B, Caroli-Bosc C, Demarquay JF, Delmont JP, Rampal P, Montet JC (2001) Role of fibrates and HMG-CoA reductase inhibitors in gallstone formation: epidemiological study in an unselected population. Dig Dis Sci 46:540–544
Chan DC, Watts GF, Barrett PH, Mamo JCL, Redgrave TG (2002) Markers of triglyceride-rich lipoprotein remnant metabolism in visceral obesity. Clin Chem 48:278–283
Chang BH, Liao W, Li L, Nakamuta M, Mack D, Chan L (1999) Liver-specific inactivation of the abetalipoproteinemia gene completely abrogates very low density lipoprotein/low density lipoprotein production in a viable conditional knockout mouse. J Biol Chem 274:6051–6055
Cherrington AD (2005) The role of hepatic insulin receptors in the regulation of glucose production. J Clin Invest 115:1136–1139
Defronzo RA (1997) Insulin resistance: a multifaceted syndrome responsible for NIDDM, obesity, hypertension, dyslipidaemia and atherosclerosis. Neth J Med 50:191–197
Dong XC, Copps KD, Guo S, Li Y, Kollipara R, DePinho RA, White MF (2008) Inactivation of hepatic Foxo1 by insulin signaling is required for adaptive nutrient homeostasis and endocrine growth regulation. Cell Metab 8:65–76
Dong X, Biswas A, Suel KE, Jackson LK, Martinez R, Gu H, Chook YM (2009) Structural basis for leucine-rich nuclear export signal recognition by CRM1. Nature 458:1136–1141
Edgerton DS, Johnson KM, Cherrington AD (2009a) Current strategies for the inhibition of hepatic glucose production in type 2 diabetes. Front Biosci 14:1169–1181
Edgerton DS, Ramnanan CJ, Grueter CA, Johnson KM, Lautz M, Neal DW, Williams PE, Cherrington AD (2009b) Effects of insulin on the metabolic control of hepatic gluconeogenesis in vivo. Diabetes 58:2766–2775

Ekberg K, Landau BR, Wajngot A, Chandramouli V, Efendic S, Brunerngraber H, Wahren J (1999) Contributions by kidney and liver to glucose production in the postabsorptive state and after 60 h of fasting. Diabetes 48:292–298

Essers MA, de Vries-Smits LM, Barker N, Polderman PE, Burgering BM, Korswagen HC (2005) Functional interaction between beta-catenin and FOXO in oxidative stress signaling. Science 308:1181–1184

Fisher SJ, Kahn CR (2003) Insulin signaling is required for insulin's direct and indirect action on hepatic glucose production. J Clin Invest 111:463–468

Foretz M, Guichard C, Ferre P, Foufelle F (1999a) Sterol regulatory element binding protein-1c is a major mediator of insulin action on the hepatic expression of glucokinase and lipogenesis-related genes. Proc Natl Acad Sci U S A 96:12737–12742

Foretz M, Pacot C, Dugail I, Lemarchand P, Guichard C, Le Liepvre X, Berthelier-Lubrano C, Spiegelman B, Kim JB, Ferre P et al (1999b) ADD1/SREBP-1c is required in the activation of hepatic lipogenic gene expression by glucose. Mol Cell Biol 19:3760–3768

Galili N, Davis RJ, Fredericks WJ, Mukhopadhyay S, Rauscher FJ III, Emanuel BS, Rovera G, Barr FG (1993) Fusion of a fork head domain gene to PAX3 in the solid tumour alveolar rhabdomyosarcoma. Nat Genet 5:230–235

Garcia-Garcia AB, Gonzalez C, Real JT, Martin de Llano JJ, Gonzalez-Albert V, Civera M, Chaves FJ, Ascaso JF, Carmena R (2005) Influence of microsomal triglyceride transfer protein promoter polymorphism -493 GT on fasting plasma triglyceride values and interaction with treatment response to atorvastatin in subjects with heterozygous familial hypercholesterolaemia. Pharmacogenet Genomics 15:211–218

Grundy SM, Ahrens EH Jr, Salen G, Schreibman PH, Nestel PJ (1972) Mechanisms of action of clofibrate on cholesterol metabolism in patients with hyperlipidemia. J Lipid Res 13:531–551

Haddad IA, Ordovas JM, Fitzpatrick T, Karathanasis SK (1986) Linkage, evolution, and expression of the rat apolipoprotein A-I, C-III, and A-IV genes. J Biol Chem 261:13268–13277

Haeusler RA, Kaestner KH, Accili D (2010) FoxOs function synergistically to promote glucose production. J Biol Chem 285:35245–35248

Hagan DL, Kienzle B, Jamil H, Hariharan N (1994) Transcriptional regulation of human and hamster microsomal triglyceride transfer protein genes. Cell type-specific expression and response to metabolic regulators. J Biol Chem 269:28737–28744

Holm C, Osterlund T, Laurell H, Contreras JA (2000) Molecular mechanisms regulating hormone-sensitive lipase and lipolysis. Annu Rev Nutr 20:365–393

Huang H, Regan KM, Wang F, Wang D, Smith DI, van Deursen JM, Tindall DJ (2005) Skp2 inhibits FOXO1 in tumor suppression through ubiquitin-mediated degradation. Proc Natl Acad Sci U S A 102:1649–1654

Hussain MM, Shi J, Dreizen P (2003) Microsomal triglyceride transfer protein and its role in apoB-lipoprotein assembly. J Lipid Res 44:22–32

Hwangbo DS, Gershman B, Tu MP, Palmer M, Tatar M (2004) Drosophila dFOXO controls lifespan and regulates insulin signalling in brain and fat body. Nature 429:562–566

Imae M, Fu Z, Yoshida A, Noguchi T, Kato H (2003) Nutritional and hormonal factors control the gene expression of FoxOs, the mammalian homologues of DAF-16. J Mol Endocrinol 30:253–262

Ishihara H, Maechler P, Gjinovci A, Herrera PL, Wollheim CB (2003) Islet beta-cell secretion determines glucagon release from neighbouring alpha-cells. Nat Cell Biol 5:330–335

Jamil H, Chu CH, Dickson JK Jr, Chen Y, Yan M, Biller SA, Gregg RE, Wetterau JR, Gordon DA (1998) Evidence that microsomal triglyceride transfer protein is limiting in the production of apolipoprotein B-containing lipoproteins in hepatic cells. J Lipid Res 39:1448–1454

Jiang G, Zhang BB (2003) Glucagon and regulation of glucose metabolism. Am J Physiol Endocrinol Metab 284:E671–E678

Jun M, Foote C, Lv J, Neal B, Patel A, Nicholls SJ, Grobbee DE, Cass A, Chalmers J, Perkovic V (2010) Effects of fibrates on cardiovascular outcomes: a systematic review and meta-analysis. Lancet 375:1875–1884

Juo SH, Han Z, Smith JD, Colangelo L, Liu K (2000) Common polymorphism in promoter of microsomal triglyceride transfer protein gene influences cholesterol, ApoB, and triglyceride levels in young African American men: results from the coronary artery risk development in young adults (CARDIA) study. Arterioscler Thromb Vasc Biol 20:1316–1322

Kamagate A, Dong HH (2008) FoxO1 integrates insulin signaling to VLDL production. Cell Cycle 7:3162–3170

Kamagate A, Qu S, Perdomo G, Su D, Kim DH, Slusher S, Meseck M, Dong HH (2008) FoxO1 mediates insulin-dependent regulation of hepatic VLDL production in mice. J Clin Invest 118:2347–2364

Kim DH, Perdomo G, Zhang T, Slusher S, Lee S, Phillips BE, Fan Y, Giannoukakis N, Gramignoli R, Strom S et al (2011) FoxO6 integrates insulin signaling with gluconeogenesis in the liver. Diabetes 60:2763–2774

Kim DH, Zhang T, Lee S, Dong HH (2013) FoxO6 in glucose metabolism. J Diabetes 5:233–240

Kissebah AH, Adams PW, Wynn V (1974) Interrelationships between insulin secretion and plasma free fatty acid and triglyceride kinetics in maturity onset diabetes and the effect of phenethybriguanide (phenformin). Diabetologia 10:119–130

Koo SH, Flechner L, Qi L, Zhang X, Screaton RA, Jeffries S, Hedrick S, Xu W, Boussouar F, Brindle P et al (2005) The CREB coactivator TORC2 is a key regulator of fasting glucose metabolism. Nature 437:1109–1111

Krauss RM (1998) Atherogenicity of triglyceride-rich lipoproteins. Am J Cardiol 81(4A):13B–17B

Ledmyr H, Karpe F, Lundahl B, McKinnon M, Skoglund-Andersson C, Ehrenborg E (2002) Variants of the microsomal triglyceride transfer protein gene are associated with plasma cholesterol levels and body mass index. J Lipid Res 43:51–58

Ledmyr H, McMahon AD, Ehrenborg E, Nielsen LB, Neville M, Lithell H, MacFarlane PW, Packard CJ, Karpe F (2004) The microsomal triglyceride transfer protein gene-493T variant lowers cholesterol but increases the risk of coronary heart disease. Circulation 109:2279–2284

Lee SS, Kennedy S, Tolonen AC, Ruvkun G (2003) DAF-16 target genes that control C. elegans life-span and metabolism. Science 300:644–647

Lee HY, Birkenfeld AL, Jornayvaz FR, Jurczak MJ, Kanda S, Popov V, Frederick DW, Zhang D, Guigni B, Bharadwaj KG et al (2011) Apolipoprotein CIII overexpressing mice are predisposed to diet-induced hepatic steatosis and hepatic insulin resistance. Hepatology 54:1650–1660

Leung GK, Veniant MM, Kim SK, Zlot CH, Raabe M, Bjorkegren J, Neese RA, Hellerstein MK, Young SG (2000) A deficiency of microsomal triglyceride transfer protein reduces apolipoprotein B secretion. J Biol Chem 275:7515–7520

Lewis GF, Steiner G (1996) Acute effects of insulin in the control of VLDL production in humans. Implications for the insulin-resistant state. Diabetes Care 19:390–393

Lewis GF, Carpentier A, Adeli K, Giacca A (2002) Disordered fat storage and mobilization in the pathogenesis of insulin resistance and type 2 diabetes. Endocr Rev 23:201–229

Liao W, Hui TY, Young SG, Davis RA (2003) Blocking microsomal triglyceride transfer protein interferes with apoB secretion without causing retention or stress in the ER. J Lipid Res 44:978–985

Lin MC, Gordon D, Wetterau JR (1995) Microsomal triglyceride transfer protein (MTP) regulation in HepG2 cells: insulin negatively regulates MTP gene expression. J Lipid Res 36:1073–1081

Lin J, Yang R, Tarr PT, Wu PH, Handschin C, Li S, Yang W, Pei L, Uldry M, Tontonoz P et al (2005) Hyperlipidemic effects of dietary saturated fats mediated through PGC-1beta coactivation of SREBP. Cell 120:261–273

Liu Y, Dentin R, Chen D, Hedrick S, Ravnskjaer K, Schenk S, Milne J, Meyers DJ, Cole P, Yates J III et al (2008) A fasting inducible switch modulates gluconeogenesis via activator/coactivator exchange. Nature 456:269–273

Maeda N, Li H, Lee D, Oliver P, Quarfordt SH, Osada J (1994) Targeted disruption of the apolipoprotein C-III gene in mice results in hypotriglyceridemia and protection from postprandial hypertriglyceridemia. J Biol Chem 269:23610–23616

Malmstrom R, Packard CJ, Watson TD, Rannikko S, Caslake M, Bedford D, Stewart P, Yki-Jarvinen H, Shepherd J, Taskinen MR (1997) Metabolic basis of hypotriglyceridemic effects of insulin in normal men. Arterioscler Thromb Vasc Biol 17:1454–1464

Malmstrom R, Packard CJ, Caslake M, Bedford D, Stewart P, Yki-Jarvinen H, Shepherd J, Taskinen MR (1998) Effects of insulin and acipimox on VLDL1 and VLDL2 apolipoprotein B production in normal subjects. Diabetes 47:779–787

Marcoux C, Tremblay M, Fredenrich A, Davignon J, Cohn JS (2001) Lipoprotein distribution of apolipoprotein C-III and its relationship to the presence in plasma of triglyceride-rich remnant lipoproteins. Metabolism 50:112–119

Massillon D, Chen W, Hawkins M, Liu R, Barzilai N, Rossetti L (1995) Quantitation of hepatic glucose fluxes and pathways of hepatic glycogen synthesis in conscious mice. Am J Physiol 269:E1037–E1043

Matsumoto M, Han S, Kitamura T, Accili D (2006) Dual role of transcription factor FoxO1 in controlling hepatic insulin sensitivity and lipid metabolism. J Clin Invest 116:2464–2472

Matsumoto M, Pocai A, Rossetti L, Depinho RA, Accili D (2007) Impaired regulation of hepatic glucose production in mice lacking the forkhead transcription factor foxo1 in liver. Cell Metab 6:208–216

Matsuzaki H, Daitoku H, Hatta M, Tanaka K, Fukamizu A (2003) Insulin-induced phosphorylation of FKHR (Foxo1) targets to proteasomal degradation. Proc Natl Acad Sci U S A 100:11285–11290

Minnich A, Tian N, Byan L, Bilder G (2001) A potent PPARalpha agonist stimulates mitochondrial fatty acid beta-oxidation in liver and skeletal muscle. Am J Physiol Endocrinol Metab 280:E270–E279

Monecke T, Guttler T, Neumann P, Dickmanns A, Gorlich D, Ficner R (2009) Crystal structure of the nuclear export receptor CRM1 in complex with Snurportin1 and RanGTP. Science 324:1087–1091

Mueller MM, Castells-Roca L, Babu V, Ermolaeva MA, Muller RU, Frommolt P, Williams AB, Greiss S, Schneider JI, Benzing T et al (2014) DAF-16/FOXO and EGL-27/GATA promote developmental growth in response to persistent somatic DNA damage. Nat Cell Biol 16:1168–1179

Nagai Y, Yonemitsu S, Erion DM, Iwasaki T, Stark R, Weismann D, Dong J, Zhang D, Jurczak MJ, Loffler MG et al (2009) The role of peroxisome proliferator-activated receptor gamma coactivator-1 beta in the pathogenesis of fructose-induced insulin resistance. Cell Metab 9:252–264

Nagashima T, Shigematsu N, Maruki R, Urano Y, Tanaka H, Shimaya A, Shimokawa T, Shibasaki M (2010) Discovery of novel forkhead box O1 inhibitors for treating type 2 diabetes: improvement of fasting glycemia in diabetic db/db mice. Mol Pharmacol 78:961–970

Nakae J, Biggs WH, Kitamura T, Cavenee W, Wright CVE, Arden KC, Accili D (2002) Regulation of insulin action and pancreatic ß-cell function by mutated alleles of the gene encoding forkhead transcription factor Foxo1. Nat Genet 32:245–253

Nakae J, Kitamura T, Kitamura Y, Biggs WH III, Arden KC, Accili D (2003) The forkhead transcription factor Foxo1 regulates adipocyte differentiation. Dev Cell 4:119–129

Nieves DJ, Cnop M, Retzlaff B, Walden CE, Brunzell JD, Knopp RH, Kahn SE (2003) The atherogenic lipoprotein profile associated with obesity and insulin resistance is largely attributable to intra-abdominal fat. Diabetes 52:172–179

Okamoto H, Nakae J, Kitamura T, Park BC, Dragatsis I, Accili D (2004) Transgenic rescue of insulin receptor-deficient mice. J Clin Invest 114:214–223

Oliveieri O, Stranieri C, Bassi A, Zaia B, Girelli D, Pizzolo F, Trabetti E, Cheng S, Grow MA, Pignatti PF et al (2002) ApoC-III gene polymorphisms and risk of coronary artery disease. J Lipid Res 43:1450–1457

Onat A, Hergenc G, Sansoy V, Fobker M, Ceyhan K, Toprak S, Assmann G (2003) Apolipoprotein C-III, a strong discriminant of coronary risk in men and a determinant of the metabolic syndrome in both genders. Atherosclerosis 168:81–89

Petersen KF, Dufour S, Hariri A, Nelson-Williams C, Foo JN, Zhang XM, Dziura J, Lifton RP, Shulman GI (2010) Apolipoprotein C3 gene variants in nonalcoholic fatty liver disease. N Engl J Med 362:1082–1089

Phillips C, Owens D, Collins P, Tomkin GH (2002) Microsomal triglyceride transfer protein: does insulin resistance play a role in the regulation of chylomicron assembly? Atherosclerosis 160:355–360

Pilkis SJ, Granner DK (1992) Molecular physiology of the regulation of hepatic gluconeogenesis and glycogenolysis. Annu Rev Physiol 54:885–909

Post SM, Duez H, Gervois PP, Staels B, Kuipers F, Princen HM (2001) Fibrates suppress bile acid synthesis via peroxisome proliferator-activated receptor-alpha-mediated downregulation of cholesterol 7alpha-hydroxylase and sterol 27-hydroxylase expression. Arterioscler Thromb Vasc Biol 21:1840–1845

Prager R, Wallace P, Olefsky JM (1987) Direct and indirect effects of insulin to inhibit hepatic glucose output in obese subjects. Diabetes 36:607–611

Puigserver P, Rhee J, Donovan J, Walkey CJ, Yoon JC, Oriente F, Kitamura Y, Altomonte J, Dong H, Accili D et al (2003) Insulin-regulated hepatic gluconeogenesis through FOXO1-PGC-1alpha interaction. Nature 423:550–555

Qu S, Altomonte J, Perdomo G, He J, Fan Y, Kamagate A, Meseck M, Dong HH (2006) Aberrant Forkhead box O1 function is associated with impaired hepatic metabolism. Endocrinology 147:5641–5652

Qu S, Su D, Altomonte J, Kamagate A, He J, Perdomo G, Tse T, Jiang Y, Dong HH (2007a) PPAR{alpha} mediates the hypolipidemic action of fibrates by antagonizing FoxO1. Am J Physiol Endocrinol Metab 292:E421–E434

Qu S, Perdomo G, Su D, D'Souza FM, Shachter NS, Dong HH (2007b) Effects of apoA-V on HDL and VLDL metabolism in APOC3 transgenic mice. J Lipid Res 48:1476–1487

Raabe M, Flynn LM, Zlot CH, Wong JS, Veniant MM, Hamilton RL, Young SG (1998) Knockout of the abetalipoproteinemia gene in mice: reduced lipoprotein secretion in heterozygotes and embryonic lethality in homozygotes. Proc Natl Acad Sci U S A 95:8686–8691

Raabe M, Veniant MM, Sullivan MA, Zlot CH, Bjorkegren J, Nielsen LB, Wong JS, Hamilton RL, Young SG (1999) Analysis of the role of microsomal triglyceride transfer protein in the liver of tissue-specific knockout mice. J Clin Invest 103:1287–1298

Rossetti L, Massillon D, Barzilai N, Vuguin P, Chen W, Hawkins M, Wu J, Wang J (1997) Short term effects of leptin on hepatic gluconeogenesis and in vivo insulin action. J Biol Chem 272:27758–27763

Samuel VT, Choi CS, Phillips TG, Romanelli AJ, Geisler JG, Bhanot S, McKay R, Monia B, Shutter JR, Lindberg RA et al (2006) Targeting foxo1 in mice using antisense oligonucleotide improves hepatic and peripheral insulin action. Diabetes 55:2042–2050

Shachter NS (2001) Apolipoproteins C-1 and C-III as important modulators of lipoprotein metabolism. Curr Opin Lipidol 12:297–304

Sparks JD, Sparks CE (1994) Insulin regulation of triacylglycerol-rich lipoprotein synthesis and secretion. Biochim Biophys Acta 1215:9–32

Stahlberg D, Reihner E, Rudling M, Berglund L, Einarsson K, Angelin B (1995) Influence of bezafibrate on hepatic cholesterol metabolism in gallstone patients: reduced activity of cholesterol 7 alpha-hydroxylase. Hepatology 21:1025–1030

Stewart MM, Laker MF, Alberti KG (1994) The contribution of lipids to coronary heart disease in diabetes mellitus. J Intern Med Suppl 736:41–46

Taghibiglou C, Carpentier A, Van Iderstine SC, Chen B, Rudy D, Aiton A, Lewis GF, Adeli K (2000) Mechanisms of hepatic very low density lipoprotein overproduction in insulin resistance. J Biol Chem 275:8416–8425

Tietge UJ, Bakillah A, Maugeais C, Tsukamoto K, Hussain M, Rader DJ (1999) Hepatic overexpression of microsomal triglyceride transfer protein (MTP) results in increased in vivo secretion of VLDL triglycerides and apolipoprotein B. J Lipid Res 40:2134–2139

Tonelli J, Kishore P, Lee DE, Hawkins M (2005) The regulation of glucose effectiveness: how glucose modulates its own production. Curr Opin Clin Nutr Metab Care 8:450–456

Tsai WC, Bhattacharyya N, Han LY, Hanover JA, Rechler MM (2003) Insulin inhibition of transcription stimulated by the forkhead protein Foxo1 is not solely due to nuclear exclusion. Endocrinology 144:5615–5622

Unger RH, Dobbs RE, Orci L (1978) Insulin, glucagon, and somatostatin secretion in the regulation of metabolism. Annu Rev Physiol 40:307–343

Valenti L, Rametta R, Dongiovanni P, Maggioni M, Ludovica Fracanzani A, Zappa M, Lattuada E, Roviaro G, Fargion S (2008) Increased expression and activity of the transcription factor Foxo1 in nonalcoholic steatohepatitis. Diabetes 57:1355–1362

Vanhanen HT, Miettinen TA (1995) Cholesterol absorption and synthesis during pravastatin, gemfibrozil and their combination. Atherosclerosis 115:135–146

Wahren J, Ekberg K (2007) Splanchnic regulation of glucose production. Annu Rev Nutr 27:329–345

Wang C, McConathy WJ, Kloer HJ, Alaupovic P (1985) Modulation of lipoprotein lipase activity by apolipoproteins: effect of apolipoprotein C-III. J Clin Invest 75:384

Yoon CF, Pulgserver P, Chen G, Donovan J, Wu Z, Rhee J, Adelmant G, Stafford J, Kahn CR, Granner DK et al (2001) Control of hepatic gluconeogenesis through the transcriptional coactivator PGC-1. Nature 413:131–138

Zhang X, Gan L, Pan H, Guo S, He X, Olson ST, Mesecar A, Adam S, Unterman TG (2002) Phosphorylation of serine 256 suppresses transactivation by FKHR (FOXO1) by multiple mechanisms. Direct and indirect effects on nuclear/cytoplasmic shuttling and DNA binding. J Biol Chem 277:45276–45284

Zhang L, Rubins NE, Ahima RS, Greenbaum LE, Kaestner KH (2005) Foxa2 integrates the transcriptional response of the hepatocyte to fasting. Cell Metab 2:141–148

Zhang W, Patil S, Chauhan B, Guo S, Powell DR, Le J, Klotsas A, Matika R, Xiao X, Franks R, Heidenreich KA, Sajan MP, Farese RV, Stolz DB, Tso P, Koo SH, Montminy M, Unterman TG (2006) FoxO1 regulates multiple metabolic pathways in the liver: effects on gluconeogenic, glycolytic, and lipogenic gene expression. J Biol Chem. 281(15):10105–17

Chapter 5
Molecular Mechanisms of Lipotoxicity in Nonalcoholic Fatty Liver Disease

Sophie C. Cazanave and Arun J. Sanyal

Abstract Nonalcoholic fatty liver disease (NAFLD) is a growing epidemic which affects up to 30 % of the American population. Nonalcoholic steatohepatitis (NASH), the more severe form of NAFLD, is associated with increased risk of cardiovascular morbidity and mortality, insulin resistance (IR), type 2 diabetes mellitus (T2DM), hepatic steatosis, elevated circulating levels of free fatty acids (FFAs) and hepatocyte apoptosis. Because lipid-induced toxicity, or lipoapoptosis, represents a key pathogenic factor for NAFLD, we will focus on the evidence connecting the accumulation of certain lipid types in the liver with the progressive induction of hepatocyte injury and secondary inflammation/fibrosis during the disease. A better understanding of the molecular and cellular pathway of lipotoxicity has both diagnostic and therapeutic implication for the treatment of NAFLD.

Keywords Hepatic lipid • NASH • ER stress • CHOP • JNK • BH3-only proteins • Death receptors • Oxidative stress • Inflammation • Fibrosis

Abbreviations

ALT	Alanine aminotransferases
AP-1	Activator protein-1
ATF	Activating transcription factor
BAs	Bile acids
CHOP	CCAAT/enhancer-binding homologous protein
CYP2E1	Cytochrome P450 2E1
DAG	Diacylglycerides
DGAT2	Diacylglycerol acyltransferase-2
DR	Death receptor

S.C. Cazanave, Ph.D. (✉) • A.J. Sanyal, M.D. (✉)
Division of Gastroenterology, Hepatology and Nutrition, Department of Internal Medicine, Virginia Commonwealth University, 1220 E. Broad Street, Richmond 23219, VA, USA
e-mail: sophie.cazanave@vcuhealth.org; arun.sanyal@vcuhealth.org

J.M. Ntambi (ed.), *Hepatic De Novo Lipogenesis and Regulation of Metabolism*,
DOI 10.1007/978-3-319-25065-6_5

ER	Endoplasmic reticulum
FAT	Fatty acid translocase
FATP	Fatty acid transport protein
FC	Free cholesterol
FFAs	Free fatty acids
FXR	Farnesoid X receptor
Hh	Hedgehog
HSCs	Hepatic stellate cells
IL	Interleukin
IRE1α	Inositol-requiring enzyme-1α
JNK	c-Jun NH2-terminal kinase
KCs	Kupffer cells
LPC	Lysophosphatidyl choline
NF-kB	Nuclear factor-kappaB
MCD	Methionine choline deficient
MCP-1	Macrophage chemotactic protein 1
NAFLD	Nonalcoholic fatty liver disease
NASH	Nonalcoholic steatohepatitis
NK	Natural killer
PERK	PKR-like ER kinase
PLA2	Phospholipase A2
ROS	Reactive oxygen species
S1P	Sphingosine-1-phosphate
SL	Sphingolipids
SphK	Sphingosine kinase
T2DM	Type 2 diabetes mellitus
TAG	Triacylglycerides
TG	Triglycerides
TNF	Tumor necrosis factor
TNF-R1	TNFα receptor 1
TRAIL	TNFα-related apoptosis-inducing ligand
VLDL	Very low-density lipoprotein particles

1 Introduction

Nonalcoholic fatty liver disease (NAFLD) affects almost a third of all adults in North America and contributes substantially to the burden of chronic liver disease (Browning et al. 2004). It has two clinical-histological phenotypes: nonalcoholic fatty liver (NAFL) and nonalcoholic steatohepatitis (NASH). NASH can progress to cirrhosis in 15–20 % of subjects and is projected to become the leading cause of hepatocellular cancer and end-stage liver disease requiring liver transplantation (Sass et al. 2005). The pathogenesis of NASH is thought to be driven by several components including hepatic steatosis along with hepatocyte damage, oxidative stress, inflammation and various degrees of fibrosis (Adams et al. 2005). NASH is now recognized as a cause of potentially progressive liver disease, posing patients at risk of developing cirrhosis, chronic liver disease with portal hypertension, liver failure and hepatocellular carcinoma (Angulo and Lindor 2002).

NAFLD associates with features of the metabolic syndrome; and insulin resistance plays a key role in the pathogenesis of both NAFLD and the metabolic syndrome (Chavez-Tapia et al. 2006; Utzschneider and Kahn 2006). Insulin resistance results in increased lipolysis in adipose tissue and augmented release of free fatty acids (FFAs) and inflammatory cytokines into the circulation. Increased FFAs flux to the other tissues results in increased triglyceride storage in the liver and directly contributes to the development of hepatic steatosis in patients with type 2 diabetes mellitus (T2DM) (Roden 2006). However, the limited capacity of non-adipose tissues to store lipids result in the induction of cellular dysfunction or cell death, a phenomenon termed lipoapoptosis (Kusminski et al. 2009; Unger and Orci 2002). According to the classical "two-hit hypothesis", lipid accumulation ("first hit") increases the vulnerability of the liver to oxidative stress ("second hit") and promotes hepatic cell death, pro-inflammatory cytokine expression, liver stellate cells activation and fibrogenesis. However recent finding suggest that FFAs (Malhi et al. 2006) and their lipotoxic intermediates such as lysophosphatidyl Choline (LPC) (Kakisaka et al. 2012a), rather than their esterified products (triglycerides) are toxic to hepatocytes; and triglyceride accumulation in the liver may be hepato-protective by decreasing the excess of FFAs in hepatocytes (Listenberger et al. 2003). This is in accordance with the fact that FFAs levels are increased in the liver of mice fed a methionine choline deficient (MCD) diet (Larter et al. 2008), and that the circulating FFAs increase in NAFLD in parallel with the increasing disease severity (Feldstein et al. 2003a; Nehra et al. 2001). Also, others lipid metabolites such as ceramides or free cholesterol (FC) also accumulate in NAFLD and contribute to hepatocyte injury and the pathogenesis of the disease. Thus, increased hepatocyte apoptosis and elevated serum caspase-cleaved cytokeratin 18 fragments are observed in patients with NAFLD (Feldstein et al. 2009). Because the amount of hepatocyte apoptosis correlates to NAFLD severity and may contribute to disease pathogenesis, understanding the cellular and molecular mechanisms of lipid-induced toxicity is relevant to the disease.

2 Altered Hepatic Lipid Metabolism in NAFLD

Lipid accumulation in the liver seems to be a crucial mechanism in NAFLD pathophysiology. Although most of the lipids accumulate as triglycerides (TG) in the liver of NAFLD patients, several other lipid metabolites such as free fatty acids (FFAs), lysophosphatidyl Choline (LPC), sphingolipids (SL) and free cholesterol (FC) also concentrate in the liver during the disease. When in excess, these lipids can act as signaling molecules and trigger apoptotic processes which contribute to disease progression.

2.1 Hepatic Free Fatty Acids and Triglycerides

Hepatic steatosis is a hallmark of NAFLD. TG accumulation in the liver can result from increased lipogenesis, reduced fat oxidation (β-oxidation), or reduced capacity to export fat in the form of very low-density lipoprotein particles (VLDL). Mostly, increased delivery of adipose-derived free fatty acid (FFAs) to the liver contributes to 60 % of hepatic fat content while augmented de novo lipogenesis is responsible for up to 30 % of stored hepatic fat (Donnelly et al. 2005; Kawano and Cohen 2013). Of note, de novo lipogenesis normally contributes to 5–7 % of hepatic TG. Decreased lipid disposal via β-oxidation or VLDL export, or increased dietary lipids only minimally contributes to hepatic steatosis (Kawano and Cohen 2013).

NASH patients have increased levels of circulating FFAs (Puri et al. 2009). FFAs enter the cells by both passive diffusion and facilitated transport by fatty acid transport protein (FATP) and fatty acid translocase (FAT/CD36). Excess FFAs are rapidly esterified within the hepatocytes to form diacylglycerides (DAG) and triacylglycerides (TAG). Thus, although no major increase in hepatic FFAs is noted in NAFLD or NASH patients when compared to normal patients, a significant increase in hepatic DAG, TAG and total lipid content is observed in patients with NASH and even more in patients with NAFLD (Puri et al. 2007). In fact, a growing body of evidence suggest that esterification of FFAs into TGs is not a toxic event but rather is part of a detoxification process for the liver. Along with this observations, in vitro and in vivo studies have demonstrated that accumulation of non-esterified saturated FFAs (*e.g.*, palmitate (C16:0) or stearate (C18:0)) is hepatotoxic because of their ability to induce hepatocyte lipoapoptosis (Malhi et al. 2006, 2007; Barreyro et al. 2007; Cazanave et al. 2009; Wei et al. 2009). In contrast, unsaturated FFAs (*e.g.*, oleate (C18:1) or palmitoleate (C16:1)) are relatively non-toxic probably because of their capacity to be rapidly esterified and incorporated into triglycerides. Thus, liver cells incubated with unsaturated FFAs exhibit little cell death but a significant increase in TG content, whereas cells treated with saturated FFAs die rapidly but minimally accumulate hepatic lipid droplets (Listenberger et al. 2003; Akazawa et al. 2010).

Also, monounsaturated FFAs can inhibit liver lipotoxicity induced by saturated FFAs by redirecting the toxic FFAs into triglyceride storage (Listenberger et al. 2003; Akazawa et al. 2010). Thus, genetic deletion of stearoyl CoA desaturase-1 (SCD1), a key enzyme in the synthesis of monounsaturated FFAs from saturated FFAs, prevents hepatic TGs accumulation but aggravates hepatocellular damage induced by a MCD diet (Li et al. 2009a). Moreover, mice deficient in diacylglycerol acyltransferase-2 (DGAT2), the enzyme that catalyzes the final step in TAG synthesis, have reduced hepatic steatosis but displayed increased oxidative stress, hepatocyte apoptosis, inflammation and fibrosis (Yamaguchi et al. 2007); in contrast liver-specific overexpression of DGAT2 promotes hepatic steatosis without inducing any liver injury or abnormalities in insulin resistance (Monetti et al. 2007). Hepatic triglyceride content declines over time in mice fed a MCD diet and inversely correlate with liver damage (Yamaguchi et al. 2007); similarly total hepatic content is greater in patients with simple steatosis than in patients with NASH (Puri et al. 2007). Decrease in fatty acid transporters could directly contributes to the modulation of hepatic fat accumulation throughout disease progression as FABP-1 expression is first overexpressed in the liver of patients with simple steatosis (NAFLD), but decreases as the disease progresses towards NASH (Charlton et al. 2009). Therefore, all these observations strongly implicate that incorporation of FFAs into neutral triglycerides represents a protective mechanism against toxic saturated FFAs, and impairment of this process might contribute to NASH pathogenesis.

2.2 Hepatic Lysophosphatidyl Choline (LPC)

Saturated FFAs can be esterified into DAG that is further converted into phosphatidylcholine (PC). Lysophosphatidylcholine (LPC) is a major phospholipid which derives from partial hydrolysis and removal of the fatty acid group of PC by the enzyme phospholipase A2 (PLA2); and recent evidence suggest that LPC also mediates FFAs-induced liver toxicity and insulin resistance in NASH (Kakisaka et al. 2012a; Han et al. 2008, 2011). Indeed, LPC levels were found to be increased in liver biopsies from patients with NAFLD in proportion to disease severity (Han et al. 2008). LPC concentrations in hepatic lipid extracts were also significantly increased in mice fed a high fat or MCD diet as compared to control mice (Pathil et al. 2012); deficiency in group IVA phospholipase A2, a cytosolic form of PLA2, protects mice against high fat diet-induced increase in alanine aminotransferases (ALT) and liver injury (Ii et al. 2009). The LPC-induced hepatic toxicity was found to be dependent on mechanisms largely similar to saturated FFAs-mediated cell death, involving endoplasmic reticulum (ER) stress, activation of the c-Jun NH2-terminal kinase (JNK) and downstream upregulation of pro-apoptotic BH3 only proteins and mitochondrial dysfunctions (Fig. 5.1) (Kakisaka et al. 2012a), and inhibitors of PLA2 prevent LPC-induced cell toxicity (Han et al. 2008). Collectively, these data suggest that increase levels of LPC mediate free fatty acid-induced hepatic toxicity.

2.3 Hepatic Sphingolipids (Ceramides and Sphingosine 1-Phosphate)

The sphingolipid (SL)signaling pathways are induced by reactive oxygen species (ROS) and pro-inflammatory molecules, both upregulated in NASH (Pagadala et al. 2012). Among the SL, ceramides and its metabolites have been described as regulator of cell stress, death ligand-induced apoptosis and insulin resistance and obesity. Ceramide can be generated through two major pathways, de novo synthesis and sphingomyelin hydrolysis. The de novo synthesis via serine palmitoyltransferase

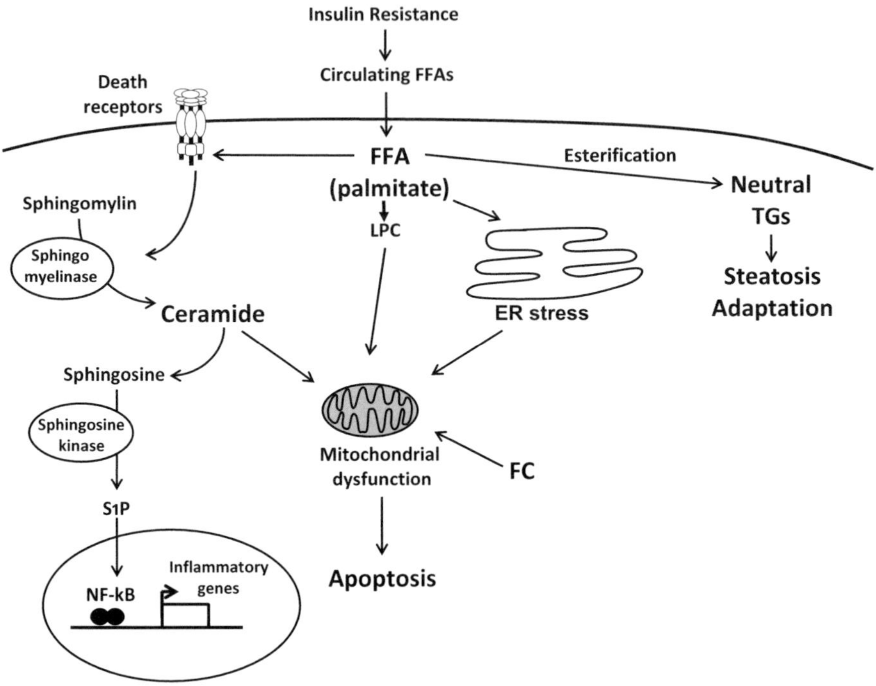

Fig. 5.1 Lipids-dependent signaling pathways. Insulin resistance, a hallmark of NAFLD, induces an increase in circulating FFAs. These circulating FFAs are transported into hepatocytes where they can be esterified to neutral triglycerides (TGs). Esterification of FFAs acts as an adaptive mechanism allowing cells to maintain viability in a context of excess saturated FFAs accumulation. Saturated FFAs (such as palmitate) are directly toxic to hepatocyte and induce several signaling pathways of apoptosis including ER stress as well as activation of death receptors cytotoxic signaling, both resulting to mitochondrial dysfunction and subsequent cellular apoptosis. Cytotoxic signaling processes triggered downstream of palmitate could be mediated by the lipid metabolite LPC. Others lipid types accumulate during NAFLD, such as free cholesterol (FC) and ceramide and may also contribute to hepatocyte toxicity by inducing the mitochondrial pathway of apoptosis. Finally, Sphingosine kinase 1/S1P signaling pathway may contribute to liver inflammation via production of nuclear factor-kappaB (NF-kB)-dependent pro-inflammatory cytokines

(STP) and ceramide synthase takes place in the ER and involves the condensation of sphingosine with a fatty acyl-CoA (normally, palmitoyl-CoA), whereas acidic or neutral sphingomyelinases (SMase) present in lysosomes, plasma membranes and cytosol hydrolyze membrane sphingomyelin to yield ceramide and phosphocholine. Ceramide synthesis depends on the levels of long chain saturated fatty acid available. Ceramide synthesis rate is increased in the liver of high fat diet-fed rats (Chocian et al. 2010); similarly ceramides levels increases after liver cells incubation with saturated FFAs (Wei et al. 2006). Activation of the death receptors Fas or tumor necrosis factor (TNF) receptor 1 (TNF-R1) trigger activation of SMases (Fig. 5.1) and release of ceramide and this process seems to contribute to Fas and TNF-R1-induced JNK-mediated apoptosis (Brenner et al. 1998; Schwandner et al. 1998).

Some studies indicate that increase levels of circulating ceramide correlate with insulin resistance and inflammation, and inhibition of de novo ceramide synthesis in obese mice ameliorates insulin action, attenuates inflammatory parameters (Holland et al. 2007) and reduces diet-induced hepatic steatosis (Kurek et al. 2014). Ceramide play a role in insulin resistance by activating JNK which inhibits insulin receptor signaling, or by preventing insulin-stimulated glucose uptake, glucose transporter 4 (GLUT4) translocation and AKT kinase activity (Pagadala et al. 2012). Therefore, ceramide represent an attractive candidate to link the excess of saturated FFAs to the activation of inflammatory pathways and the development of an insulin resistance state during fatty liver disease. However, conflicting data on the role of ceramide in contributing to the pathogenesis of NAFLD have been published. First, ceramide concentrations are not increased in the liver of patients with NAFLD when compared to normal patients (Kotronen et al. 2009). Second, saturated fat-induced insulin resistance in rats is independent of ceramide generation (Galbo et al. 2013). Finally, some studies demonstrated that saturated FFAs-induced ER stress and hepatocyte toxicity, a key event in NAFLD progression, was also ceramide independent (Wei et al. 2006; Feldstein et al. 2006).

Ceramide are further hydrolyzed to generate sphingosine, which can be phosphorylated by sphingosine kinase (SphK) to form sphingosine-1-phosphate (S1P). Bioactive S1P have been primarily studied in lung and skeletal muscles disease in which they have been implicated in the regulation of inflammation, cancer and angiogenesis. Recent data have demonstrated that circulating levels of S1P are increased in obese humans or mice (Kowalski et al. 2013); and inhibition of sphingolipid synthesis restores insulin sensitivity and decreases hepatic steatosis in obese rodents (Cowart 2009; Fayyaz et al. 2014). Also, hepatic Sphk1 protein levels and activity, as well as S1P hepatic levels are increased in high fructose-fed rats (Wang et al. 2013a). In this model, inhibition of hepatic SphK1/ S1P signaling pathway decreased liver inflammation via the suppression of nuclear factor-kappaB (NF-kB)-dependent production of pro-inflammatory cytokines, such as interleukin (IL)-1β, IL-6 and TNFα. Also, incubation of rat or human hepatocytes with palmitate results in elevation of S1P intracellular levels which counteract insulin signaling (Fayyaz et al. 2014). Finally, some recent data implicate S1P in liver fibrogenesis as they play a role in the migration of myofibroblast into damage liver areas (Li et al. 2009b, 2011). Therefore, impairment of the SphK1/ S1P signaling pathway is implicated in the pathological development of inflammation, obesity and diabetes and its role in NAFLD remains to be further elucidated.

2.4 Cholesterol and Bile Acids

Several studies have provided evidence for widespread abnormalities in cholesterol homeostatic pathways in NAFLD. Comparative hepatic and serum lipidomic analysis of subjects with fatty liver disease has demonstrated a progressive increase in free cholesterol (FC), without a similar increase in liver cholesterol ester in patients with NAFLD and NASH as compared to controls (Puri et al. 2007, 2009). Further changes in the expression status of genes involved in cholesterol metabolism were identified in NAFLD patients (Min et al. 2012). Indeed, increase expression of the rate-limiting enzyme in cholesterol synthesis HMC CoA reductase (HMGCR) is observed in individuals with NAFLD and it correlates with FC levels and the severity of the liver disease. Also, both sterol-regulatory-element-binding protein 2 (SREBP-2, a transcription factor that regulates several genes implicated in cholesterol synthesis) and steroidogenic acute regulatory protein (StAR, a mitochondrial-cholesterol transporting polypeptide) were overexpressed in NASH patients as compared to those with simple steatosis (Min et al. 2012; Caballero et al. 2009). Further evidence of FC implication in NASH pathogenesis was provided by experimental animal models of NASH (Van Rooyen et al. 2011; Simon et al. 2014). Obese hyper-insulinemic mice have increased hepatic FC levels which correlate with histologic severity of NASH (Van Rooyen et al. 2011). In this model, increased dietary cholesterol content results in increased ALT values, hepatocyte apoptosis, macrophage accumulation and fibrosis. In another study, a high cholesterol diet which results in mitochondrial FC loading in rat hepatocytes, depleted the mitochondrial glutathione pool, and induced microvesicular steatosis and cell sensitization to TNFα- and Fas-induced hepatocellular damage (Mari et al. 2006). More recent data from Farrell's group show that FC accumulates into the hepatocyte plasma membrane, the endoplasmic reticulum (ER) and mitochondria in the liver of obese diabetic mice fed a high cholesterol diet; and accumulation of FC in hepatocyte triggers JNK1-dependent apoptotic cascade involving mitochondrial dysfunction and culminating in cellular demise (Fig. 5.1) (Gan et al. 2014). Although FC deposits in hepatocyte ER in this animal model of NASH, FC-mediated lipotoxicity was independent of an ER stress, a result in accord with a previous study (Mari et al. 2006). Therefore the role of FC is relevant to NASH pathogenesis.

Bile acids (BAs) are end products of cholesterol catabolism generated by the key enzyme cholesterol 7 alpha-hydroxylase (CYP7A1) in the liver. Recent evidence indicate that bile acids are increased in serum and liver tissue of NASH patients (Bechmann et al. 2013; Dasarathy et al. 2011; Aranha et al. 2008), and dysregulation in BAs homeostasis can result in liver injury and inflammation. Hydrophobic BAs are toxic to hepatocyte in vitro by inducing cell surface expression of death receptors Fas or TNFα-related apoptosis-inducing ligand (TRAIL), with ligand-independent oligomerization and induction of these death receptors and activation of downstream cytotoxic signaling (Faubion et al. 1999; Higuchi et al. 2004). BAs-dependent toxicity also involve mitochondrial dysfunction and caspases activation (Rodrigues et al. 1998). Endogenous BAs are ligands of the farnesoid X receptor (FXR), an orphan nuclear receptor highly expressed in the liver. The BAs that activate FXR are the primary and secondary bile acids: chenodeoxycholic acid, deoxycholic acid, lithocholic acid, and cholic acid, with decreasing potency (Wang et al. 1999). FXR activation plays a pivotal role in decreasing BA synthesis and gluconeogenesis, enhancing triglyceride clearance, and inhibiting inflammation and fibrosis (Anderson and Borlak 2008; Contos et al. 2004). NASH is associated with increased gluconeogenesis and lipogenesis, and a apoptotic and pro-inflammatory/

fibrogenic state, a readout indicative of decreased FXR activity. Interestingly, the hepatic expression of small heterodimer partner (SHP), a downstream FXR target, is decreased in NAFLD patients (Bechmann et al. 2013). *FXR* deficiency in mice represents a significant risk factor for the progression of NAFLD to NASH with increased hepatic steatosis, hepatocellular injury, inflammation, and fibrosis in an experimental model of NASH (Kong et al. 2009), whereas FXR activation protects human islets from lipotoxicity (Popescu et al. 2010). Although, it is unclear why in NAFLD, FXR activity decreases in the face of increased bile salts including those that are considered to be endogenous ligands for FXR, it seems that both of the processes contributes to NAFLD progression. Recently, a major clinical trial has confirmed the ability of FXR agonists to reduce steatosis, inflammation, hepatocyte ballooning and fibrosis in NASH patients (Neuschwander-Tetri et al. 2015).

3 Mechanisms of Hepatocyte Lipoapoptosis

Apoptosis or programmed cell death is a cardinal pathogenic feature of NASH (Feldstein et al. 2003a). Within the context of NAFLD, hepatic lipid deposition overload and failure of the hepatocyte to dispose of excess FFAs by converting them into triglyceride is associated with increased risk for hepatocyte apoptosis, or lipo-apoptosis. Thus, in patients with NAFLD, the amount of apoptosis correlates with disease severity and the stage of fibrosis (Feldstein et al. 2003a; Kalsch et al. 2011). In line with this concept, elevated serum caspase-cleaved cytokeratin 18 fragments, an apoptotic marker, discriminate simple human hepatic steatosis from NASH (Feldstein et al. 2009).

Apoptosis is a highly regulated form of cell death. Hepatocytes can undergo apoptosis by either an extrinsic death receptor pathway or by an intrinsic intracellular pathway converging onto the mitochondria to mediate cell death. The extrinsic pathway is initiated by the ligation of death ligands, such as Fas or TRAIL, to their respective receptors; the intrinsic pathway of apoptosis is activated by intracellular stress such as oxidative stress or by membrane-bound organelles dysfunctions including ER stress, and lysosomal and mitochondrial permeabilization. In hepatocyte, both of these pathways induces mitochondrial dysfunctions and downstream activation of effector caspases (cysteine-aspartic proteases) for efficient apoptosis. The mitochondrial pathway of cell death is tightly controlled by interactions between members of the Bcl-2 family of proteins (Youle and Strasser 2008). These proteins can be grouped into three distinct categories: the anti-apoptotic members including Bcl-2, Bcl-xL and Mcl-1, the multidomain pro-apoptotic members such as Bax and Bak, and the BH3-only proteins or biosensors of cell death that includes BIM and PUMA. These biosensors disrupt mitochondrial integrity by directly activating the executioners Bax and/or Bak, whereas the anti-apoptotic members of this family hold Bax and Bak in check thereby preserving mitochondrial integrity. Thus, the Bcl-2 proteins play an essential role in regulating hepatic injury.

3.1 Endoplasmic Reticulum (ER) Stress

Recent evidence indicate that disturbances in the ER seems to be implicated in both lipoapoptotic pathways and the development of steatosis and disease progression in NASH. The ER is a multifunctional intracellular membranous network supporting many vital cellular functions, including protein synthesis, posttranslational modification including *N*-linked glycosylation, folding and transport to the membrane, lipid synthesis and packaging, carbohydrate metabolism and the regulation of calcium homeostasis.

In conditions of physiologic changes impacting the rate of these processes, or ER stress, an adaptive and protective signaling network, termed the unfolded protein response (UPR), is activated by ER-membrane sensors, which serve to re-establish ER homeostasis and promote survival. However, when faced with a prolonged ER stress, apoptotic pathways are activated causing cell demise. Three transmembrane biosensors embedded in the ER orchestrate the UPR, namely activating transcription factor (ATF) 6, PKR-like ER kinase (PERK) and inositol-requiring enzyme-1α (IRE1α), which are kept inactive as long as they are bound to the intraluminal chaperone, glucose regulated protein GRP78/BIP (Malhi and Kaufman 2011). Once activated, these kinases will trigger several mechanisms to (1) reduce protein translation and avoid further accumulation of misfolded proteins in the ER; (2) upregulate transcription of protein folding genes; and (3) activate ER-assisted degradation to limit the accumulation of misfolded protein. If the disruption to ER function cannot be resolved, a programmed cell death pathway is initiated mainly downstream of c-Jun N-terminal kinase (JNK) and the transcription factor CCAAT/enhancer binding homologous protein (CHOP) (Fig. 5.2). PERK activation induces the phosphorylation of the eukaryotic translation initiation factor 2A (eIF2α) resulting in general protein translation. PERK activation also results in the selective translation of the transcription factor ATF4 which regulates GRP78 promoter and also transcriptionally upregulates *CHOP* expression. Active IRE1α cleaves *X-box-binding protein (XBP)*-1 mRNA, and the resultant spliced protein sXBP-1 promotes expression of genes implicated in the degradation of misfolded ER glycoproteins. Also, when activated IRE1α can recruit the adaptor molecule TNF-receptor associated factor (TRAF)2, which activates the apoptosis-signal-regulating kinase (ASK)1/JNK pathway (Ozcan et al. 2004). ATF6α mainly upregulates the transcription of chaperones, which are components of the ER-associated degradation (ERAD) pathway, but can also transcriptionally induce CHOP expression (Tabas and Ron 2011).

Compelling evidence have demonstrated that ER stress is activated in both human and animal models of obesity and NASH (Ozcan et al. 2004; Puri et al. 2008; Wang et al. 2006a). Indeed, Ozcan *et al.* have first observed that in dietary and genetic models of obesity, enhanced ER stress-activated hyperactivation of JNK mediates insulin resistance by phosphorylating and inhibiting insulin receptor substrate 1 (IRS-1). Furthermore, in a nutritional murine model of steatohepatitis, increased in ER stress-induced CHOP expression and *XBP-1* mRNA splicing asso-

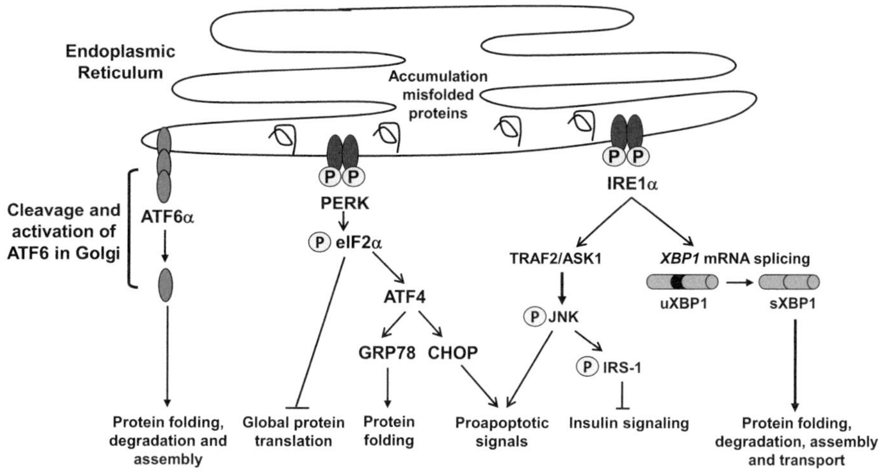

Fig. 5.2 ER-mediated adaptive and pro-apoptotic responses. Excessive accumulation of unfolded proteins in the ER results, at first, in the induction of adaptive rescuer responses or UPR by three parallel signaling branches: activating transcription factor (ATF) 6α, PKR-like ER kinase (PERK)-eIf2α and inositol-requiring enzyme-1α (IRE1α)-XBP1. The UPR acts to alleviate ER stress by inhibiting global protein translation, by increasing folding capacity by up-regulating chaperone protein GRP78, and by promoting the degradation, assembly and transport of misfolded proteins. However, sustained UPR activation in response to prolonged ER stress, will lead to the induction of pro-apoptotic signals mediated mainly by JNK and CHOP. Indeed, ER stress can activate JNK through IRE1a-TRAF2-ASK1 complex. Also, PERK-mediated preferential translation of the transcription factor ATF4 induces the expression of the pro-apoptotic transcription factor CHOP. *UPR* unfolded protein response, *eIf2α* eukaryotic translation initiation factor 2A, *uXBP1* unspliced X-box-binding protein 1, *sXBP1* spliced X-box-binding protein 1, *TRAF2* TNF-receptor associated factor 2, *ASK1* apoptosis-signal-regulating kinase 1

ciated with liver injury (Wang et al. 2006a). In obese humans undergoing bariatric surgery, ER stress is detected in liver tissues and is highly regulated by weight loss (Gregor et al. 2009). Finally, Puri *et al.* observed a strong activation of PERK in the liver of patients with NAFLD and NASH; and human NASH was associated with increased JNK activity (Puri et al. 2008).

It appears that, more than the steatosis itself, it is the composition of fatty acids in the steatotic liver that drives ER stress-mediated liver damage. Hence, induction of ER stress markers was observed only in male Wistar rats fed a diet enriched in saturated fatty acid as compared to rat fed a polyunsaturated fatty acid diet, despite similar accumulation of hepatic triglycerides with these two diets (Wang et al. 2006a). Also, deletion in mice of the enzyme that catalyzes the synthesis of monounsaturated fatty acids from saturated fatty acids, stearoyl-coA desaturase 1 (SCD1), results in unsaturated fatty acid insufficiency and induces a marked increase in the abundance of ER stress markers (e.g. sXBP-1 and CHOP) in the liver (Flowers et al. 2008). In line with this concept, numerous in vitro studies have demonstrated that over-accumulation of saturated FFAs, but not unsaturated FFAs, in liver cells trigger an ER stress which was mitochondrial dependent (Cazanave et al. 2009; Akazawa et al. 2010). Also, liver cells co-incubation with unsaturated FFA palmitoleate or oleate could rescue the cells from saturated FFAs-mediated ER stress and toxicity (Akazawa et al. 2010; Wang et al. 2013b). ER stress-associated JNK and CHOP activation mediates apoptosis in part by up-regulating the TRAIL receptor 2 (TRAIL-R2 or death receptor 5 [DR5]) (Malhi et al. 2007; Cazanave et al. 2011) and enhancing the expression and function of the pro-apoptotic BH3-only proteins BIM and PUMA, causing mitochondrial dysfunction and cellular demise (Barreyro et al. 2007; Cazanave et al. 2009); and these mechanisms will be discussed in greater detail in the forthcoming sections. Saturated FFAs also deplete ER-calcium stores, and the excess calcium in the cytosol can be absorb by the mitochondria, resulting in dysfunction of the organelle and inducing subsequent cell death (Wei et al. 2009).

3.2 CHOP-Dependent Apoptosis

CCAAT/enhancer-binding homologous protein (CHOP) is the most well characterized inducible leucine zipper transcription factor that emanates from the stressed ER. Both ATF4 and ATF6α contributes to *CHOP* transcription; and CHOP is linked to the lipoapoptotic process in several ways. CHOP expression is increased in human NASH (Puri et al. 2008), and numerous in vitro studies have demonstrated that saturated FFAs induces CHOP expression in liver cells (Akazawa et al. 2010; Cazanave et al. 2010). Although it was demonstrated that *CHOP* deletion has minimal effect on hepatocytes toxicity induced by a low dose of palmitate (Pfaffenbach et al. 2010), *CHOP* knockdown protects liver cells to high concentrations of saturated FFAs (Cazanave et al. 2010). Several CHOP downstream target genes have been proposed to promote apoptosis. CHOP can transcriptionally upregulate the expression of TRAIL-R2 or DR5 (Yamaguchi and Wang 2004) which subsequently promote apoptosis by the extrinsic pathway of cell death (Fig. 5.3) (Guicciardi and Gores 2009). Another CHOP-dependent targets implicate several members of the Bcl-2 family of proteins that regulate the mitochondrial pathway of cell death. On the one hand, CHOP sensitizes cells to ER stress-dependent toxicity by down regulating the expression of the anti-apoptotic protein Bcl-2 (McCullough et al. 2001). On the other hand, the transcription complexes CHOP/C/EBPα or CHOP/activator protein-1 (AP-1) heterodimers directly binds, respectively, to the promoter of the pro-apoptotic BH3-only protein *BIM* or *PUMA* and up-regulates the expression of the BIM and PUMA (Fig. 5.3) (Cazanave et al. 2010; Puthalakath et al. 2007). CHOP also regulates the expression of the pro-apoptotic protein tribbles related protein 3 (TRB3), which indirectly induces PUMA expression and compromises mitochondrial function (Ohoka et al. 2005). Interestingly, the anti-apoptotic Bcl-xL, which inhibits BIM-dependent mitochondrial dysfunctions, prevents ER stress-induced apoptosis (Gaudette et al. 2014). Thus, CHOP-dependent apoptosis likely relies on disrupting the fine balance between pro- and anti-apoptotic Bcl-2 proteins therefore causing toxicity by the mitochondrial pathway. Also, CHOP mediates

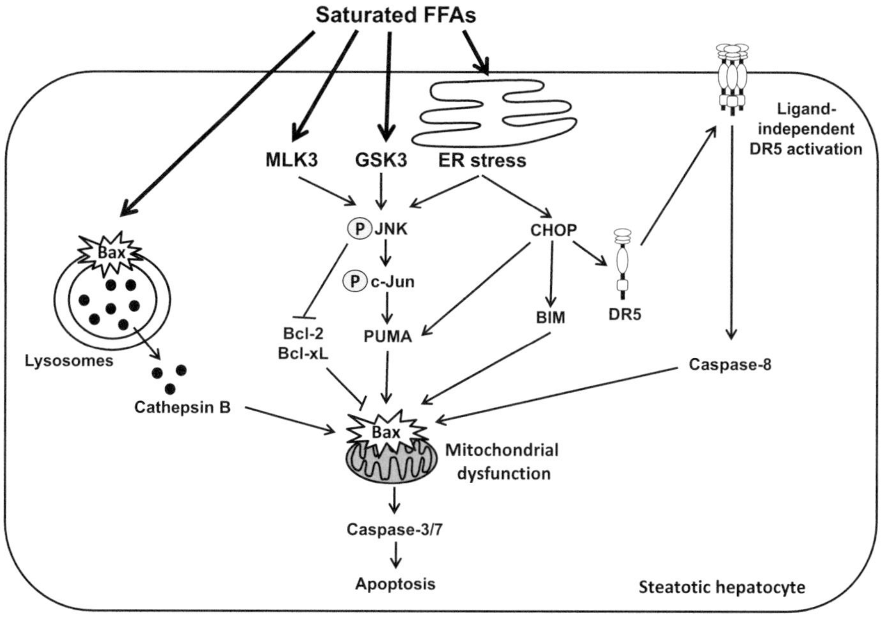

Fig. 5.3 Molecular mechanisms of hepatocyte lipoapoptosis. Saturated FFAs accumulate in the endoplasmic reticulum (ER) and induce an ER stress, which in turn induce the transcription factor CHOP and induce JNK activity. Saturated FFAs also activate JNK downstream of the mixed-lineage protein kinase 3 (MLK3) and the glycogen synthase kinase (GSK)-3 signaling cascades. Active JNK phosphorylates the transcription factor c-Jun, which cooperates with CHOP to upregulate the transcription of the pro-apoptotic BH3-only protein PUMA. CHOP also mediate the upregulation of another BH3-only protein BIM; and BIM and PUMA cooperate in activating the executioner pro-apoptotic protein Bax, causing mitochondrial dysfunction, activation of the effector caspases 3/7 and cellular apoptosis. JNK further induces the mitochondrial pathway of apoptosis by inhibiting the anti-apoptotic functions of the anti-apoptotic proteins Bcl-2 and Bcl-xL. CHOP also upregulates the expression of the death receptor DR5, resulting in increased DR5 cell surface expression which clustering of the receptor leading to recruitment and activation of the executioner caspase-8, which ultimately induces Bax. Finally, saturated FFAs induce Bax-dependent permeabilization of lysosomes and release of cathepsin B in the cytosol resulting in downstream mitochondrial dysfunctions and apoptosis

apoptosis by activation the ER oxidase 1α leading to calcium release from the ER and mitochondrial dysfunctions (Li et al. 2009c).

Although some line of evidence implicate CHOP in ER stress-induced lipotoxicity in vitro, a paradoxical protective role of CHOP is observed in in vivo murine model of NASH (Soon et al. 2010; Malhi et al. 2013). Indeed, in the absence of CHOP, mice develop a worsened NASH phenotype using an MCD or high fat diet; and it was found that, in that model, CHOP has a pro-inflammatory effect in fatty liver injury, by favoring survival of activated macrophages and accumulation in the liver which aggravates liver injury (Malhi et al. 2013). Hepatocyte-specific *CHOP* knockout mice will be necessary to delineate the exact contribution of CHOP in the onset and progression of fatty liver injury.

3.3 c-Jun NH2-Terminal Kinase Signaling Pathway

c-Jun N-terminal kinase (JNK) is a stress-activated protein kinase, and only the two isoforms JNK1 and JNK2 are expressed in liver tissue. JNK activity has been implicated as a central player in regulating insulin action and it represents a critical mechanism in the pathogenesis of obesity, T2DM and fatty liver disease. In both experimental animal models of obesity (Schattenberg et al. 2006; Singh et al. 2009) and in human NASH (Cazanave et al. 2009; Puri et al. 2008), JNK activity is increased in the liver. In the context of obesity and fatty liver disease, saturated FFAs activate JNK downstream of the mixed-lineage protein kinase 3 (MLK3) (Kant et al. 2013; Jaeschke and Davis 2007) and the glycogen synthase kinase (GSK)-3 signaling (Ibrahim et al. 2011), and/or secondary to FFAs-induced ER stress (Fig. 5.3) (Wei et al. 2009; Urano et al. 2000). Loss of *jnk1* prevents the development of insulin resistance and obesity in murine model of NASH (Schattenberg et al. 2006; Singh et al. 2009) as JNK can mediate the insulin resistance state by phosphorylating the insulin receptor substrate (IRS)1 and therefore suppressing insulin receptor signaling (Li and Yu 2013). Both JNK1 and JNK2 have been implicated in liver injury (Wang et al. 2006b); however JNK1 is more strongly associated with steatohepatitis (Schattenberg et al. 2006; Singh et al. 2009). *Jnk1* null mice are protected against hepatocyte injury and apoptosis induced by a MCD diet (Schattenberg et al. 2006). JNK1, but not JNK2, can regulate genes transcription through its ability to phosphorylate and activate the transcription factor c-Jun, a member of the activator protein-1 (AP-1) complex. In both human and animal models of NASH, increased phosphorylation of c-Jun and/or increased DNA-binding activity of AP-1 complexes containing c-Jun is observed in the liver (Schattenberg et al. 2006; Singh et al. 2009; Videla et al. 2009). JNK1-dependent phosphorylation of c-Jun mediates, in part, hepatocyte lipoapoptosis by transcriptionally upregulating the expression of the pro-apoptotic BH3-only protein PUMA (Cazanave et al. 2009) which induces mitochondrial dysfunction (Fig. 5.3). JNK further induces the mitochondrial pathway of apoptosis by regulating the function of other Bcl-2 family members through post-transcriptional mechanisms: JNK phosphorylates the anti-apoptotic proteins Bcl-2 and Bcl-xL thereby inhibiting their anti-apoptotic properties (Yamamoto et al. 1999); whereas JNK-mediated phosphorylation of the pro-apoptotic proteins BIM, BAD and Bax increases the ability of these proteins to trigger mitochondrial-dependent cell death (Lei and Davis 2003; Donovan et al. 2002; Kim et al. 2006). Finally, JNK also sensitizes liver cells to death receptor-mediated pathway of apoptosis by upregulating the expression of DR5 (Malhi et al. 2007) and Fas ligand (Eichhorst et al. 2000).

3.4 Bcl-2 Family Proteins and Mitochondrial-Lysosomal Pathway

Accumulating data indicate that the Bcl-2 family proteins are key regulators of lipoapoptosis, as they orchestrate the activation of the mitochondrial pathway of cell death. As stated before, the BH3-only proteins BIM and/or PUMA are upregulated by saturated FFAs and in human NASH, and they cooperate to induce liver cell death (Barreyro et al. 2007; Cazanave et al. 2009). Thus, liver cells knockdown for *BIM* or *PUMA* are partially protected against palmitate toxicity. In that context, *BIM* expression is transcriptionally regulated by the transcription factor Forkhead box O3a (FoxO3a), activated upon dephosphorylation by the protein phosphatase 2A (PP2A) (Barreyro et al. 2007); whereas CHOP and JNK1-activated AP-1 cooperatively mediate the transcriptional upregulation of *PUMA* (Cazanave et al. 2010). The expression of the anti-apoptotic Bcl-2 proteins Bcl-xL and Mcl-1 is also altered during lipoapoptosis (Feldstein et al. 2006; Masuoka et al. 2009) and contributes to the apoptotic process. Indeed, Bcl-xL expression is decreased in palmitate-treated liver cells (Feldstein et al. 2006), and in the liver of MCD-fed mice (Farrell et al. 2009); similarly, saturated FFAs-induced toxicity implicate protein kinase C-dependent proteasomal degradation of Mcl-1 (Masuoka et al. 2009). Preventing the decrease of these anti-apoptotic proteins by overexpressing Bcl-xL or a non-degradable form of Mcl-1 could attenuate the lipotoxic insult (Feldstein et al. 2006; Masuoka et al. 2009). BIM and PUMA directly or indirectly activates the pro-apoptotic executioner protein Bax (Ren et al. 2010); and the pro-survival activity of Mcl-1 and Bcl-xL proceeds via inhibition of Bax function at the mitochondria by direct interaction with Bax (Vela et al. 2013; Germain et al. 2008) or indirectly by sequestration of activator BH3-only proteins. Once activated, Bax translocates from the cytosol to insert into the outer mitochondrial membrane, where it oligomerizes and forms pores which induce membrane permeabilization. Following this event, proteins that participate in the execution phase of apoptosis are activated, namely caspases 3, 6 and 7 that trigger DNA damage and cell death; and *caspase 3* knock-out mice are protected against diet-induced liver injury (Thapaliya et al. 2014). Several line of evidence suggest a major role of mitochondrial dysfunction in NAFLD. Structural and functional mitochondrial dysfunctions are observed in human and animal models of NASH (Sanyal et al. 2001; Wei et al. 2008) and lead to an overproduction of reactive oxygen species (ROS) that trigger lipid peroxidation, pro-inflammatory cytokines production and cell death (Begriche et al. 2006).

Lysosomal permeabilization with subsequent release of the major lysosomal cysteine protease cathepsin B into the cytosol has been reported in human liver tissues from patients with NAFLD (Feldstein et al. 2004) and could also contribute to mitochondrial dysfunctions and caspases activation associated with the disease. Indeed, it was reported that saturated FFAs activates the lysosomal pathway of apoptosis through Bax-dependent permeabilization of lysosomes and release of cathepsin B in the cytosol (Fig. 5.3) (Guicciardi et al. 2004); and this event was preceding mitochondrial permeabilization and cell death (Feldstein et al. 2004). Genetic loss of *cathepsin B* protected mice from diet induced-mitochondrial dysfunctions and liver injury (Feldstein et al. 2004; Li et al. 2008); and genetic or chemical inhibition of cathepsin B activity could attenuate palmitate-induced mitochondrial dysfunction and liver cell death (Feldstein et al. 2004), highlighting the contribution of lysosomal permeabilization in mediating lipoapoptosis.

3.5 Death Receptors

Hepatocytes express death receptors that belongs to the tumor necrosis factor (TNF) receptor superfamily, namely Fas, TNF receptor 1 (TNF-R1) and TRAIL-R1 and 2 (or DR4 and DR5), which have all been implicated in lipotoxicity and NASH. When engaged to their natural ligands, these receptors activate caspase cascades, such as the initiator caspase-8 and the executor caspase-3 that mediate cell death. In the hepatocyte, the extrinsic death receptor-mediated pathway of cell death also requires mitochondrial dysfunction for efficient apoptosis (Guicciardi and Gores 2009).

Compelling data have demonstrated a crucial role for DR5-dependent signaling as a key mechanism of hepatocyte lipoapoptosis (Cazanave et al. 2011). Increased expression of DR5 is observed in liver biopsies from patients with NASH (Malhi et al. 2007; Kahraman et al. 2010), and in liver cells after treatment with saturated FFAs (Malhi et al. 2007; Cazanave et al. 2011). More specifically, saturated FFAs/ER stress-induced accumulation and aggregation of DR5 within the plasma membrane, activates DR5 signaling in a ligand independent manner (Fig. 5.3) (Cazanave et al. 2011; Lu et al. 2014), causing hepatocyte apoptosis (Cazanave et al. 2011). Thus, FFAs-induced steatosis sensitizes hepatocytes to TRAIL cytotoxicity (Malhi et al. 2007). Also, FFAs-induced proteasomal elimination of cellular inhibitor of apoptosis protein-1 (cIAP-1), a member of the antiapoptotic IAP family that negatively regulate death receptor signaling, may also contribute to the sensitization of steatotic hepatocyte to TRAIL cytotoxicity (Akazawa et al. 2013). In contrast, silencing DR5 expression protected in vitro against FFAs-mediated hepatocyte apoptosis (Cazanave et al. 2011). More recently, Idrissova *et al.* has shown that mice deficient in *TRAIL-R* were protected against key features of NASH, including steatosis, hepatocyte apoptosis inflammation and fibrosis induced by a high fat, fructose and cholesterol diet (Idrissova et al. 2015). These results further highlight the determinant role of TRAIL signaling in mediating hepatocyte apoptosis during fatty liver injury.

The expression of Fas is also increased in livers of patients with NASH (Feldstein et al. 2003a) and in experimental models of NASH (Farrell et al. 2009), rendering steatotic liver highly susceptible to Fas-induced hepatocyte cell death (Feldstein et al. 2003b). However, no data directly implicate Fas-dependent signaling in the development of NASH in vivo. TNF-R1 and its ligands TNF-α are also both increased in NASH, but could not discriminate steatohepatitis from steatosis (Ribeiro et al. 2004; Crespo et al. 2001; Younossi et al. 2008). Some studies indicate that *TNF-R1*$^{-/-}$ mice are protected against diet-induced obesity (Romanatto et al. 2009), and that anti-TNF-α therapy improves hepatic histology in animal models of NASH (Li et al. 2003). Conversely, diet-induced steatohepatitis is not prevented in both TNF-α and TNF-R1 knockout mice, indicating that the TNF-α-dependent signaling might not be primordial in NASH pathogenesis (Deng et al. 2005; Schnyder-Candrian et al. 2005). Therefore, the role of TNF-R1 signaling in NASH is still controversial and merits to be further explored.

4 Cross-Talk between Lipoapoptosis, Oxidative Stress and Inflammatory Processes

The pathogenesis of NASH is multifactorial, and oxidative stress has been implicated in disease progression, as its drives both hepatocyte cell death and inflammation leading to T2DM and steatohepatitis. As mentioned earlier, NASH patients exhibit mitochondrial dysfunctions (Sanyal et al. 2001) resulting in an substantial accumulation of ROS that trigger oxidative degradation of lipids, pro-inflammatory cytokines release and cell death.

4.1 NAFLD and Oxidative Stress

Oxidative stress is observed in both human and experimental model of NASH (Kojima et al. 2007; Matsunami et al. 2010) as a result of an increase in the generation of ROS by the mitochondria and a decrease in anti-oxidant defenses. FFAs accumulation in hepatocyte favors excess generation of oxidized fatty acid which are in turn capable to trigger ROS production. Also, Bax-dependent permeabilization of the mitochondrial membranes by saturated FFAs, may disrupt the electron flow in the mitochondrial respiratory chain causing a substantial accumulation of ROS. In vitro, palmitate treatment induces ROS generation in liver cells (Gao et al. 2010). Cytochrome P450 2E1 (CYP2E1), a member of the oxido-reductase cytochrome family, represents an additional source of oxidative stress as this enzyme can oxidize fatty acids and generate ROS. Higher CYP2E1 expression and activity is observed in the liver of human and animal models of NAFLD (Bell et al. 2011; Abdelmegeed et al. 2012); gene silencing of CYP2E1 expression prevents high fat diet-induced lipid peroxidation and increased in phosphorylated JNK (Abdelmegeed et al. 2012). Consequently, increased production of ROS further induces lipid peroxidation and DNA damage; and both 4-hydroxy-2-noneral (4-HNE, a highly toxic aldehyde product of lipid peroxidation) and 8-hydroxydeoxyguanosine (8-OH-dG, the oxidized derivative of deoxyguanosine) increase in NAFLD as disease progresses from simple steatosis to steatohepatitis (Seki et al. 2002). Also, serum malondialdehyde (MDA) levels, a marker of lipid oxidation were significantly higher in patients with NASH (Koruk et al. 2004).

The Keap1-Nrf2 pathway is also activated in response to oxidative stress, and seems to contribute to NASH pathogenesis. On the one hand, Keap1 is inactivated by oxidative stress causing the stabilization of the transcription factor Nrf2 and increased expression of Nrf2-downstream target detoxifying and antioxidant genes; and Nrf2 expression is increased in human NAFLD and NASH (Xu et al. 2011). On the other hand, cellular degradation of Keap1 through autophagy-dependent mechanisms induces JNK activation and upregulates BIM and PUMA expression and contributes to hepatocyte lipoapoptosis in an Nrf2-independent manner (Cazanave et al. 2014). Some in vivo studies indicate that *Keap1* deletion increases insulin resistance and hepatocyte steatosis in murine models of steatohepatitis (More et al. 2013). Conversely, genetic activation of Nrf2 signaling by *Keap1* knockdown decreases the onset of T2DM in a genetic or diet-induced murine model of NASH (Uruno et al. 2013). Therefore the role of Keap1-Nrf2 in the context of fatty liver disease is still controversial, possibly due to the divergent Nrf2-dependent and Nrf2-independent Keap1-mediated signaling, and remains to be clarified.

Continued oxidative stress can also result in chronic inflammation, which can further induce liver fibrogenesis leading to cirrhosis and tumor development. Oxidative stress can induce a variety of transcription factors including NF-kB and JNK-activated AP-1 which mediates the upregulation of inflammatory cytokines and chemokines, and the recruitment of activated immune cells to the liver with direct implications in NASH pathogenesis.

Anti-oxidant vitamin E supplementation has been proven to attenuate oxidative stress, mitochondrial lipid peroxidation, necro-inflammation and hepatic apoptosis in several animal models of NAFLD [as reviewed by Al-Busafi et al. (2012)]. As well, vitamin E in combination with the insulin-sensitizer pioglitazone improves liver enzymes and histology in patients with NASH (Al-Busafi et al. 2012; Sanyal et al. 2004).

4.2 NAFLD and the Hepatic Inflammatory Responses

Liver inflammation, an essential response to tissue damage, is coordinated by secreted mediators such as cytokines and chemokines which coordinate cellular defenses and the scarring of the liver (fibrosis). The origin of liver inflammation in NASH is multiple but lead to the recruitment and/or activation of distinct immune cells.

Visceral adipose tissue of obese patients is inflamed and releases macrophage chemokines and cytokines such as macrophage chemotactic protein 1 (MCP-1), interleukin (IL)-6, IL-8 and TNFα (Makki et al. 2013); and therefore adipose tissue could substantially contributes to increase systemic levels of these pro-inflammatory mediators which would alter hepatic inflammatory system in patients with NAFLD. Hepatocyte cell death may also trigger hepatic inflammation, by inducing NF-kB and AP-1-dependent upregulation of several cytokines and/or by releasing pro-inflammatory extracellular vesicles which could recruit and activate immune cells (Hirsova and Gores 2015). Both NF-kB and JNK are activated in NASH; NF-kB is important for hepatic inflammatory recruitment in steatohepatitis (Dela Pena et al. 2005) and *jnk1* knockout mice are protected against MCD-induced liver inflammation (Schattenberg et al. 2006). NF-kB and/or AP-1 transcriptionally upregulates MCP-1, IL-8 and IL-6, which are consistently found to be upregulated in human and animal models of NASH (Peverill et al. 2014).

Several cells of the innate immune system have been implicated in eliciting liver inflammation during NASH, including the resident liver macrophages or Kupffer cells (KCs), natural killer (NK) T cells and hepatic stellate cells (HSCs). Indeed, pro-inflammatory activated KCs are critical for the production of TNFα and the recruitment of blood derived-monocytes which further exacerbate liver damage and induce liver fibrogenesis as a feed-forward loop. Indeed, engulfment of hepatocyte apoptotic bodies by KCs induces the expression of death receptor ligands in KCs (Canbay et al. 2003) which may aggravate hepatocyte cell death. Sequential depletion of KCs decreases MCD-induced steatosis, liver injury and pro-inflammatory monocyte infiltration (Tosello-Trampont et al. 2012). NKT cells may have a protective role at the early stage of NAFLD (simple steatosis), but they accumulate and promote fibrosis with the progression toward NASH (Tajiri and Shimizu 2012). Type I NKT cells seems to the predominant NKT cell subtype that promote liver inflammation and fibrosis, and a recent elegant study highlights the importance of activated NKT cells in mediating fibrosis and cancer development associated with the NASH phenotype in a murine model of NASH (Wolf et al. 2014). Finally, HSCs are the main collagen-producing cells in the liver. While quiescent in normal liver, these cells become activated during liver damage and therefore express a range of chemokines that recruit leukocytes, and also produce components of the extracellular matrix (e.g., collagen α1(i) and smooth muscle actin (SMA)-α) which contributes to fibrosis in NASH (Washington et al. 2000).

Liver fibrosis results from excessive hepatocyte damage caused by a chronic expression of pro-inflammatory cytokines and recruitment of activated immune cells to the liver which provide a favourable environment to the development of hepatocarcinoma (HCC). As mentioned earlier, liver fibrosis in NASH is mediated by activated HSCs. Of the different signaling pathways that induce liver fibrosis, the hedgehog (Hh) pathway has gained a considerable interest into the mechanisms that activate HSCs and promote fibrogenesis during NASH. Indeed, Hh pathway activation correlates with disease severity in NASH (Guy et al. 2012); and ballooned hepatocytes have been identified as an important source of Hh ligands in NASH (Rangwala et al. 2011). A recent study indicate that ballooned hepatocytes in NASH display diminished levels of caspase-9 (an essential initiator caspase required for the mitochondrial pathway of apoptosis), and represent an "undead" state of the hepatocyte in which the cell death program is initiated upon palmitate treatment but cannot be executed and secrete Hh ligands (sonic hedgehog, SHH) in a JNK-dependent manner (Fig. 5.4) (Kakisaka et al. 2012b). Hepatic NKT cells also drive the production of Hh ligands during NASH (Syn et al. 2012), and Hh pathway activation in HSC promotes the state transition of quiescent HSCs to activated myofibroblastic HSCs. Inhibitor of Hh signaling can convert activated myofibroblasts back to quiescent HSCs (Chen et al. 2012); and inhibition of Hh signaling in a murine model of NASH attenuates liver injury, macrophage-associated inflammation and fibrosis (Hirsova et al. 2013).

Hence, inflammation in NASH is intricately linked to hepatocyte cell death and drives hepatic fibrosis.

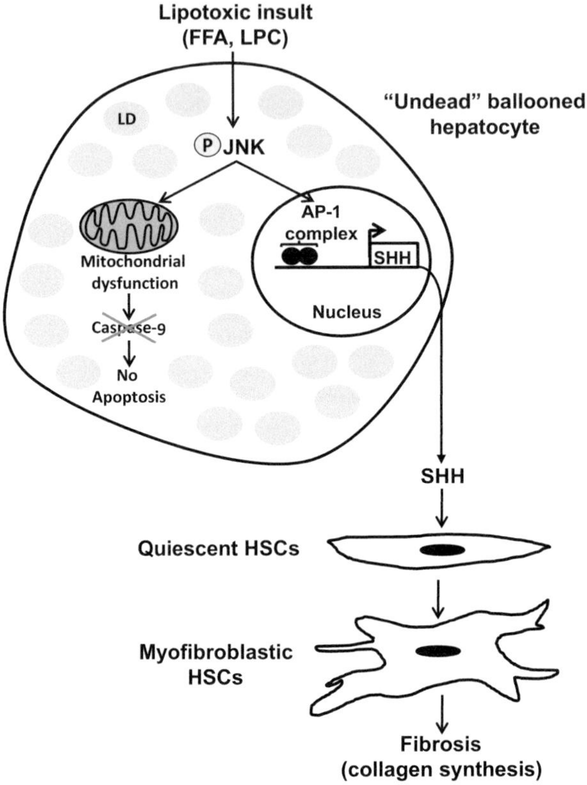

Fig. 5.4 Secretion of the hedgehog ligand (Hh), sonic hedgehog (SHH), by ballooned hepatocytes contributes to liver fibrosis in NASH. Ballooned hepatocytes in NASH display diminished levels of caspase-9 and represent an "undead" state of the hepatocyte in which the cell death program is initiated upon lipotoxic insult by saturated FFAs or LPC but cannot be executed. These "undead" hepatocytes secrete the hedgehog ligand SHH, via a JNK/AP-1 complex-dependent transcriptional upregulation of SHH expression. Increased release of SHH by ballooned hepatocytes results in Hh pathway activation in hepatic stellate cells (HSCs) promoting the transition of HSCs from a quiescent state to an activated, myofibroblastic phenotype which produces extracellular matrix and promotes liver fibrosis. *LD* lipid droplet

5 Conclusion

NAFLD is the most common cause of chronic liver disease in North America and is rapidly becoming a leading cause of liver-related morbidity and mortality. NAFLD is associated with insulin resistance which leads to the accumulation of fat, mainly triglycerides, resulting in hepatic steatosis. Although steatosis is not toxic, accumulation of certain lipid types in the liver, such as FFAs, ceramides or free cholesterol and certain of their intermediates, induce hepatocyte cell death. The lipotoxic signaling pathways implicate impairment of proper function of liver cell organelles such as mitochondria, endoplasmic reticulum and lysosomes, with activation of JNK and CHOP, dysregulated expression of Bcl2 family members and death receptors and subsequent activation of effector caspases. Although an important amount of evidence has been accumulated concerning the role of specific lipid types in NAFLD pathogenesis, the precise mechanisms by which each of these lipids contributes to liver injury, inflammation and fibrosis remain incompletely defined. There is a major unmet need for development of therapeutics to prevent and treat NAFLD; and a better understanding of the cellular and molecular events linking toxic lipid accumulation to hepatocyte injury and NAFLD progression is of public health interest to help identify novel therapeutic approaches for treating this disease.

Acknowledgements This work was supported, in whole or in part, by NIH Grants R01 DK081450 (A.J.S) and T32 07150 (A.J.S. and S.C.C.).

References

Abdelmegeed MA, Banerjee A, Yoo SH, Jang S, Gonzalez FJ, Song BJ (2012) Critical role of cytochrome P450 2E1 (CYP2E1) in the development of high fat-induced non-alcoholic steatohepatitis. J Hepatol 57:860–866

Adams LA, Lymp JF, St Sauver J, Sanderson SO, Lindor KD, Feldstein A, Angulo P (2005) The natural history of nonalcoholic fatty liver disease: a population-based cohort study. Gastroenterology 129:113–121

Akazawa Y, Cazanave S, Mott JL, Elmi N, Bronk SF, Kohno S, Charlton MR, Gores GJ (2010) Palmitoleate attenuates palmitate-induced Bim and PUMA up-regulation and hepatocyte lipoapoptosis. J Hepatol 52:586–593

Akazawa Y, Guicciardi ME, Cazanave SC, Bronk SF, Werneburg NW, Kakisaka K, Nakao K, Gores GJ (2013) Degradation of cIAPs contributes to hepatocyte lipoapoptosis. Am J Physiol Gastrointest Liver Physiol 305:G611–G619

Al-Busafi SA, Bhat M, Wong P, Ghali P, Deschenes M (2012) Antioxidant therapy in nonalcoholic steatohepatitis. Hepat Res Treat 2012:947575

Anderson N, Borlak J (2008) Molecular mechanisms and therapeutic targets in steatosis and steatohepatitis. Pharmacol Rev 60:311–357

Angulo P, Lindor KD (2002) Non-alcoholic fatty liver disease. J Gastroenterol Hepatol 17(Suppl):S186–S190

Aranha MM, Cortez-Pinto H, Costa A, da Silva IB, Camilo ME, de Moura MC, Rodrigues CM (2008) Bile acid levels are increased in the liver of patients with steatohepatitis. Eur J Gastroenterol Hepatol 20:519–525

Barreyro FJ, Kobayashi S, Bronk SF, Werneburg NW, Malhi H, Gores GJ (2007) Transcriptional regulation of Bim by FoxO3A mediates hepatocyte lipoapoptosis. J Biol Chem 282:27141–27154

Bechmann LP, Kocabayoglu P, Sowa JP, Sydor S, Best J, Schlattjan M, Beilfuss A, Schmitt J, Hannivoort RA, Kilicarslan A et al (2013) Free fatty acids repress small heterodimer partner (SHP) activation and adiponectin counteracts bile acid-induced liver injury in superobese patients with nonalcoholic steatohepatitis. Hepatology 57:1394–1406

Begriche K, Igoudjil A, Pessayre D, Fromenty B (2006) Mitochondrial dysfunction in NASH: causes, consequences and possible means to prevent it. Mitochondrion 6:1–28

Bell LN, Molleston JP, Morton MJ, Klipsch A, Saxena R, Vuppalanchi R, Chalasani N (2011) Hepatic lipid peroxidation and cytochrome P-450 2E1 in pediatric nonalcoholic fatty liver disease and its subtypes. J Clin Gastroenterol 45:800–807

Brenner B, Ferlinz K, Grassme H, Weller M, Koppenhoefer U, Dichgans J, Sandhoff K, Lang F, Gulbins E (1998) Fas/CD95/Apo-I activates the acidic sphingomyelinase via caspases. Cell Death Differ 5:29–37

Browning JD, Szczepaniak LS, Dobbins R, Nuremberg P, Horton JD, Cohen JC, Grundy SM, Hobbs HH (2004) Prevalence of hepatic steatosis in an urban population in the United States: impact of ethnicity. Hepatology 40:1387–1395

Caballero F, Fernandez A, De Lacy AM, Fernandez-Checa JC, Caballeria J, Garcia-Ruiz C (2009) Enhanced free cholesterol, SREBP-2 and StAR expression in human NASH. J Hepatol 50:789–796

Canbay A, Feldstein AE, Higuchi H, Werneburg N, Grambihler A, Bronk SF, Gores GJ (2003) Kupffer cell engulfment of apoptotic bodies stimulates death ligand and cytokine expression. Hepatology 38:1188–1198

Cazanave SC, Mott JL, Elmi NA, Bronk SF, Werneburg NW, Akazawa Y, Kahraman A, Garrison SP, Zambetti GP, Charlton MR et al (2009) JNK1-dependent PUMA expression contributes to hepatocyte lipoapoptosis. J Biol Chem 284:26591–26602

Cazanave SC, Elmi NA, Akazawa Y, Bronk SF, Mott JL, Gores GJ (2010) CHOP and AP-1 cooperatively mediate PUMA expression during lipoapoptosis. Am J Physiol Gastrointest Liver Physiol 299:G236–G243

Cazanave SC, Mott JL, Bronk SF, Werneburg NW, Fingas CD, Meng XW, Finnberg N, El-Deiry WS, Kaufmann SH, Gores GJ (2011) Death receptor 5 signaling promotes hepatocyte lipoapoptosis. J Biol Chem 286:39336–39348

Cazanave SC, Wang X, Zhou H, Rahmani M, Grant S, Durrant DE, Klaassen CD, Yamamoto M, Sanyal AJ (2014) Degradation of Keap1 activates BH3-only proteins Bim and PUMA during hepatocyte lipoapoptosis. Cell Death Differ 21:1303–1312

Charlton M, Viker K, Krishnan A, Sanderson S, Veldt B, Kaalsbeek AJ, Kendrick M, Thompson G, Que F, Swain J et al (2009) Differential expression of lumican and fatty acid binding protein-1: new insights into the histologic spectrum of nonalcoholic fatty liver disease. Hepatology 49:1375–1384

Chavez-Tapia NC, Mendez-Sanchez N, Uribe M (2006) The metabolic syndrome as a predictor of nonalcoholic fatty liver disease. Ann Intern Med 144:379, author reply 80

Chen Y, Choi SS, Michelotti GA, Chan IS, Swiderska-Syn M, Karaca GF, Xie G, Moylan CA, Garibaldi F, Premont R, et al (2012) Hedgehog controls hepatic stellate cell fate by regulating metabolism. Gastroenterology 143:1319–1329.e1–e11

Chocian G, Chabowski A, Zendzian-Piotrowska M, Harasim E, Lukaszuk B, Gorski J (2010) High fat diet induces ceramide and sphingomyelin formation in rat's liver nuclei. Mol Cell Biochem 340:125–131

Contos MJ, Choudhury J, Mills AS, Sanyal AJ (2004) The histologic spectrum of nonalcoholic fatty liver disease. Clin Liver Dis 8:481–500, vii

Cowart LA (2009) Sphingolipids: players in the pathology of metabolic disease. Trends Endocrinol Metab 20:34–42

Crespo J, Cayon A, Fernandez-Gil P, Hernandez-Guerra M, Mayorga M, Dominguez-Diez A, Fernandez-Escalante JC, Pons-Romero F (2001) Gene expression of tumor necrosis factor alpha and TNF-receptors, p55 and p75, in nonalcoholic steatohepatitis patients. Hepatology 34:1158–1163

Dasarathy S, Yang Y, McCullough AJ, Marczewski S, Bennett C, Kalhan SC (2011) Elevated hepatic fatty acid oxidation, high plasma fibroblast growth factor 21, and fasting bile acids in nonalcoholic steatohepatitis. Eur J Gastroenterol Hepatol 23:382–388

Dela Pena A, Leclercq I, Field J, George J, Jones B, Farrell G (2005) NF-kappaB activation, rather than TNF, mediates hepatic inflammation in a murine dietary model of steatohepatitis. Gastroenterology 129:1663–1674

Deng QG, She H, Cheng JH, French SW, Koop DR, Xiong S, Tsukamoto H (2005) Steatohepatitis induced by intragastric overfeeding in mice. Hepatology 42:905–914
Donnelly KL, Smith CI, Schwarzenberg SJ, Jessurun J, Boldt MD, Parks EJ (2005) Sources of fatty acids stored in liver and secreted via lipoproteins in patients with nonalcoholic fatty liver disease. J Clin Invest 115:1343–1351
Donovan N, Becker EB, Konishi Y, Bonni A (2002) JNK phosphorylation and activation of BAD couples the stress-activated signaling pathway to the cell death machinery. J Biol Chem 277:40944–40949
Eichhorst ST, Muller M, Li-Weber M, Schulze-Bergkamen H, Angel P, Krammer PH (2000) A novel AP-1 element in the CD95 ligand promoter is required for induction of apoptosis in hepatocellular carcinoma cells upon treatment with anticancer drugs. Mol Cell Biol 20:7826–7837
Farrell GC, Larter CZ, Hou JY, Zhang RH, Yeh MM, Williams J, dela Pena A, Francisco R, Osvath SR, Brooling J et al (2009) Apoptosis in experimental NASH is associated with p53 activation and TRAIL receptor expression. J Gastroenterol Hepatol 24:443–452
Faubion WA, Guicciardi ME, Miyoshi H, Bronk SF, Roberts PJ, Svingen PA, Kaufmann SH, Gores GJ (1999) Toxic bile salts induce rodent hepatocyte apoptosis via direct activation of Fas. J Clin Invest 103:137–145
Fayyaz S, Henkel J, Japtok L, Kramer S, Damm G, Seehofer D, Puschel GP, Kleuser B (2014) Involvement of sphingosine 1-phosphate in palmitate-induced insulin resistance of hepatocytes via the S1P2 receptor subtype. Diabetologia 57:373–382
Feldstein AE, Canbay A, Angulo P, Taniai M, Burgart LJ, Lindor KD, Gores GJ (2003a) Hepatocyte apoptosis and fas expression are prominent features of human nonalcoholic steatohepatitis. Gastroenterology 125:437–443
Feldstein AE, Canbay A, Guicciardi ME, Higuchi H, Bronk SF, Gores GJ (2003b) Diet associated hepatic steatosis sensitizes to Fas mediated liver injury in mice. J Hepatol 39:978–983
Feldstein AE, Werneburg NW, Canbay A, Guicciardi ME, Bronk SF, Rydzewski R, Burgart LJ, Gores GJ (2004) Free fatty acids promote hepatic lipotoxicity by stimulating TNF-alpha expression via a lysosomal pathway. Hepatology 40:185–194
Feldstein AE, Werneburg NW, Li Z, Bronk SF, Gores GJ (2006) Bax inhibition protects against free fatty acid-induced lysosomal permeabilization. Am J Physiol Gastrointest Liver Physiol 290:G1339–G1346
Feldstein AE, Wieckowska A, Lopez AR, Liu YC, Zein NN, McCullough AJ (2009) Cytokeratin-18 fragment levels as noninvasive biomarkers for nonalcoholic steatohepatitis: a multicenter validation study. Hepatology 50:1072–1078
Flowers MT, Keller MP, Choi Y, Lan H, Kendziorski C, Ntambi JM, Attie AD (2008) Liver gene expression analysis reveals endoplasmic reticulum stress and metabolic dysfunction in SCD1-deficient mice fed a very low-fat diet. Physiol Genomics 33:361–372
Galbo T, Perry RJ, Jurczak MJ, Camporez JP, Alves TC, Kahn M, Guigni BA, Serr J, Zhang D, Bhanot S et al (2013) Saturated and unsaturated fat induce hepatic insulin resistance independently of TLR-4 signaling and ceramide synthesis in vivo. Proc Natl Acad Sci U S A 110:12780–12785
Gan LT, Van Rooyen DM, Koina ME, McCuskey RS, Teoh NC, Farrell GC (2014) Hepatocyte free cholesterol lipotoxicity results from JNK1-mediated mitochondrial injury and is HMGB1 and TLR4-dependent. J Hepatol 61:1376–1384
Gao D, Nong S, Huang X, Lu Y, Zhao H, Lin Y, Man Y, Wang S, Yang J, Li J (2010) The effects of palmitate on hepatic insulin resistance are mediated by NADPH oxidase 3-derived reactive oxygen species through JNK and p38MAPK pathways. J Biol Chem 285:29965–29973
Gaudette BT, Iwakoshi NN, Boise LH (2014) Bcl-xL protein protects from C/EBP homologous protein (CHOP)-dependent apoptosis during plasma cell differentiation. J Biol Chem 289:23629–23640
Germain M, Milburn J, Duronio V (2008) MCL-1 inhibits BAX in the absence of MCL-1/BAX interaction. J Biol Chem 283:6384–6392

Gregor MF, Yang L, Fabbrini E, Mohammed BS, Eagon JC, Hotamisligil GS, Klein S (2009) Endoplasmic reticulum stress is reduced in tissues of obese subjects after weight loss. Diabetes 58:693–700
Guicciardi ME, Gores GJ (2009) Life and death by death receptors. FASEB J 23:1625–1637
Guicciardi ME, Leist M, Gores GJ (2004) Lysosomes in cell death. Oncogene 23:2881–2890
Guy CD, Suzuki A, Zdanowicz M, Abdelmalek MF, Burchette J, Unalp A, Diehl AM, Nash CRN (2012) Hedgehog pathway activation parallels histologic severity of injury and fibrosis in human nonalcoholic fatty liver disease. Hepatology 55:1711–1721
Han MS, Park SY, Shinzawa K, Kim S, Chung KW, Lee JH, Kwon CH, Lee KW, Lee JH, Park CK et al (2008) Lysophosphatidylcholine as a death effector in the lipoapoptosis of hepatocytes. J Lipid Res 49:84–97
Han MS, Lim YM, Quan W, Kim JR, Chung KW, Kang M, Kim S, Park SY, Han JS, Park SY et al (2011) Lysophosphatidylcholine as an effector of fatty acid-induced insulin resistance. J Lipid Res 52:1234–1246
Higuchi H, Grambihler A, Canbay A, Bronk SF, Gores GJ (2004) Bile acids up-regulate death receptor 5/TRAIL-receptor 2 expression via a c-Jun N-terminal kinase-dependent pathway involving Sp1. J Biol Chem 279:51–60
Hirsova P, Gores GJ (2015) Death receptor-mediated cell death and proinflammatory signaling in nonalcoholic steatohepatitis. Cell Mol Gastroenterol Hepatol 1:17–27
Hirsova P, Ibrahim SH, Bronk SF, Yagita H, Gores GJ (2013) Vismodegib suppresses TRAIL-mediated liver injury in a mouse model of nonalcoholic steatohepatitis. PLoS One 8:e70599
Holland WL, Brozinick JT, Wang LP, Hawkins ED, Sargent KM, Liu Y, Narra K, Hoehn KL, Knotts TA, Siesky A et al (2007) Inhibition of ceramide synthesis ameliorates glucocorticoid-, saturated-fat-, and obesity-induced insulin resistance. Cell Metab 5:167–179
Ibrahim SH, Akazawa Y, Cazanave SC, Bronk SF, Elmi NA, Werneburg NW, Billadeau DD, Gores GJ (2011) Glycogen synthase kinase-3 (GSK-3) inhibition attenuates hepatocyte lipoapoptosis. J Hepatol 54:765–772
Idrissova L, Malhi H, Werneburg NW, LeBrasseur NK, Bronk SF, Fingas C, Tchkonia T, Pirtskhalava T, White TA, Stout MB et al (2015) Trail receptor deletion in mice suppresses the inflammation of nutrient excess. J Hepatol 62(5):1156–1163
Ii H, Yokoyama N, Yoshida S, Tsutsumi K, Hatakeyama S, Sato T, Ishihara K, Akiba S (2009) Alleviation of high-fat diet-induced fatty liver damage in group IVA phospholipase A2-knockout mice. PLoS One 4:e8089
Jaeschke A, Davis RJ (2007) Metabolic stress signaling mediated by mixed-lineage kinases. Mol Cell 27:498–508
Kahraman A, Schlattjan M, Kocabayoglu P, Yildiz-Mezileto glu S, Schlensak M, Fingas CD, Wedemeyer I, Marquitan G, Gieseler RK, Baba HA et al (2010) Major histocompatibility complex class I-related chains A and B (MIC A/B): a novel role in nonalcoholic steatohepatitis. Hepatology 51:92–102
Kakisaka K, Cazanave SC, Fingas CD, Guicciardi ME, Bronk SF, Werneburg NW, Mott JL, Gores GJ (2012a) Mechanisms of lysophosphatidylcholine-induced hepatocyte lipoapoptosis. Am J Physiol Gastrointest Liver Physiol 302:G77–G84
Kakisaka K, Cazanave SC, Werneburg NW, Razumilava N, Mertens JC, Bronk SF, Gores GJ (2012b) A hedgehog survival pathway in 'undead' lipotoxic hepatocytes. J Hepatol 57:844–851
Kalsch J, Bechmann LP, Kalsch H, Schlattjan M, Erhard J, Gerken G, Canbay A (2011) Evaluation of biomarkers of NAFLD in a cohort of morbidly obese patients. J Nutr Metab 2011:369168
Kant S, Barrett T, Vertii A, Noh YH, Jung DY, Kim JK, Davis RJ (2013) Role of the mixed-lineage protein kinase pathway in the metabolic stress response to obesity. Cell Rep 4:681–688
Kawano Y, Cohen DE (2013) Mechanisms of hepatic triglyceride accumulation in non-alcoholic fatty liver disease. J Gastroenterol 48:434–441
Kim BJ, Ryu SW, Song BJ (2006) JNK- and p38 kinase-mediated phosphorylation of Bax leads to its activation and mitochondrial translocation and to apoptosis of human hepatoma HepG2 cells. J Biol Chem 281:21256–21265

Kojima H, Sakurai S, Uemura M, Fukui H, Morimoto H, Tamagawa Y (2007) Mitochondrial abnormality and oxidative stress in nonalcoholic steatohepatitis. Alcohol Clin Exp Res 31:S61–S66

Kong B, Luyendyk JP, Tawfik O, Guo GL (2009) Farnesoid X receptor deficiency induces nonalcoholic steatohepatitis in low-density lipoprotein receptor-knockout mice fed a high-fat diet. J Pharmacol Exp Ther 328:116–122

Koruk M, Taysi S, Savas MC, Yilmaz O, Akcay F, Karakok M (2004) Oxidative stress and enzymatic antioxidant status in patients with nonalcoholic steatohepatitis. Ann Clin Lab Sci 34:57–62

Kotronen A, Seppanen-Laakso T, Westerbacka J, Kiviluoto T, Arola J, Ruskeepaa AL, Oresic M, Yki-Jarvinen H (2009) Hepatic stearoyl-CoA desaturase (SCD)-1 activity and diacylglycerol but not ceramide concentrations are increased in the nonalcoholic human fatty liver. Diabetes 58:203–208

Kowalski GM, Carey AL, Selathurai A, Kingwell BA, Bruce CR (2013) Plasma sphingosine-1-phosphate is elevated in obesity. PLoS One 8:e72449

Kurek K, Piotrowska DM, Wiesiolek-Kurek P, Lukaszuk B, Chabowski A, Gorski J, Zendzian-Piotrowska M (2014) Inhibition of ceramide de novo synthesis reduces liver lipid accumulation in rats with nonalcoholic fatty liver disease. Liver Int 34:1074–1083

Kusminski CM, Shetty S, Orci L, Unger RH, Scherer PE (2009) Diabetes and apoptosis: lipotoxicity. Apoptosis 14(12):1484–1495

Larter CZ, Yeh MM, Haigh WG, Williams J, Brown S, Bell-Anderson KS, Lee SP, Farrell GC (2008) Hepatic free fatty acids accumulate in experimental steatohepatitis: role of adaptive pathways. J Hepatol 48:638–647

Lei K, Davis RJ (2003) JNK phosphorylation of Bim-related members of the Bcl2 family induces Bax-dependent apoptosis. Proc Natl Acad Sci U S A 100:2432–2437

Li H, Yu X (2013) Emerging role of JNK in insulin resistance. Curr Diabetes Rev 9:422–428

Li Z, Yang S, Lin H, Huang J, Watkins PA, Moser AB, Desimone C, Song XY, Diehl AM (2003) Probiotics and antibodies to TNF inhibit inflammatory activity and improve nonalcoholic fatty liver disease. Hepatology 37:343–350

Li Z, Berk M, McIntyre TM, Gores GJ, Feldstein AE (2008) The lysosomal-mitochondrial axis in free fatty acid-induced hepatic lipotoxicity. Hepatology 47:1495–1503

Li ZZ, Berk M, McIntyre TM, Feldstein AE (2009a) Hepatic lipid partitioning and liver damage in nonalcoholic fatty liver disease: role of stearoyl-CoA desaturase. J Biol Chem 284:5637–5644

Li C, Jiang X, Yang L, Liu X, Yue S, Li L (2009b) Involvement of sphingosine 1-phosphate (SIP)/S1P3 signaling in cholestasis-induced liver fibrosis. Am J Pathol 175:1464–1472

Li G, Mongillo M, Chin KT, Harding H, Ron D, Marks AR, Tabas I (2009c) Role of ERO1-alpha-mediated stimulation of inositol 1,4,5-triphosphate receptor activity in endoplasmic reticulum stress-induced apoptosis. J Cell Biol 186:783–792

Li C, Zheng S, You H, Liu X, Lin M, Yang L, Li L (2011) Sphingosine 1-phosphate (S1P)/S1P receptors are involved in human liver fibrosis by action on hepatic myofibroblasts motility. J Hepatol 54:1205–1213

Listenberger LL, Han X, Lewis SE, Cases S, Farese RV Jr, Ory DS, Schaffer JE (2003) Triglyceride accumulation protects against fatty acid-induced lipotoxicity. Proc Natl Acad Sci U S A 100:3077–3082

Lu M, Lawrence DA, Marsters S, Acosta-Alvear D, Kimmig P, Mendez AS, Paton AW, Paton JC, Walter P, Ashkenazi A (2014) Cell death. Opposing unfolded-protein-response signals converge on death receptor 5 to control apoptosis. Science 345:98–101

Makki K, Froguel P, Wolowczuk I (2013) Adipose tissue in obesity-related inflammation and insulin resistance: cells, cytokines, and chemokines. ISRN Inflamm 2013:139239

Malhi H, Kaufman RJ (2011) Endoplasmic reticulum stress in liver disease. J Hepatol 54:795–809

Malhi H, Bronk SF, Werneburg NW, Gores GJ (2006) Free fatty acids induce JNK-dependent hepatocyte lipoapoptosis. J Biol Chem 281:12093–12101

Malhi H, Barreyro FJ, Isomoto H, Bronk SF, Gores GJ (2007) Free fatty acids sensitise hepatocytes to TRAIL mediated cytotoxicity. Gut 56:1124–1131
Malhi H, Kropp EM, Clavo VF, Kobrossi CR, Han J, Mauer AS, Yong J, Kaufman RJ (2013) C/EBP homologous protein-induced macrophage apoptosis protects mice from steatohepatitis. J Biol Chem 288:18624–18642
Mari M, Caballero F, Colell A, Morales A, Caballeria J, Fernandez A, Enrich C, Fernandez-Checa JC, Garcia-Ruiz C (2006) Mitochondrial free cholesterol loading sensitizes to TNF- and Fas-mediated steatohepatitis. Cell Metab 4:185–198
Masuoka HC, Mott J, Bronk SF, Werneburg NW, Akazawa Y, Kaufmann SH, Gores GJ (2009) Mcl-1 degradation during hepatocyte lipoapoptosis. J Biol Chem 284:30039–30048
Matsunami T, Sato Y, Ariga S, Sato T, Kashimura H, Hasegawa Y, Yukawa M (2010) Regulation of oxidative stress and inflammation by hepatic adiponectin receptor 2 in an animal model of nonalcoholic steatohepatitis. Int J Clin Exp Pathol 3:472–481
McCullough KD, Martindale JL, Klotz LO, Aw TY, Holbrook NJ (2001) Gadd153 sensitizes cells to endoplasmic reticulum stress by down-regulating Bcl2 and perturbing the cellular redox state. Mol Cell Biol 21:1249–1259
Min HK, Kapoor A, Fuchs M, Mirshahi F, Zhou H, Maher J, Kellum J, Warnick R, Contos MJ, Sanyal AJ (2012) Increased hepatic synthesis and dysregulation of cholesterol metabolism is associated with the severity of nonalcoholic fatty liver disease. Cell Metab 15:665–674
Monetti M, Levin MC, Watt MJ, Sajan MP, Marmor S, Hubbard BK, Stevens RD, Bain JR, Newgard CB, Farese RV Sr et al (2007) Dissociation of hepatic steatosis and insulin resistance in mice overexpressing DGAT in the liver. Cell Metab 6:69–78
More VR, Xu J, Shimpi PC, Belgrave C, Luyendyk JP, Yamamoto M, Slitt AL (2013) Keap1 knockdown increases markers of metabolic syndrome after long-term high fat diet feeding. Free Radic Biol Med 61:85–94
Nehra V, Angulo P, Buchman AL, Lindor KD (2001) Nutritional and metabolic considerations in the etiology of nonalcoholic steatohepatitis. Dig Dis Sci 46:2347–2352
Neuschwander-Tetri BA, Loomba R, Sanyal AJ, Lavine JE, Van Natta ML, Abdelmalek MF, Chalasani N, Dasarathy S, Diehl AM, Hameed B et al (2015) Farnesoid X nuclear receptor ligand obeticholic acid for non-cirrhotic, non-alcoholic steatohepatitis (FLINT): a multicentre, randomised, placebo-controlled trial. Lancet 385:956–965
Ohoka N, Yoshii S, Hattori T, Onozaki K, Hayashi H (2005) TRB3, a novel ER stress-inducible gene, is induced via ATF4-CHOP pathway and is involved in cell death. EMBO J 24:1243–1255
Ozcan U, Cao Q, Yilmaz E, Lee AH, Iwakoshi NN, Ozdelen E, Tuncman G, Gorgun C, Glimcher LH, Hotamisligil GS (2004) Endoplasmic reticulum stress links obesity, insulin action, and type 2 diabetes. Science 306:457–461
Pagadala M, Kasumov T, McCullough AJ, Zein NN, Kirwan JP (2012) Role of ceramides in nonalcoholic fatty liver disease. Trends Endocrinol Metab 23:365–371
Pathil A, Mueller J, Warth A, Chamulitrat W, Stremmel W (2012) Ursodeoxycholyl lysophosphatidylethanolamide improves steatosis and inflammation in murine models of nonalcoholic fatty liver disease. Hepatology 55:1369–1378
Peverill W, Powell LW, Skoien R (2014) Evolving concepts in the pathogenesis of NASH: beyond steatosis and inflammation. Int J Mol Sci 15:8591–8638
Pfaffenbach KT, Gentile CL, Nivala AM, Wang D, Wei Y, Pagliassotti MJ (2010) Linking endoplasmic reticulum stress to cell death in hepatocytes: roles of C/EBP homologous protein and chemical chaperones in palmitate-mediated cell death. Am J Physiol Endocrinol Metab 298:E1027–E1035
Popescu IR, Helleboid-Chapman A, Lucas A, Vandewalle B, Dumont J, Bouchaert E, Derudas B, Kerr-Conte J, Caron S, Pattou F et al (2010) The nuclear receptor FXR is expressed in pancreatic beta-cells and protects human islets from lipotoxicity. FEBS Lett 584:2845–2851
Puri P, Baillie RA, Wiest MM, Mirshahi F, Choudhury J, Cheung O, Sargeant C, Contos MJ, Sanyal AJ (2007) A lipidomic analysis of nonalcoholic fatty liver disease. Hepatology 46:1081–1090

Puri P, Mirshahi F, Cheung O, Natarajan R, Maher JW, Kellum JM, Sanyal AJ (2008) Activation and dysregulation of the unfolded protein response in nonalcoholic fatty liver disease. Gastroenterology 134:568–576

Puri P, Wiest MM, Cheung O, Mirshahi F, Sargeant C, Min HK, Contos MJ, Sterling RK, Fuchs M, Zhou H et al (2009) The plasma lipidomic signature of nonalcoholic steatohepatitis. Hepatology 50:1827–1838

Puthalakath H, O'Reilly LA, Gunn P, Lee L, Kelly PN, Huntington ND, Hughes PD, Michalak EM, McKimm-Breschkin J, Motoyama N et al (2007) ER stress triggers apoptosis by activating BH3-only protein Bim. Cell 129:1337–1349

Rangwala F, Guy CD, Lu J, Suzuki A, Burchette JL, Abdelmalek MF, Chen W, Diehl AM (2011) Increased production of sonic hedgehog by ballooned hepatocytes. J Pathol 224:401–410

Ren D, Tu HC, Kim H, Wang GX, Bean GR, Takeuchi O, Jeffers JR, Zambetti GP, Hsieh JJ, Cheng EH (2010) BID, BIM, and PUMA are essential for activation of the BAX- and BAK-dependent cell death program. Science 330:1390–1393

Ribeiro PS, Cortez-Pinto H, Sola S, Castro RE, Ramalho RM, Baptista A, Moura MC, Camilo ME, Rodrigues CM (2004) Hepatocyte apoptosis, expression of death receptors, and activation of NF-kappaB in the liver of nonalcoholic and alcoholic steatohepatitis patients. Am J Gastroenterol 99:1708–1717

Roden M (2006) Mechanisms of disease: hepatic steatosis in type 2 diabetes—pathogenesis and clinical relevance. Nat Clin Pract Endocrinol Metab 2:335–348

Rodrigues CM, Fan G, Wong PY, Kren BT, Steer CJ (1998) Ursodeoxycholic acid may inhibit deoxycholic acid-induced apoptosis by modulating mitochondrial transmembrane potential and reactive oxygen species production. Mol Med 4:165–178

Romanatto T, Roman EA, Arruda AP, Denis RG, Solon C, Milanski M, Moraes JC, Bonfleur ML, Degasperi GR, Picardi PK et al (2009) Deletion of tumor necrosis factor-alpha receptor 1 (TNFR1) protects against diet-induced obesity by means of increased thermogenesis. J Biol Chem 284:36213–36222

Sanyal AJ, Campbell-Sargent C, Mirshahi F, Rizzo WB, Contos MJ, Sterling RK, Luketic VA, Shiffman ML, Clore JN (2001) Nonalcoholic steatohepatitis: association of insulin resistance and mitochondrial abnormalities. Gastroenterology 120:1183–1192

Sanyal AJ, Mofrad PS, Contos MJ, Sargeant C, Luketic VA, Sterling RK, Stravitz RT, Shiffman ML, Clore J, Mills AS (2004) A pilot study of vitamin E versus vitamin E and pioglitazone for the treatment of nonalcoholic steatohepatitis. Clin Gastroenterol Hepatol 2:1107–1115

Sass DA, Chang P, Chopra KB (2005) Nonalcoholic fatty liver disease: a clinical review. Dig Dis Sci 50:171–180

Schattenberg JM, Singh R, Wang Y, Lefkowitch JH, Rigoli RM, Scherer PE, Czaja MJ (2006) JNK1 but not JNK2 promotes the development of steatohepatitis in mice. Hepatology 43:163–172

Schnyder-Candrian S, Czarniecki J, Lerondel S, Corpataux J, Ryffel B, Schnyder B (2005) Hepatic steatosis in the absence of tumor necrosis factor in mice. Cytokine 32:287–295

Schwandner R, Wiegmann K, Bernardo K, Kreder D, Kronke M (1998) TNF receptor death domain-associated proteins TRADD and FADD signal activation of acid sphingomyelinase. J Biol Chem 273:5916–5922

Seki S, Kitada T, Yamada T, Sakaguchi H, Nakatani K, Wakasa K (2002) In situ detection of lipid peroxidation and oxidative DNA damage in non-alcoholic fatty liver diseases. J Hepatol 37:56–62

Simon Y, Kessler SM, Gemperlein K, Bohle RM, Muller R, Haybaeck J, Kiemer AK (2014) Elevated free cholesterol in a p62 overexpression model of non-alcoholic steatohepatitis. World J Gastroenterol 20:17839–17850

Singh R, Wang Y, Xiang Y, Tanaka KE, Gaarde WA, Czaja MJ (2009) Differential effects of JNK1 and JNK2 inhibition on murine steatohepatitis and insulin resistance. Hepatology 49:87–96

Soon RK Jr, Yan JS, Grenert JP, Maher JJ (2010) Stress signaling in the methionine-choline-deficient model of murine fatty liver disease. Gastroenterology 139:1730–1739, 1739.e1

Syn WK, Agboola KM, Swiderska M, Michelotti GA, Liaskou E, Pang H, Xie G, Philips G, Chan IS, Karaca GF et al (2012) NKT-associated hedgehog and osteopontin drive fibrogenesis in non-alcoholic fatty liver disease. Gut 61:1323–1329

Tabas I, Ron D (2011) Integrating the mechanisms of apoptosis induced by endoplasmic reticulum stress. Nat Cell Biol 13:184–190

Tajiri K, Shimizu Y (2012) Role of NKT cells in the pathogenesis of NAFLD. Int J Hepatol 2012:850836

Thapaliya S, Wree A, Povero D, Inzaugarat ME, Berk M, Dixon L, Papouchado BG, Feldstein AE (2014) Caspase 3 inactivation protects against hepatic cell death and ameliorates fibrogenesis in a diet-induced NASH model. Dig Dis Sci 59:1197–1206

Tosello-Trampont AC, Landes SG, Nguyen V, Novobrantseva TI, Hahn YS (2012) Kuppfer cells trigger nonalcoholic steatohepatitis development in diet-induced mouse model through tumor necrosis factor-alpha production. J Biol Chem 287:40161–40172

Unger RH, Orci L (2002) Lipoapoptosis: its mechanism and its diseases. Biochim Biophys Acta 1585:202–212

Urano F, Wang X, Bertolotti A, Zhang Y, Chung P, Harding HP, Ron D (2000) Coupling of stress in the ER to activation of JNK protein kinases by transmembrane protein kinase IRE1. Science 287:664–666

Uruno A, Furusawa Y, Yagishita Y, Fukutomi T, Muramatsu H, Negishi T, Sugawara A, Kensler TW, Yamamoto M (2013) The Keap1-Nrf2 system prevents onset of diabetes mellitus. Mol Cell Biol 33:2996–3010

Utzschneider KM, Kahn SE (2006) Review: the role of insulin resistance in nonalcoholic fatty liver disease. J Clin Endocrinol Metab 91:4753–4761

Van Rooyen DM, Larter CZ, Haigh WG, Yeh MM, Ioannou G, Kuver R, Lee SP, Teoh NC, Farrell GC (2011) Hepatic free cholesterol accumulates in obese, diabetic mice and causes nonalcoholic steatohepatitis. Gastroenterology 141:1393–1403, 403.e1–e5

Vela L, Gonzalo O, Naval J, Marzo I (2013) Direct interaction of Bax and Bak proteins with Bcl-2 homology domain 3 (BH3)-only proteins in living cells revealed by fluorescence complementation. J Biol Chem 288:4935–4946

Videla LA, Tapia G, Rodrigo R, Pettinelli P, Haim D, Santibanez C, Araya AV, Smok G, Csendes A, Gutierrez L et al (2009) Liver NF-kappaB and AP-1 DNA binding in obese patients. Obesity 17:973–979

Wang H, Chen J, Hollister K, Sowers LC, Forman BM (1999) Endogenous bile acids are ligands for the nuclear receptor FXR/BAR. Mol Cell 3:543–553

Wang D, Wei Y, Pagliassotti MJ (2006a) Saturated fatty acids promote endoplasmic reticulum stress and liver injury in rats with hepatic steatosis. Endocrinology 147:943–951

Wang Y, Singh R, Lefkowitch JH, Rigoli RM, Czaja MJ (2006b) Tumor necrosis factor-induced toxic liver injury results from JNK2-dependent activation of caspase-8 and the mitochondrial death pathway. J Biol Chem 281:15258–15267

Wang X, Zhang DM, Gu TT, Ding XQ, Fan CY, Zhu Q, Shi YW, Hong Y, Kong LD (2013a) Morin reduces hepatic inflammation-associated lipid accumulation in high fructose-fed rats via inhibiting sphingosine kinase 1/sphingosine 1-phosphate signaling pathway. Biochem Pharmacol 86:1791–1804

Wang D, Wei Y, Frye M, Gentile CL, Pagliassotti MJ (2013b) Saturated fatty acid-induced cytotoxicity in liver cells does not involve phosphatase and tensin homologue deleted on chromosome 10. J Nutr Metab 2013:514206

Washington K, Wright K, Shyr Y, Hunter EB, Olson S, Raiford DS (2000) Hepatic stellate cell activation in nonalcoholic steatohepatitis and fatty liver. Hum Pathol 31:822–828

Wei Y, Wang D, Topczewski F, Pagliassotti MJ (2006) Saturated fatty acids induce endoplasmic reticulum stress and apoptosis independently of ceramide in liver cells. Am J Physiol Endocrinol Metab 291:E275–E281

Wei Y, Rector RS, Thyfault JP, Ibdah JA (2008) Nonalcoholic fatty liver disease and mitochondrial dysfunction. World J Gastroenterol 14:193–199

Wei Y, Wang D, Gentile CL, Pagliassotti MJ (2009) Reduced endoplasmic reticulum luminal calcium links saturated fatty acid-mediated endoplasmic reticulum stress and cell death in liver cells. Mol Cell Biochem 331:31–40

Wolf MJ, Adili A, Piotrowitz K, Abdullah Z, Boege Y, Stemmer K, Ringelhan M, Simonavicius N, Egger M, Wohlleber D et al (2014) Metabolic activation of intrahepatic CD8+ T cells and NKT cells causes nonalcoholic steatohepatitis and liver cancer via cross-talk with hepatocytes. Cancer Cell 26:549–564

Xu W, Shao L, Zhou C, Wang H, Guo J (2011) Upregulation of Nrf2 expression in non-alcoholic fatty liver and steatohepatitis. Hepatogastroenterology 58:2077–2080

Yamaguchi H, Wang HG (2004) CHOP is involved in endoplasmic reticulum stress-induced apoptosis by enhancing DR5 expression in human carcinoma cells. J Biol Chem 279:45495–45502

Yamaguchi K, Yang L, McCall S, Huang J, Yu XX, Pandey SK, Bhanot S, Monia BP, Li YX, Diehl AM (2007) Inhibiting triglyceride synthesis improves hepatic steatosis but exacerbates liver damage and fibrosis in obese mice with nonalcoholic steatohepatitis. Hepatology 45:1366–1374

Yamamoto K, Ichijo H, Korsmeyer SJ (1999) BCL-2 is phosphorylated and inactivated by an ASK1/Jun N-terminal protein kinase pathway normally activated at G(2)/M. Mol Cell Biol 19:8469–8478

Youle RJ, Strasser A (2008) The BCL-2 protein family: opposing activities that mediate cell death. Nat Rev Mol Cell Biol 9:47–59

Younossi ZM, Jarrar M, Nugent C, Randhawa M, Afendy M, Stepanova M, Rafiq N, Goodman Z, Chandhoke V, Baranova A (2008) A novel diagnostic biomarker panel for obesity-related nonalcoholic steatohepatitis (NASH). Obes Surg 18:1430–1437

Chapter 6
Role of the Endocannabinoid System in Hepatic Lipogenesis

Joseph Tam

Abstract Endocannabinoids are endogenous lipid mediators recognized by the cannabinoid-1 and 2 (CB_1 and CB_2, respectively) receptors that mediate the effects of marijuana. The endocannabinoid system, consisting of endocannabinoids (eCBs), their receptors, and the enzymes involved in their biosynthesis and degradation, is present in many tissues including the liver. The present chapter reviews accumulating evidence that the eCB system is involved in the regulation of de novo hepatic lipogenesis.

Keywords Endocannabinoids • CB_1 receptor • CB_2 receptor • Hepatic lipogenesis • Fatty liver • Diet-induced obesity • Monounsaturated fatty acids

1 The Endocannabinoid System

The psychoactive and medicinal effects of *Cannabis sativa* (marijuana) have been recognized for millennia (Abel 1975). Yet, it is only in the last few decades that the insights into its mechanism of action began to emerge due to some landmark discoveries in the field of cannabinoid research. While more than 60 plant-derived cannabinoid molecules have been identified in marijuana (Pertwee 2006), only Δ^9-tetrahydrocannabinol (THC), which was first isolated, identified, and synthesized in 1964 (Gaoni and Mechoulam 1964), is responsible for its psychoactive properties. After this milestone discovery in cannabinoid research, it took more than two decades to identify the THC binding site in the brain (Devane et al. 1988), which was later cloned and named cannabinoid-1 (CB_1) receptor (Matsuda et al. 1990). In addition to the brain-type CB_1 receptor, a second cannabinoid receptor was identified in lymphoid tissue and was named CB_2 (Munro et al. 1993). These cannabinoid receptors are G protein-coupled receptors that mainly signal via G_i/G_o proteins, even though they can also activate G_s,

J. Tam (✉)
Obesity and Metabolism Laboratory, Faculty of Medicine, The Institute for Drug Research, School of Pharmacy, The Hebrew University of Jerusalem, Jerusalem 91120, Israel
e-mail: yossit@ekmd.huji.ac.il

J.M. Ntambi (ed.), *Hepatic De Novo Lipogenesis and Regulation of Metabolism*,
DOI 10.1007/978-3-319-25065-6_6

$G_{q/11}$, and G protein-independent signaling pathways (Howlett 2005). Although the CB_1 receptor is the most abundant G protein-coupled receptor in the mammalian brain, where it is almost exclusively localized on excitatory and inhibitory presynaptic nerve terminals (Freund et al. 2003), it is also expressed at much lower, yet functionally relevant, levels in many peripheral tissues, such as adipose tissue (Bensaid et al. 2003), liver (Osei-Hyiaman et al. 2005), skeletal muscle (Pagotto et al. 2006), kidney (Larrinaga et al. 2010), and pancreatic β-cells (Nakata and Yada 2008). In contrast, the CB_2 receptor is expressed primarily on cells of the immune and hematopoietic systems, but the receptor is also expressed in many other tissues, including the liver (Julien et al. 2005), skeleton (Ofek et al. 2006), and the brain (Van Sickle et al. 2005; Gong et al. 2006). The CB_1 and CB_2 receptors share low (44 %) sequence homology and a similar ligand binding profile (Munro et al. 1993).

The cloning of CB receptors in mammalian cells led to the identification of their endogenous ligands. Arachidonoyl ethanolamide (AEA, anandamide) was the first CB receptor ligand to be identified (Devane et al. 1992), followed by 2-arachidonoyl glycerol (2-AG) (Sugiura et al. 1995; Mechoulam et al. 1995). While a few related ligands that can activate CB receptors have been identified (Pacher et al. 2006), AEA and 2-AG are the most extensively studied. These endocannabinoids (eCBs) are derived "on demand" from arachidonic acid via multiple pathways and are released from cells after their production to activate CB receptors. AEA and 2-AG are produced in response to an elevation in intracellular calcium levels or metabotropic receptor activation (Pacher et al. 2006). The eCBs are different from classical neurotransmitters because they are not stored in vesicles. Once released, they remain attached to the membrane due to their lipophilicity and can be taken back up by cells through a high affinity transport mechanism (Fowler 2007). Their clearance depends on cellular uptake and enzymatic degradation. AEA is degraded mainly by the membrane-associated fatty-acid amide hydrolase (FAAH) (McKinney and Cravatt 2005), while 2-AG is primarily degraded by monoglyceride lipase (MAGL) (Dinh et al. 2002). Endogenous cannabinoids, their receptors, and the enzymes/proteins involved in their biosynthesis, transport, and degradation jointly make up the 'eCB system'.

2 The Hepatic Endocannabinoid System

While increasing evidence indicates that the eCB system plays a key role in the pathogenesis of many liver diseases, the initial report on the presence and functional importance of this system in the liver (Osei-Hyiaman et al. 2005) was completely unexpected. In fact, for many years the liver was used as a negative control to study the function of neuronal CB_1 receptors (Galiegue et al. 1995). Normally, the expression of CB_1 and CB_2 receptors in whole liver is fairly low (Xu et al. 2006; Osei-Hyiaman et al. 2005; Jourdan et al. 2010; Galiegue et al. 1995; Quarta et al. 2010). However, in liver pathologies, the expression of both CB receptors is upregulated (Mendez-Sanchez et al. 2007; Buckley et al. 1998; Julien et al. 2005; Xu et al. 2006; Floreani et al. 2010). CB_1 receptors are expressed on hepatocytes

(Michalopoulos et al. 2003; Jeong et al. 2008), stellate cells (Teixeira-Clerc et al. 2006; Siegmund et al. 2007), hepatic myofibroblasts (Julien et al. 2005), and hepatic vascular endothelial cells (Batkai et al. 2001; Biecker et al. 2004; Domenicali et al. 2005), whereas CB_2 receptors are expressed in Kupffer cells and hepatic myofibroblasts (Julien et al. 2005; Teixeira-Clerc et al. 2010) but are not expressed by hepatocytes (Mallat et al. 2013). In normal liver, eCBs are present at levels that are comparable to those found in the brain (Osei-Hyiaman et al. 2005; Siegmund et al. 2007). While mounting evidence suggests that the eCB system is involved in several liver pathologies, such as hepatic fibrosis, cancer, cirrhosis, and ischemia-reperfusion injury, the main focus of this chapter will be its involvement in nonalcoholic fatty liver disease (NAFLD), lipid metabolism, and hepatic de novo lipogenesis.

3 Regulation of Hepatic Lipogenesis by the Endocannabinoid System

3.1 *Role of the CB_1 Receptor in Hepatic Lipogenesis*

It is widely recognized that energy intake and fat utilization are regulated in an organized manner and that stimulating CB_1 receptors on adipocytes affects lipid metabolism through the regulation of adiponectin synthesis (Bensaid et al. 2003), an increased lipoprotein lipase activity (Cota et al. 2003) and/or controlling lipid hydroperoxide levels (Nam et al. 2012). Yet, the role of adipose tissue in lipogenesis is fairly minor compared to that of the liver (Diraison et al. 2003). Recently, emerging evidence indicates that high fat diet (HFD)-induced obesity is associated with an increase in de novo lipogenesis despite the increased availability of dietary fat (Diraison et al. 2003; Lin et al. 2005; Biddinger et al. 2005). The possibility that the eCB system is involved in the regulation of hepatic de novo lipogenesis was the focus of many recent studies. The initial evidence for such involvement came from Osei-Hyiaman and colleagues, who showed the complete resistance of CB_1-deficient mice to HFD-induced hepatic steatosis, whereas wild-type animals on the same diet developed hepatic steatosis (Osei-Hyiaman et al. 2005). These striking differences suggest a role for the CB_1 receptor in the hepatic de novo lipogenic response to a HFD. In fact, when wild-type mice were checked 3 weeks after the initiation of the HFD, before a significant effect on body weight could be detected, the net level of de novo fatty acid synthesis was dramatically increased compared to that in the standard diet (STD) lean controls (Osei-Hyiaman et al. 2005). Mechanistically, stimulation of CB_1 receptors with the potent CB_1 agonist HU-210 (20 μg/kg) in mice increases the hepatic mRNA expression levels of the lipogenic transcription factor sterol regulatory element-binding protein-1c (SREBP-1c) and its target genes acetyl-CoA carboxylase-1 (ACC1), fatty acid synthase (FAS), and stearoyl-CoA desaturase-1 (SCD1). In accordance with these findings, HU-210 treatment increases the hepatic incorporation of tritium into fatty acids in mice injected with intrahepatic ^3H_2O, an effect that is absent in CB_1-deficient mice (Osei-Hyiaman et al. 2005).

Following these findings, and that activation of hepatic CB_1 receptors with AEA induces the expression of SREBP-1c and lipogenesis in isolated primary hepatocytes (Osei-Hyiaman et al. 2005), others have also shown that high-dose of AEA (100 nM) administration via water upregulates the transcription of CB_1 receptor and the expression of SREBP-1c in livers of zebrafish larvae at 7 and 15 days post-fertilization (Migliarini and Carnevali 2008). In addition, hyperactivation of the eCB system by AEA treatment in cultured liver explants increases the expression of SREBP-1c and its related lipogenic genes (ACC1, FAS and glucokinase), effect that is inhibited by the presence of the potent and selective CB_1 receptor antagonist, rimonabant (Jourdan et al. 2012). Moreover, CB_1-mediated hepatic de novo lipogenesis may explain the finding that daily cannabis smoking by chronic hepatitis C patients is an independent risk factor for increased hepatic steatosis but not for obesity (Hezode et al. 2008). Additional support for the steatogenic role of eCBs in humans comes from a study demonstrating that increases in the arterial and hepatic venous concentrations of 2-AG, as well as splanchnic production of triglycerides containing saturated fatty acids are positively correlated with liver fat content (Westerbacka et al. 2010). This data suggests that the human fatty liver takes up 2-AG and overproduces triacylglycerols containing saturated fatty acids, which might reflect increased de novo lipogenesis.

In HFD-induced obese mice both hepatic levels of AEA and the expression of CB_1 receptors are increased compared to animals fed a STD (Osei-Hyiaman et al. 2005; Jourdan et al. 2010; Liu et al. 2013). This finding suggests that intake of a HFD activates the hepatic eCB system, which could contribute to increased hepatic de novo lipogenesis. Even though hepatic AEA levels were found to be higher under HFD conditions, its biosynthetic rate remained unchanged. On the other hand, the rate of AEA catabolism was reduced due to decreased activity of its degrading enzyme, FAAH (Osei-Hyiaman et al. 2005). This could explain why a HFD specifically increases AEA levels and not 2-AG levels, as the latter is not an in vivo substrate of FAAH. This finding, together with earlier observations that lipids present in tissue extracts inhibit FAAH activity (Katayama et al. 1997), suggest that a HFD may induce the generation of endogenous FAAH inhibitors that would then contribute to activating the eCB system by reducing the levels of the degrading enzyme of AEA. In fact, a recent study by Liu and colleagues proves that activation of CB_1 receptors in the liver promote the synthesis of monounsaturated fatty acids (MUFAs) by increasing the activity of SCD1. These endogenously produced MUFAs act as potent inhibitors of FAAH and are therefore responsible for the increased hepatic levels of AEA in HFD-induced obese mice (Liu et al. 2013). Unlike wild-type mice, SCD1-deficient animals are resistant to the HFD-induced increase in hepatic AEA and the corresponding decrease in FAAH activity in the liver. An in vivo treatment with an SCD1 inhibitor improves HFD-induced metabolic impairments in wild-type mice (Liu et al. 2013). Taken together, these findings reveal a positive feedback loop between the hepatic eCB/CB_1 receptor system and SCD1 that contributes to increased de novo hepatic lipogenesis in obesity.

Using a chemical approach to test the effects of increased hepatic eCB signaling in mice, Ruby and colleagues treated wild-type mice with an inhibitor of MAGL, the enzyme responsible for the degradation of 2-AG, and found that elevated levels of 2-AG result in increased expression of SREBP-1c and FAS in the liver and accumulation of triglycerides and apoE-depleted, triglyceride-rich apolipoproteins in the plasma. These changes were absent in both CB_1-deficient and ApoE-deficient mice and could be prevented by CB_1 blockade (Ruby et al. 2008). Moreover, a recent study by the same group reveals that inhibiting both MAGL and FAAH by isopropyl dodecylfluorophosphonate (IDFP), the most potent pharmacological inhibitor of eCB degradation, acutely increases hepatic triglycerides content (Ruby et al. 2011). Taken together, these findings implicate a hyperactive eCB/CB_1 system as the cause of hepatic de novo lipogenesis.

In agreement with the above findings, recent evidence suggests that blockade of CB_1 receptors reduces hepatic de novo lipogenesis. Treatment of mice on a HFD with rimonabant, markedly reduces the rate of hepatic fatty acid synthesis (Osei-Hyiaman et al. 2005). Using an in vitro model of NAFLD in which HepG2 cells were exposed to saturated and unsaturated fatty acids, Shi and colleagues demonstrated that activation of CB_1 receptors in these cells increases the expression of SREBP-1c and its downstream targets ACC1 and FAS, an effect that is completely inhibited in cells pretreated with rimonabant (Shi et al. 2014). The mechanism by which rimonabant alleviates hepatic de novo lipogenesis has been recently demonstrated in hepatocytes (Wu et al. 2011). Briefly, CB_1 blockade by rimonabant decreases lipogenesis through AMP-activated protein kinase (AMPK) via the cyclic adenosine monophosphate (cAMP)-dependent protein kinase A (PKA)-liver kinase B1 (LKB1) axis and downstream of $G\alpha_{i/o}$ inhibition. AMPK, in turn, inactivates the liver X receptor α (LXRα), a lipid sensor that stimulates fatty acid synthesis and triglyceride accumulation in hepatocytes (Wu et al. 2011). Most probably, the stimulation of CB_1 and $G\alpha_{i/o}$ works via a similar mechanism but with the opposite effects.

Like a HFD, chronic alcohol use has been shown to contribute to enhanced hepatic de novo lipogenesis and hepatic steatosis (Lieber and Schmid 1961; You et al. 2002; Muramatsu et al. 1981). Chronic alcohol use (Basavarajappa et al. 2000, 2003), but not acute use (Rubio et al. 2007; Ferrer et al. 2007), can increase eCB levels. Thus, the role of chronic alcohol use in regulating hepatic de novo lipogenesis via the eCB/CB_1 receptor system was tested in several recent studies. Exposure of mice to an ethanol diet increases the hepatic expression level of the CB_1 receptor (Jeong et al. 2008; Mukhopadhyay et al. 2010; Patsenker et al. 2011; Trebicka et al. 2011) and the amount of 2-AG in hepatic stellate cells. These results suggest a paracrine mechanism whereby hepatic stellate cell-derived 2-AG activates CB_1 receptors on adjacent hepatocytes to stimulate hepatic de novo lipogenesis (Jeong et al. 2008; Chanda et al. 2013). In fact, ethanol exposure increases the hepatic expression of SREBP-1c and its target, FAS, an effect that is blunted or absent in both global CB_1-deficient and hepatocyte-specific CB_1-deficient ($LCB_1^{-/-}$) mice (Jeong et al. 2008; Trebicka et al. 2011), which are also resistant to ethanol-induced hepatic steatosis. This paracrine effect, together with high levels of retinoic acid in hepatic stellate cells and its established role in the control of gene expression, suggest a possible role of retinoic acid and its receptors in regulating hepatic CB_1 expression. Indeed, the expression level of the CB_1 receptor in primary human or mouse hepatocytes is increased by retinoic acid receptor-gamma (RARγ) and

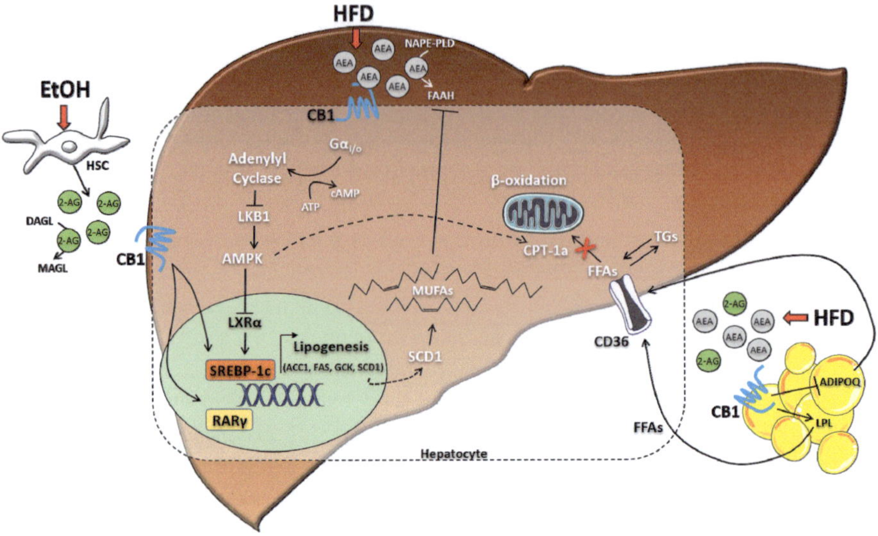

Fig. 6.1 Role of the CB_1 receptor in hepatic lipogenesis. Increased hepatic eCB tone by HFD (either by increased amount of AEA and/or elevated expression level of the CB_1 receptor) upregulates the hepatic mRNA expression levels of the lipogenic transcription factor, SREBP-1c, and its targets genes, including SCD1. In turn, increased activity of SCD1 promotes the synthesis of MUFAs that act as potent inhibitors of FAAH that is responsible for the increased hepatic levels of AEA during HFD feeding. Like a HFD, chronic alcohol use increases the hepatic expression level of the CB_1 receptor and 2-AG levels in HSCs, which in turn activate hepatocyte CB_1 receptors to stimulate hepatic de novo lipogenesis. In addition, stimulation of CB_1 receptor in extrahepatic sites influences hepatic lipogenesis indirectly, by promoting the activation of LPL, which in turn induces the transfer of FFAs from adipose tissue to the liver, and increasing CD36 expression in ADIOPQ-dependent manner. *HFD* high-fat diet, *cAMP* cyclic adenosine monophosphate, *LKB1* liver kinase B1, *AMPK* AMP-activated protein kinase, *LXRα* liver X receptor α, *SREBP-1c* sterol regulatory element-binding protein-1c, *ACC1* acetyl-CoA carboxylase-1, *FAS* fatty acid synthase, *SCD1* stearoyl-CoA desaturase-1, *GCK* glucokinase, *RARγ* retinoic acid receptor-gamma, *DAGL* diacylglycerol lipase, *MAGL* monoacylglycerol lipase, *FAAH* fatty-acid amide hydrolase, *NAPE-PLD* N-acyl phosphatidylethanolamine phospholipase D, *FFAs* free fatty acids, *TGs* triglycerides, *LPL* lipoprotein lipase, *ADIPOQ* adiponectin, *CPT-1a* carnitine palmitoyltransferase 1a, *AEA* anandamide, *2-AG* 2-arachidonoylglycerol

panRAR agonists and the effect is attenuated by siRNA knockdown of RARγ (Mukhopadhyay et al. 2010). In addition, both the CB_1 receptor and RARγ are upregulated in hepatocytes isolated from mice fed a HFD or chronic alcohol diet, providing further evidence that induction of the eCB system involves retinoic acid (Mukhopadhyay et al. 2010). These findings support the idea that hepatic de novo lipogenesis is increased via CB_1 activation in alcoholic fatty liver disease.

Taken together, all of the findings described above strongly suggest a role for hepatic CB_1 receptors in hepatic de novo lipogenesis, which may contribute to fat accumulation in the liver (Fig. 6.1). Yet, the findings do not rule out the possibility that eCBs activate CB_1 receptors in extrahepatic sites and influence hepatic lipogenesis indirectly. Two novel mouse models, $LCB_1^{-/-}$ mice (Osei-Hyiaman et al. 2008) and mice with transgenic expression of hepatic CB_1 receptors on a global CB_1-deficient background (htg$CB_1^{-/-}$) (Tam et al. 2010; Liu et al. 2012) have been used to address a direct versus indirect role for hepatic CB_1 receptors in the development of HFD-induced hepatic steatosis. When placed on a HFD, $LCB_1^{-/-}$ mice developed the same degree of obesity as wild-type mice, but unlike global $CB_1^{-/-}$ mice they were only partially protected from hepatic steatosis (Ravinet Trillou et al. 2004; Osei-Hyiaman et al. 2008). Similarly, htg$CB_1^{-/-}$ mice, like global $CB_1^{-/-}$ animals, remained largely resistant to the steatotic effect of a HFD (Tam et al. 2010; Liu et al. 2012). Therefore, the major source of liver fat may be adipose tissue because stimulating CB_1 receptors in adipocytes promotes the activation of lipoprotein lipase (Cota et al. 2003), which in turn induces the transfer of fatty acids from adipose tissue to the liver (Jourdan et al. 2010). Blocking CB_1 receptors either globally with rimonabant (Tam et al. 2014) or peripherally with JD5037, a peripherally-restricted cannabinoid inverse agonist (Cinar et al. 2014; Tam et al. 2012), reverses the HFD-induced upregulation of fatty acid translocase/CD36, which mediates the uptake of free fatty acids from the circulation to the liver. This reversal was found to be mediated indirectly by adiponectin (Tam et al. 2014). These findings are also in agreement with the observation that the specific deletion of CB_1 receptors in adipocytes protects mice from HFD-induced hepatic steatosis possibly through increased adiponectin signaling (Mancini et al. 2010).

3.2 *Role of the CB_2 Receptor in Hepatic Lipogenesis*

In contrast to the CB_1 receptor, the role of the CB_2 receptor in the regulation of hepatic de novo lipogenesis is not fully elucidated. CB_2 receptor expression is undetectable in normal liver, but is induced during embryonic development (Buckley et al. 1998) and in pathological conditions, such as NAFLD (Mendez-Sanchez et al. 2007), liver fibrosis (Julien et al. 2005), the regenerating liver (Teixeira-Clerc et al. 2010), and hepatocellular carcinoma (Xu et al. 2006). The involvement of CB_2 receptors in hepatic fat metabolism is suggested by the observation that CB_2 receptor expression is modestly increased in hepatocytes from both leptin-deficient *ob/ob* and HFD-induced obese mice (Mendez-Sanchez et al. 2007). Moreover, CB_2-deficient mice are resistant to HFD-induced hepatic steatosis (Deveaux et al. 2009; Agudo et al. 2010), and CB_2 agonism enhances liver triglyceride accumulation in wild-type mice fed with a HFD (Deveaux et al. 2009). Treatment of immortalized human hepatocytes and human hepatoma HepG2 cells with the CB_2 agonist, AM1241, dose-dependently induces lipid accumulation in the presence of oleic acid possibly as a result of increased FAS expression (De Gottardi et al. 2010). Because CB_1 expression is also upregulated by AM1241, there may be a reciprocal regulation between the CB_1 and CB_2 receptors that governs the lipogenic effect of CB_2 agonism. Further studies are needed to delineate the specific steatogenic role of CB_2 receptors.

4 Concluding Remarks

The eCB system is present in the liver, is involved in the control of hepatic de novo lipogenesis, and has important therapeutic implications specifically in relation to targeting CB_1 receptors. Increased CB_1 receptor activity contributes to the development of diet-induced and alcohol-induced fatty liver, whereas either globally or peripherally restricted CB_1 antagonism attenuates or prevents these changes. These findings support the idea that CB_1 antagonists may be useful therapeutics for hepatic steatosis and NAFLD. While neuropsychiatric side effects limit the therapeutic usefulness of brain-penetrant CB_1 receptor antagonists, second generation, peripherally-restricted CB_1 receptor antagonists (Tam et al. 2010, 2012) may alleviate this problem and offer superior therapeutic benefit for these conditions.

Acknowledgment This research was supported by an Israel Science Foundation grant (grant No. 617/14) to J.T.

References

Abel EL (1975) Cannabis: effects on hunger and thirst. Behav Biol 15(3):255–281

Agudo J, Martin M, Roca C, Molas M, Bura AS, Zimmer A, Bosch F, Maldonado R (2010) Deficiency of CB2 cannabinoid receptor in mice improves insulin sensitivity but increases food intake and obesity with age. Diabetologia 53(12):2629–2640

Basavarajappa BS, Saito M, Cooper TB, Hungund BL (2000) Stimulation of cannabinoid receptor agonist 2-arachidonylglycerol by chronic ethanol and its modulation by specific neuromodulators in cerebellar granule neurons. Biochim Biophys Acta 1535(1):78–86

Basavarajappa BS, Saito M, Cooper TB, Hungund BL (2003) Chronic ethanol inhibits the anandamide transport and increases extracellular anandamide levels in cerebellar granule neurons. Eur J Pharmacol 466(1–2):73–83

Batkai S, Jarai Z, Wagner JA, Goparaju SK, Varga K, Liu J, Wang L, Mirshahi F, Khanolkar AD, Makriyannis A, Urbaschek R, Garcia N Jr, Sanyal AJ, Kunos G (2001) Endocannabinoids acting at vascular CB1 receptors mediate the vasodilated state in advanced liver cirrhosis. Nat Med 7(7):827–832

Bensaid M, Gary-Bobo M, Esclangon A, Maffrand JP, Le Fur G, Oury-Donat F, Soubrie P (2003) The cannabinoid CB1 receptor antagonist SR141716 increases Acrp30 mRNA expression in adipose tissue of obese fa/fa rats and in cultured adipocyte cells. Mol Pharmacol 63(4):908–914

Biddinger SB, Almind K, Miyazaki M, Kokkotou E, Ntambi JM, Kahn CR (2005) Effects of diet and genetic background on sterol regulatory element-binding protein-1c, stearoyl-CoA desaturase 1, and the development of the metabolic syndrome. Diabetes 54(5):1314–1323

Biecker E, Sagesser H, Reichen J (2004) Vasodilator mRNA levels are increased in the livers of portal hypertensive NO-synthase 3-deficient mice. Eur J Clin Invest 34(4):283–289

Buckley NE, Hansson S, Harta G, Mezey E (1998) Expression of the CB1 and CB2 receptor messenger RNAs during embryonic development in the rat. Neuroscience 82(4):1131–1149

Chanda D, Kim YH, Li T, Misra J, Kim DK, Kim JR, Kwon J, Jeong WI, Ahn SH, Park TS, Koo SH, Chiang JY, Lee CH, Choi HS (2013) Hepatic cannabinoid receptor type 1 mediates alcohol-induced regulation of bile acid enzyme genes expression via CREBH. PLoS One 8(7):e68845

Cinar R, Godlewski G, Liu J, Tam J, Jourdan T, Mukhopadhyay B, Harvey-White J, Kunos G (2014) Hepatic cannabinoid-1 receptors mediate diet-induced insulin resistance by increasing de novo synthesis of long-chain ceramides. Hepatology 59(1):143–153

Cota D, Marsicano G, Tschop M, Grubler Y, Flachskamm C, Schubert M, Auer D, Yassouridis A, Thone-Reineke C, Ortmann S, Tomassoni F, Cervino C, Nisoli E, Linthorst AC, Pasquali R, Lutz B, Stalla GK, Pagotto U (2003) The endogenous cannabinoid system affects energy balance via central orexigenic drive and peripheral lipogenesis. J Clin Invest 112(3):423–431

De Gottardi A, Spahr L, Ravier-Dall'Antonia F, Hadengue A (2010) Cannabinoid receptor 1 and 2 agonists increase lipid accumulation in hepatocytes. Liver Int 30(10):1482–1489

Devane WA, Dysarz FA III, Johnson MR, Melvin LS, Howlett AC (1988) Determination and characterization of a cannabinoid receptor in rat brain. Mol Pharmacol 34(5):605–613

Devane WA, Hanus L, Breuer A, Pertwee RG, Stevenson LA, Griffin G, Gibson D, Mandelbaum A, Etinger A, Mechoulam R (1992) Isolation and structure of a brain constituent that binds to the cannabinoid receptor. Science 258(5090):1946–1949

Deveaux V, Cadoudal T, Ichigotani Y, Teixeira-Clerc F, Louvet A, Manin S, Nhieu JT, Belot MP, Zimmer A, Even P, Cani PD, Knauf C, Burcelin R, Bertola A, Le Marchand-Brustel Y, Gual P, Mallat A, Lotersztajn S (2009) Cannabinoid CB2 receptor potentiates obesity-associated inflammation, insulin resistance and hepatic steatosis. PLoS One 4(6):e5844

Dinh TP, Carpenter D, Leslie FM, Freund TF, Katona I, Sensi SL, Kathuria S, Piomelli D (2002) Brain monoglyceride lipase participating in endocannabinoid inactivation. Proc Natl Acad Sci U S A 99(16):10819–10824

Diraison F, Yankah V, Letexier D, Dusserre E, Jones P, Beylot M (2003) Differences in the regulation of adipose tissue and liver lipogenesis by carbohydrates in humans. J Lipid Res 44(4):846–853

Domenicali M, Ros J, Fernandez-Varo G, Cejudo-Martin P, Crespo M, Morales-Ruiz M, Briones AM, Campistol JM, Arroyo V, Vila E, Rodes J, Jimenez W (2005) Increased anandamide induced relaxation in mesenteric arteries of cirrhotic rats: role of cannabinoid and vanilloid receptors. Gut 54(4):522–527

Ferrer B, Bermudez-Silva FJ, Bilbao A, Alvarez-Jaimes L, Sanchez-Vera I, Giuffrida A, Serrano A, Baixeras E, Khaturia S, Navarro M, Parsons LH, Piomelli D, Rodriguez de Fonseca F (2007) Regulation of brain anandamide by acute administration of ethanol. Biochem J 404(1):97–104

Floreani A, Lazzari R, Macchi V, Porzionato A, Variola A, Colavito D, Leon A, Guido M, Baldo V, De Caro R, Bergasa NV (2010) Hepatic expression of endocannabinoid receptors and their novel polymorphisms in primary biliary cirrhosis. J Gastroenterol 45(1):68–76

Fowler CJ (2007) The pharmacology of the cannabinoid system—a question of efficacy and selectivity. Mol Neurobiol 36(1):15–25

Freund TF, Katona I, Piomelli D (2003) Role of endogenous cannabinoids in synaptic signaling. Physiol Rev 83(3):1017–1066

Galiegue S, Mary S, Marchand J, Dussossoy D, Carriere D, Carayon P, Bouaboula M, Shire D, Le Fur G, Casellas P (1995) Expression of central and peripheral cannabinoid receptors in human immune tissues and leukocyte subpopulations. Eur J Biochem 232(1):54–61

Gaoni Y, Mechoulam R (1964) Isolation, structure and partial synthesis of an active constituent of hashish. J Am Chem Soc 86:1646–1647

Gong JP, Onaivi ES, Ishiguro H, Liu QR, Tagliaferro PA, Brusco A, Uhl GR (2006) Cannabinoid CB2 receptors: immunohistochemical localization in rat brain. Brain Res 1071(1):10–23

Hezode C, Zafrani ES, Roudot-Thoraval F, Costentin C, Hessami A, Bouvier-Alias M, Medkour F, Pawlostky JM, Lotersztajn S, Mallat A (2008) Daily cannabis use: a novel risk factor of steatosis severity in patients with chronic hepatitis C. Gastroenterology 134(2):432–439

Howlett AC (2005) Cannabinoid receptor signaling. Handb Exp Pharmacol 168:53–79

Jeong WI, Osei-Hyiaman D, Park O, Liu J, Batkai S, Mukhopadhyay P, Horiguchi N, Harvey-White J, Marsicano G, Lutz B, Gao B, Kunos G (2008) Paracrine activation of hepatic CB1 receptors by stellate cell-derived endocannabinoids mediates alcoholic fatty liver. Cell Metab 7(3):227–235

Jourdan T, Djaouti L, Demizieux L, Gresti J, Verges B, Degrace P (2010) CB1 antagonism exerts specific molecular effects on visceral and subcutaneous fat and reverses liver steatosis in diet-induced obese mice. Diabetes 59(4):926–934

Jourdan T, Demizieux L, Gresti J, Djaouti L, Gaba L, Verges B, Degrace P (2012) Antagonism of peripheral hepatic cannabinoid receptor-1 improves liver lipid metabolism in mice: evidence from cultured explants. Hepatology 55(3):790–799

Julien B, Grenard P, Teixeira-Clerc F, Van Nhieu JT, Li L, Karsak M, Zimmer A, Mallat A, Lotersztajn S (2005) Antifibrogenic role of the cannabinoid receptor CB2 in the liver. Gastroenterology 128(3):742–755

Katayama K, Ueda N, Kurahashi Y, Suzuki H, Yamamoto S, Kato I (1997) Distribution of anandamide amidohydrolase in rat tissues with special reference to small intestine. Biochim Biophys Acta 1347(2–3):212–218

Larrinaga G, Varona A, Perez I, Sanz B, Ugalde A, Candenas ML, Pinto FM, Gil J, Lopez JI (2010) Expression of cannabinoid receptors in human kidney. Histol Histopathol 25(9): 1133–1138

Lieber CS, Schmid R (1961) The effect of ethanol on fatty acid metabolism; stimulation of hepatic fatty acid synthesis in vitro. J Clin Invest 40:394–399

Lin X, Yue P, Xie Y, Davidson NO, Sakata N, Ostlund RE Jr, Chen Z, Schonfeld G (2005) Reduced intestinal fat absorptive capacity but enhanced susceptibility to diet-induced fatty liver in mice heterozygous for ApoB38.9 truncation. Am J Physiol Gastrointest Liver Physiol 289(1): G146–G152

Liu J, Zhou L, Xiong K, Godlewski G, Mukhopadhyay B, Tam J, Yin S, Gao P, Shan X, Pickel J, Bataller R, O'Hare J, Scherer T, Buettner C, Kunos G (2012) Hepatic cannabinoid receptor-1 mediates diet-induced insulin resistance via inhibition of insulin signaling and clearance in mice. Gastroenterology 142(5):1218–1228, e1211

Liu J, Cinar R, Xiong K, Godlewski G, Jourdan T, Lin Y, Ntambi JM, Kunos G (2013) Monounsaturated fatty acids generated via stearoyl CoA desaturase-1 are endogenous inhibitors of fatty acid amide hydrolase. Proc Natl Acad Sci U S A 110(47):18832–18837

Mallat A, Teixeira-Clerc F, Lotersztajn S (2013) Cannabinoid signaling and liver therapeutics. J Hepatol 59(4):891–896

Mancini G, Quarta C, Srivastava RK, Klaus S, Pagotto U, Lutz B (2010) Adipocyte-specific CB1 conditional knock-out mice: new insights in the study of obesity and metabolic syndrome. In: 20th annual symposium on the cannabinoids, Research Triangle Park, NC International Cannabinoid Research Society, 17

Matsuda LA, Lolait SJ, Brownstein MJ, Young AC, Bonner TI (1990) Structure of a cannabinoid receptor and functional expression of the cloned cDNA. Nature 346(6284):561–564

McKinney MK, Cravatt BF (2005) Structure and function of fatty acid amide hydrolase. Annu Rev Biochem 74:411–432

Mechoulam R, Ben-Shabat S, Hanus L, Ligumsky M, Kaminski NE, Schatz AR, Gopher A, Almog S, Martin BR, Compton DR et al (1995) Identification of an endogenous 2-monoglyceride, present in canine gut, that binds to cannabinoid receptors. Biochem Pharmacol 50(1):83–90

Mendez-Sanchez N, Zamora-Valdes D, Pichardo-Bahena R, Barredo-Prieto B, Ponciano-Rodriguez G, Bermejo-Martinez L, Chavez-Tapia NC, Baptista-Gonzalez HA, Uribe M (2007) Endocannabinoid receptor CB2 in nonalcoholic fatty liver disease. Liver Int 27(2):215–219

Michalopoulos GK, Bowen WC, Mule K, Luo J (2003) HGF-, EGF-, and dexamethasone-induced gene expression patterns during formation of tissue in hepatic organoid cultures. Gene Expr 11(2):55–75

Migliarini B, Carnevali O (2008) Anandamide modulates growth and lipid metabolism in the zebrafish Danio rerio. Mol Cell Endocrinol 286(1–2 Suppl 1):S12–S16

Mukhopadhyay B, Liu J, Osei-Hyiaman D, Godlewski G, Mukhopadhyay P, Wang L, Jeong WI, Gao B, Duester G, Mackie K, Kojima S, Kunos G (2010) Transcriptional regulation of cannabinoid receptor-1 expression in the liver by retinoic acid acting via retinoic acid receptor-gamma. J Biol Chem 285(25):19002–19011

Munro S, Thomas KL, Abu-Shaar M (1993) Molecular characterization of a peripheral receptor for cannabinoids. Nature 365(6441):61–65

Muramatsu M, Kuriyama K, Yuki T, Ohkuma S (1981) Hepatic lipogenesis and mobilization of peripheral fats in the formation of alcoholic fatty liver. Jpn J Pharmacol 31(6):931–940

Nakata M, Yada T (2008) Cannabinoids inhibit insulin secretion and cytosolic Ca^{2+} oscillation in islet beta-cells via CB1 receptors. Regul Pept 145(1–3):49–53

Nam DH, Lee MH, Kim JE, Song HK, Kang YS, Lee JE, Kim HW, Cha JJ, Hyun YY, Kim SH, Han SY, Han KH, Han JY, Cha DR (2012) Blockade of cannabinoid receptor 1 improves insulin resistance, lipid metabolism, and diabetic nephropathy in db/db mice. Endocrinology 153(3):1387–1396

Ofek O, Karsak M, Leclerc N, Fogel M, Frenkel B, Wright K, Tam J, Attar-Namdar M, Kram V, Shohami E, Mechoulam R, Zimmer A, Bab I (2006) Peripheral cannabinoid receptor, CB2, regulates bone mass. Proc Natl Acad Sci U S A 103(3):696–701

Osei-Hyiaman D, DePetrillo M, Pacher P, Liu J, Radaeva S, Batkai S, Harvey-White J, Mackie K, Offertaler L, Wang L, Kunos G (2005) Endocannabinoid activation at hepatic CB1 receptors stimulates fatty acid synthesis and contributes to diet-induced obesity. J Clin Invest 115(5):1298–1305

Osei-Hyiaman D, Liu J, Zhou L, Godlewski G, Harvey-White J, Jeong WI, Batkai S, Marsicano G, Lutz B, Buettner C, Kunos G (2008) Hepatic CB1 receptor is required for development of diet-induced steatosis, dyslipidemia, and insulin and leptin resistance in mice. J Clin Invest 118(9):3160–3169

Pacher P, Batkai S, Kunos G (2006) The endocannabinoid system as an emerging target of pharmacotherapy. Pharmacol Rev 58(3):389–462

Pagotto U, Marsicano G, Cota D, Lutz B, Pasquali R (2006) The emerging role of the endocannabinoid system in endocrine regulation and energy balance. Endocr Rev 27(1):73–100

Patsenker E, Stoll M, Millonig G, Agaimy A, Wissniowski T, Schneider V, Mueller S, Brenneisen R, Seitz HK, Ocker M, Stickel F (2011) Cannabinoid receptor type I modulates alcohol-induced liver fibrosis. Mol Med 17(11–12):1285–1294

Pertwee RG (2006) Cannabinoid pharmacology: the first 66 years. Br J Pharmacol 147(Suppl 1):S163–S171

Quarta C, Bellocchio L, Mancini G, Mazza R, Cervino C, Braulke LJ, Fekete C, Latorre R, Nanni C, Bucci M, Clemens LE, Heldmaier G, Watanabe M, Leste-Lassere T, Maitre M, Tedesco L, Fanelli F, Reuss S, Klaus S, Srivastava RK, Monory K, Valerio A, Grandis A, De Giorgio R, Pasquali R, Nisoli E, Cota D, Lutz B, Marsicano G, Pagotto U (2010) CB(1) signaling in forebrain and sympathetic neurons is a key determinant of endocannabinoid actions on energy balance. Cell Metab 11(4):273–285

Ravinet Trillou C, Delgorge C, Menet C, Arnone M, Soubrie P (2004) CB1 cannabinoid receptor knockout in mice leads to leanness, resistance to diet-induced obesity and enhanced leptin sensitivity. Int J Obes Relat Metab Disord 28(4):640–648

Rubio M, McHugh D, Fernandez-Ruiz J, Bradshaw H, Walker JM (2007) Short-term exposure to alcohol in rats affects brain levels of anandamide, other N-acylethanolamines and 2-arachidonoyl-glycerol. Neurosci Lett 421(3):270–274

Ruby MA, Nomura DK, Hudak CS, Mangravite LM, Chiu S, Casida JE, Krauss RM (2008) Overactive endocannabinoid signaling impairs apolipoprotein E-mediated clearance of triglyceride-rich lipoproteins. Proc Natl Acad Sci U S A 105(38):14561–14566

Ruby MA, Nomura DK, Hudak CS, Barber A, Casida JE, Krauss RM (2011) Acute overactive endocannabinoid signaling induces glucose intolerance, hepatic steatosis, and novel cannabinoid receptor 1 responsive genes. PLoS One 6(11):e26415

Shi D, Zhan X, Yu X, Jia M, Zhang Y, Yao J, Hu X, Bao Z (2014) Inhibiting CB1 receptors improves lipogenesis in an in vitro non-alcoholic fatty liver disease model. Lipids Health Dis 13(1):173

Siegmund SV, Qian T, de Minicis S, Harvey-White J, Kunos G, Vinod KY, Hungund B, Schwabe RF (2007) The endocannabinoid 2-arachidonoyl glycerol induces death of hepatic stellate cells via mitochondrial reactive oxygen species. FASEB J 21(11):2798–2806

Sugiura T, Kondo S, Sukagawa A, Nakane S, Shinoda A, Itoh K, Yamashita A, Waku K (1995) 2-Arachidonoylglycerol: a possible endogenous cannabinoid receptor ligand in brain. Biochem Biophys Res Commun 215(1):89–97

Tam J, Vemuri VK, Liu J, Batkai S, Mukhopadhyay B, Godlewski G, Osei-Hyiaman D, Ohnuma S, Ambudkar SV, Pickel J, Makriyannis A, Kunos G (2010) Peripheral CB1 cannabinoid receptor blockade improves cardiometabolic risk in mouse models of obesity. J Clin Invest 120(8):2953–2966

Tam J, Cinar R, Liu J, Godlewski G, Wesley D, Jourdan T, Szanda G, Mukhopadhyay B, Chedester L, Liow JS, Innis RB, Cheng K, Rice KC, Deschamps JR, Chorvat RJ, McElroy JF, Kunos G (2012) Peripheral cannabinoid-1 receptor inverse agonism reduces obesity by reversing leptin resistance. Cell Metab 16(2):167–179

Tam J, Godlewski G, Earley BJ, Zhou L, Jourdan T, Szanda G, Cinar R, Kunos G (2014) Role of adiponectin in the metabolic effects of cannabinoid type 1 receptor blockade in mice with diet-induced obesity. Am J Physiol Endocrinol Metab 306(4):E457–E468

Teixeira-Clerc F, Julien B, Grenard P, Tran Van Nhieu J, Deveaux V, Li L, Serriere-Lanneau V, Ledent C, Mallat A, Lotersztajn S (2006) CB1 cannabinoid receptor antagonism: a new strategy for the treatment of liver fibrosis. Nat Med 12(6):671–676

Teixeira-Clerc F, Belot MP, Manin S, Deveaux V, Cadoudal T, Chobert MN, Louvet A, Zimmer A, Tordjmann T, Mallat A, Lotersztajn S (2010) Beneficial paracrine effects of cannabinoid receptor 2 on liver injury and regeneration. Hepatology 52(3):1046–1059

Trebicka J, Racz I, Siegmund SV, Cara E, Granzow M, Schierwagen R, Klein S, Wojtalla A, Hennenberg M, Huss S, Fischer HP, Heller J, Zimmer A, Sauerbruch T (2011) Role of cannabinoid receptors in alcoholic hepatic injury: steatosis and fibrogenesis are increased in CB2 receptor-deficient mice and decreased in CB1 receptor knockouts. Liver Int 31(6):860–870

Van Sickle MD, Duncan M, Kingsley PJ, Mouihate A, Urbani P, Mackie K, Stella N, Makriyannis A, Piomelli D, Davison JS, Marnett LJ, Di Marzo V, Pittman QJ, Patel KD, Sharkey KA (2005) Identification and functional characterization of brainstem cannabinoid CB2 receptors. Science 310(5746):329–332

Westerbacka J, Kotronen A, Fielding BA, Wahren J, Hodson L, Perttila J, Seppanen-Laakso T, Suortti T, Arola J, Hultcrantz R, Castillo S, Olkkonen VM, Frayn KN, Oresic M, Yki-Jarvinen H (2010) Splanchnic balance of free fatty acids, endocannabinoids, and lipids in subjects with nonalcoholic fatty liver disease. Gastroenterology 139(6):1961–1971, e1961

Wu HM, Yang YM, Kim SG (2011) Rimonabant, a cannabinoid receptor type 1 inverse agonist, inhibits hepatocyte lipogenesis by activating liver kinase B1 and AMP-activated protein kinase axis downstream of Galpha i/o inhibition. Mol Pharmacol 80(5):859–869

Xu X, Liu Y, Huang S, Liu G, Xie C, Zhou J, Fan W, Li Q, Wang Q, Zhong D, Miao X (2006) Overexpression of cannabinoid receptors CB1 and CB2 correlates with improved prognosis of patients with hepatocellular carcinoma. Cancer Genet Cytogenet 171(1):31–38

You M, Fischer M, Deeg MA, Crabb DW (2002) Ethanol induces fatty acid synthesis pathways by activation of sterol regulatory element-binding protein (SREBP). J Biol Chem 277(32): 29342–29347

Chapter 7
Role of Lipogenesis and Lipid Desaturases in Non-alcoholic Fatty Liver Disease

Mohamed Amine Lounis, Sabri Rial, James M. Ntambi, and Catherine Mounier

Abstract Non-alcoholic fatty liver disease (NAFLD) is a clinico-pathological change characterized by the accumulation of triacylglycerol (TG) in hepatic lipid droplets (LD). NAFLD can range from a simple steatosis to Non-alcoholic steatohepatitis (NASH) characterized by hepatic injury, inflammation, and eventually fibrosis. NAFLD can also be associated with insulin resistance (IR), ER stress, oxidative stress and inflammation. The cause of NAFLD is due to modification of various metabolic pathways including increased fatty acid (FA) uptake and/or reduced FA oxidation, decreased VLDL secretion and increased de novo lipogenesis (DNL). NAFLD is often observed in obese patients where energy is in excess and energy expenditure is low. In these conditions, most of carbohydrates are converted into TG through DNL. We aim here to present the most recent studies demonstrating the key role of DNL in NAFLD development. A special focus will be made on desaturases especially the stearoyl CoA desaturase 1, a central enzyme implicated in fatty liver disease as well as on the delta 5 and delta 6 desaturases, two enzymes involved in the synthesis of omega 3 and omega 6 fatty acids.

Keywords Non-alcoholic fatty liver disease (NAFLD) • Hepatic steatosis • Insulin resistance • Lipogenesis • Lipid droplet • Stearoyl CoA desaturase 1 (SCD1) • Delta 5 desaturase • Delta 6 desaturase

M.A. Lounis • S. Rial • C. Mounier (✉)
Département des Sciences Biologiques, Centre de recherche BioMed, Université du Québec à Montréal, Case Postale 8888 Succursale Centre-ville, Montréal, QC, Canada, H3C 3P8
e-mail: mounier.catherine@uqam.ca

J.M. Ntambi
Department of Biochemistry, University of Wisconsin-Madison, Madison, WI 53706, USA

Department of Nutritional Sciences, University of Wisconsin-Madison, Madison, WI 53706, USA

J.M. Ntambi (ed.), *Hepatic De Novo Lipogenesis and Regulation of Metabolism*,
DOI 10.1007/978-3-319-25065-6_7

1 Non Alcoholic Fatty Liver Disease (NAFLD) and Associated Pathologies

NAFLD is a major public health issue due to its high prevalence worldwide estimated to be 20–30 % of the total population (Milic and Stimac 2012), and increasing to 57–74 % among obese patients (Angulo 2002). It is a clinico-pathological change characterized by the accumulation of TG in hepatic LD (Teli et al. 1995). NAFLD is characterized by the presence of cytoplasmic LD in more than 5 % of the hepatocytes or by hepatic TG content exceeding the 95th percentile for lean and healthy who do not have any history of alcohol consumption, no viral infection or autoimmune liver disease (Cohen et al. 2011; Fabbrini et al. 2010; Stefan and Haring 2011). TG accumulation in NAFLD is probably due to the modification of various metabolic pathways including increase in fatty acid (FA) uptake (Donnelly et al. 2005), reduction in FA oxidation (Fromenty et al. 2004) or VLDL secretion (Donnelly et al. 2005) or finally increase in DNL (Postic and Girard 2008).

Liver histology can range the disease from simple steatosis (NAFLD) (>5 % fat infiltration, with or without minimal inflammation) to non-alcoholic steatohepatitis (NASH) characterized by hepatocyte injury (ballooning degeneration in presence or not of Mallory bodies), inflammation and eventually fibrosis (Neuschwander-Tetri and Caldwell 2003) (Fig. 7.1). Simple steatosis is thought to be a relatively benign

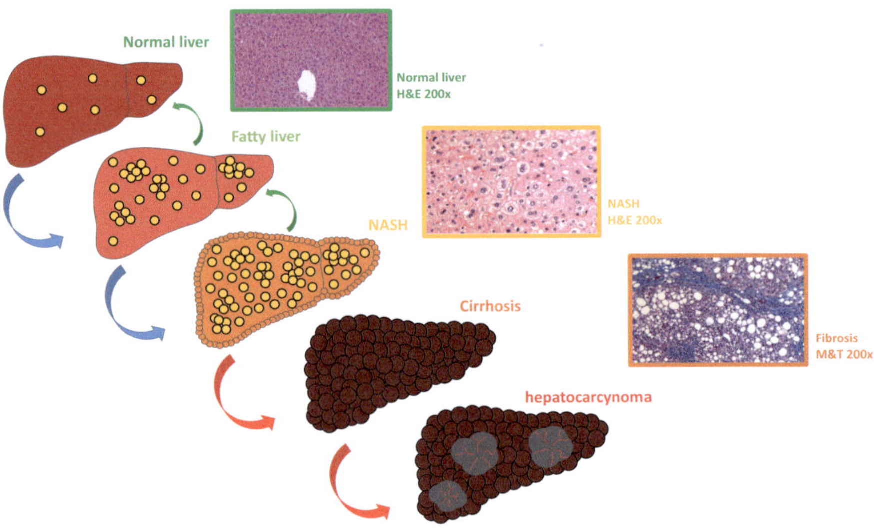

Fig. 7.1 Schematic representation of pathological conditions spectrum characterizing the NAFLD progression. In comparison with a normal liver, four steps of the NAFLD progression are schematized here from simple fatty liver to hepatocarcinoma through NASH and non-alcoholic cirrhosis. The three histologic pictures from microscopic imaging are incorporated to compare normal hepatocytes with hepatocytes in fibrosis characterized by perivenular/pericellular step and with NASH hepatocytes presenting ballooned morphology (Kanuri and Bergheim 2013). *NASH* non-alcoholic steatohepatitis, *H&E* hematoxylin-Eosin stain, *MT* Masson-trichrome stain

state, whereas NASH represents the form of NAFLD that could potentially progress to cirrhosis and the following complications such as hepatocellular carcinoma (HCC). Day and collaborators were the first to describe the "two hit hypothesis" as a model for NAFLD-progression. The "first hit" was defined as a hepatocellular lipid accumulation (steatosis) resulting from an imbalance of cellular uptake and combustion. The "second hit" was defined as an additional inflammation (NASH) resulting from an imbalance of pro- and anti-inflammatory factors (Day and James 1998). For many years it was considered that patients with "simple" steatosis (NAFLD) show very slow or no histological progression, while NASH patients can exhibit histological and clinical progression to cirrhotic-stage disease associated with all the known complications of liver insufficiency (Musso et al. 2011). However, a recent study provides evidences that all forms of NAFLD significantly increased the risk of cirrhosis and HCC probably as the result of various allelic variants of genes involved in hepatic LD formation (Yki-Jarvinen 2014).

Accumulation of lipids (especially diacylglycerols (DAG) and ceramides) in liver is associated with the installation of hepatic IR (Samuel et al. 2010). The relationship between hepatic DAG accumulation and hepatic IR could be attributed to the activation of PKCε (Samuel et al. 2004, 2007; Frangioudakis et al. 2009). Increased PKCε directly reduced the insulin-induced tyrosine phosphorylation of the insulin receptor substrate (IRS-2) by the insulin receptor kinase, leading to the reduction in insulin-stimulated hepatic glycogen synthesis and insulin-inhibition of neoglucogenesis. Supporting the key roles of PKCε and DAG, hepatic knockdown expression of PKCε (Samuel et al. 2007) or DGAT2 (diacylglycerol acyltransferase-2) (Choi et al. 2007) using antisense oligonucleotides (ASO) protects mice from hepatic IR induces by high fat diet (HFD). A role for ceramides in hepatic IR was also suggested as reduced levels of ceramides improves IR in animal models of diet-induced obesity. In this study, the authors suggested that the effect of ceramides on insulin signaling is mediated by a direct interaction of these lipids with the protein kinase B (Chavez and Summers 2012). However, a more recent study showed an opposite result. The TLR-4 knockout mice showed that despite previous evidences, the TLR-4 toll like receptor signaling is not involved in ceramides synthesis and that IR is not dependent of ceramides accumulation but on DAG implicating the activation of a PKCε/IRS2-dependent signaling pathway (Galbo et al. 2013).

Another feature associated with NAFLD is the apparition of ER stress. The ER plays an important role in the synthesis, folding and trafficking of proteins. Accumulation of lipid in the liver induces the dysfunction of ER, causing the accumulation of unfolded proteins therein, triggering an unfolded protein response (UPR) (Puri et al. 2008). Recent data indicates that UPR is activated in NAFLD playing an important role in the development and the progression of the disease (Feng et al. 2003; Boden et al. 2011). Saturated fatty acids such as palmitate activate the UPR in liver characterized by a preferential induction of PERK signaling (Wei et al. 2006; Cao et al. 2012). Conversely, short-term exposure to unsaturated fatty acids attenuates ER stress in setting the excess of palmitate (Fu et al. 2012; Listenberger et al. 2003). UPR is characterized by the arrest of protein synthesis through the phosphorylation of eIF-2α, by the activation of proteosomal protein degradation and by the activation of the ATF4-CHOP-GADD34 axis. The latter allows cells to adapt to ER dysfunction. When cells fail to adapt, JNK is activated leading to inflammation and apoptosis (Kaplowitz and Ji 2006). In both NAFLD and NASH, an increased eIF-2α phosphorylation is observed. However, in NASH, the downstream recovery pathway (ATF4-CHOP-GADD34) is not activated. Consequently, JNK phosphorylation is increased. This may explain at least in part the apparition of inflammation in the NASH state (Puri et al. 2008).

Oxidative stress is also activated in NAFLD. It is the result of a serious imbalance between limited antioxidant defenses and an excessive formation of both reactive oxygen species (ROS) and reactive nitrogen species (RNS) (Robertson et al. 2001). Free fatty acids (FFA) are metabolized via the mitochondrial β-oxidation pathway and the tricarboxylic acid (TCA) cycle. Accelerated β-oxidation, due to an excess of lipid in the liver as observed in NAFLD, induces an excessive electron flux in the electron transport chain of the mitochondria increasing ROS production and leading to mitochondrial dysfunction (Rolo et al. 2012). NASH is also characterized by an elevated hepatic DNL resulting in an increased production of malonyl-COA. In turn, malonyl-CoA inhibits the carnitine palmitoyltransferase-1 (CPT-1) leading to a decrease of long chain fatty acids import into the mitochondria, and consequently a reduction in β-oxidation. Consequently, TG accumulate in the liver aggravating the steatosis (Rolo et al. 2012; Vega et al. 2000).

The severity of NAFLD correlates with the expression of tumor necrosis factor-α (TNFα) receptor while the production of TNFα is dependent on FA concentration (Crespo et al. 2001). In addition, activation of the TNF receptor increases the expression of SREBP1c, inducing hepatic DNL and lipid accumulation (Endo et al. 2007). In *ob/ob* mice, secretion of TNFα and IL6 activates the hepatic Kupffer cells promoting hepatotoxicity, IR and steatohepatitis (Odegaard et al. 2008; Li and Diehl 2003).

2 Characteristics of the Hepatic Lipid Droplets

NAFLD is characterized by the presence of LD in hepatocytes. LD are dynamic cellular organelles found in most cells and particularly in cells specialized in lipids storage such as white and brown adipose tissue, hepatocytes and enterocytes. LD cores contain neutral lipids predominantly TG or sterol esters. These organelles may also include retinyl esters, waxes and ether lipids (Farese and Walther 2009; Martin and Parton 2006; Brasaemle and Wolins 2012). These lipids are surrounded by a phospholipid monolayer coated with a large number of proteins. They regulate many aspects of LD biology including the synthesis and the mobilization of lipids as well as their interaction with other cellular organelles. LD could either be formed de novo or derived from existing LD by fusion (Long et al. 2012). De novo formation of LD in eukaryotes occurs from the ER (Jacquier et al. 2011, 2013) where neutral lipids are synthesized (Buhman et al. 2001).

The precise mechanism of LD formation remains however mostly unanswered. The most widely accepted model proposed for LD biogenesis lasts in three different steps (Wilfling et al. 2014) (Fig. 7.2).

(a) Neutral lipids are synthesized in the ER and accumulate within the bilayer. Neutral lipids are highly mobile in the bilayer and may spontaneously aggregate on the basis of their thermal fluctuations and interactions with membrane proteins or other lipids.
(b) Once the local concentration of neutral lipid reaches a crucial threshold, an intra-membrane lipid accumulation occurs leading to the formation of lens.
(c) As lens accumulates additional neutral lipids, the bilayer deforms and a nascent LD buds into the cytoplasm. The nascent droplet might remain attached to the ER or separate completely.

LD can growth by two general mechanisms: LD expansion or LD fusion (Wilfling et al. 2014). The expansion is triggered by the relocalization of TG synthesis

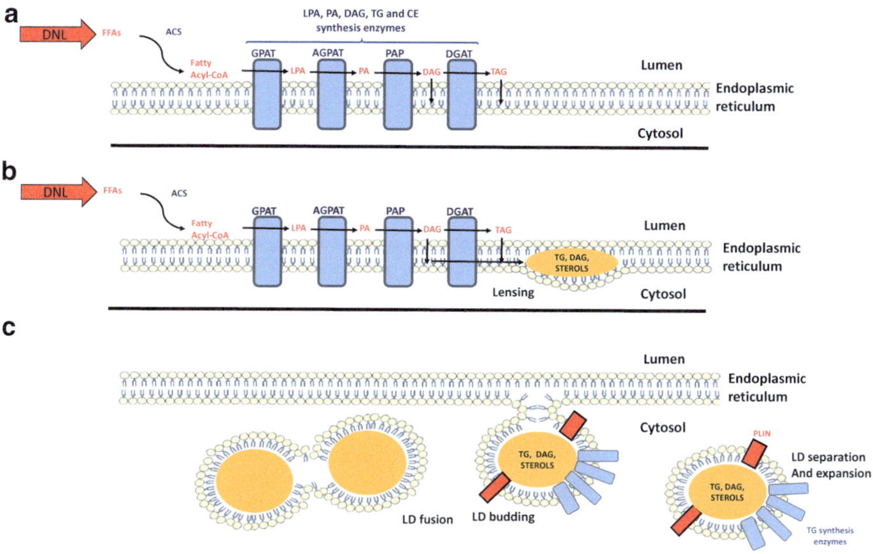

Fig. 7.2 The lensing model of lipids droplets (LD) biogenesis. This schematic representation depicts the three steps of the "lensing" model, the most widely accepted model for LD biogenesis. (**a**) Newly synthesized FFA are acetylated by ACBP and esterified by DAG and TG synthesis enzymes. (**b**) LPA, PA, DAG and TG accumulate within the phospholipidic bilayer inducing a membrane lensing in the cytosolic surface of the ER. (**c**) Formation of a prolonged lensing induces a sufficient bilayer deformation allowing the nascent of LD. The LD can separate completely, fuse together or remain attached to the ER. Associated proteins such as PLIN ensure the stabilization of the LD (Guo et al. 2009; Kohlwein et al. 2013). *DNL* de novo lipogenesis, *FFA* free fatty acids, *ACBP* acetyl-CoA binding protein, *GPAT* glycerol-3-phosphate acyltransferase, *AGPAT* 1-acylglycerol-3-phosphate O-acyltransferase, *PAP* phosphatidic acid phosphohydrolase, *DGAT* diacylglycerol acyltransferase, *LPA* lysophosphatidic acid, *PA* phosphatidic acid, *DAG* diacylglycerol, *TG* triacylglycerol, *PLIN* perilipin

enzymes from the ER to the surface of LD (Wilfling et al. 2014; Beller et al. 2010). When phosphatidylcholine is limited and the tension at the surface is relatively high, large LD can be formed by fusion/coalescence of two or more LD (Beller et al. 2010; Wilfling et al. 2014). During steatogenesis, the expression pattern of several LD associated proteins (proteins belonging to the perilipins family) changes. This is probably associated with the increased expression of PPARγ, a transcription factor that targets several genes implicated in LD formation (Inoue et al. 2005; Matsusue et al. 2008). Notably, perilipin 1, only expressed in adipose tissue in normal condition is also expressed in fatty liver (Fujii et al. 2009; Straub et al. 2008). ADRP levels are also elevated in fatty liver (Motomura et al. 2006). High expression of hepatic lipase (Yamada et al. 2011) and lysophospholipase-like1 (Speliotes et al. 2011), two enzymes with TG lipase activity, as well as DGAT2, an enzyme involved in TG synthesis (Kantartzis et al. 2009), have been associated with the risk of developing hepatic steatosis. Interestingly, several studies showed that genetic variations of genes implicated in hepatic lipid accumulation and in LD such as ATGL and CGI58, are not associated with IR (Cohen et al. 2011; Hooper et al. 2011). This is consistent with the hypothesis that the sequestration of lipids into hepatic LD protects the liver from free fatty acid induced lipotoxicity that promotes IR.

3 Role of Lipogenesis in NAFLD

Numerous studieshave implicated DNL in the development of NAFLD. DNL consists in the conversion of carbohydrates into lipids. These newly synthesized lipids are essentially esterified into TG and then either shipped *via* the very low density lipoproteins (VLDL) to the peripheral tissues including lipid-storing tissues like muscles and adipose tissues or accumulated in LD (Strable and Ntambi 2010; Rui 2014). The classical DNL pathway is widely documented and can be summarized as follow (Fig. 7.3).

In pro-lipogenic conditions, blood glucose is internalized and pyruvate is produced by the glycolysis. The pyruvate is converted into acetyl-CoA by the ATP-citrate lyase (Rui 2014; Ameer et al. 2014) and converted in malonyl-CoA by the acetyl-CoA carboxylase (ACC). The malonyl-CoA is in turn converted into palmitate (C16:0) by the fatty acid synthase (FAS) from the malonyl and acetyl-CoA (Strable and Ntambi 2010; Rui 2014; Lodhi et al. 2011). Subsequently, the palmitate can be modified by elongation, desaturation or esterification.

Elongation is mainly catalyzed by the ELOVL6 enzyme (Strable and Ntambi 2010). The substrates of ELOVL6 also include saturated and monounsaturated fatty acids from the DNL or from the diet (Moon et al. 2001; Kessler et al. 2014; Matsuzaka and Shimano 2011). The stearoyl CoA desaturase 1 (SCD1) is responsible for the Δ-9 desaturation of fatty acids produced by DNL and coming from the diet generating the palmitoleate (C16:1n-9) and oleate (C18:1n-9). The monounsaturated fatty acids (MUFA) are then preferentially mobilized for the formation of phospholipids, cholesteryl esters and TG (Mauvoisin and Mounier 2011; Strable

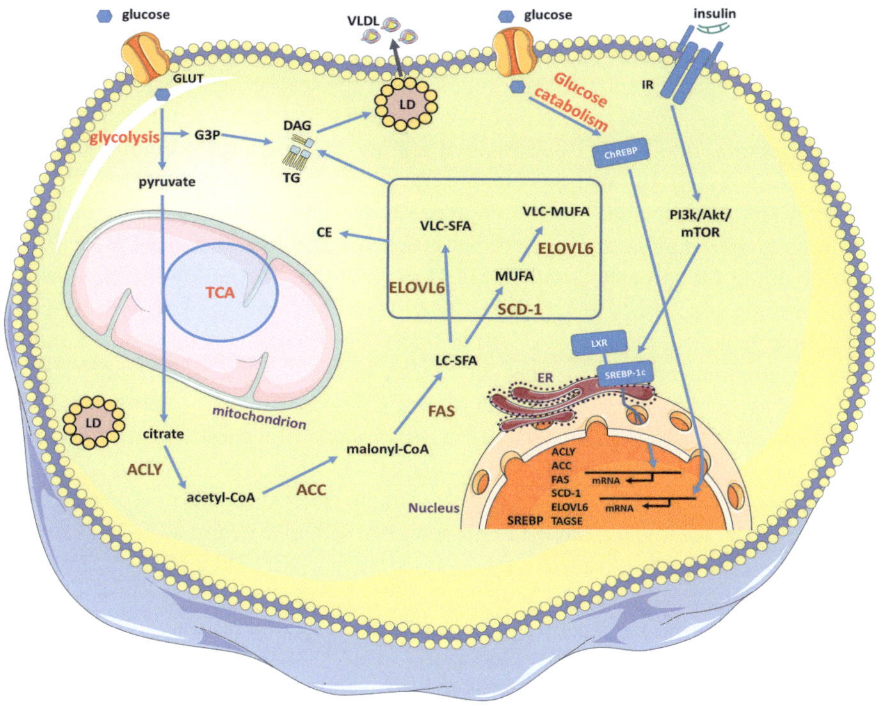

Fig. 7.3 De novo lipogenesis pathway. This schematic representation depicts the lipogenic process and its positive regulators from the glucose absorption to the fatty acids and cholesterol esters production. Insulin and glucose are represented as the two main activators of the process. Insulin activates the PI3k/Akt/mTOR pathway leading to the cleavage of SREBP while glucose serves at a lipogenic precursor and an activator for ChREBP and LXR (Strable and Ntambi 2010; Kersten 2001)

and Ntambi 2010). These syntheses occur in the ER involving GPAT (glycerol-3-phosphate acyltransferase), PAP (phosphatidic acid phosphohydrolase) and DGAT (diacylglycerol acyltransferase) (Nye et al. 2008; Coleman and Mashek 2011).

DNL is considered a great contributor to NAFLD and obesity progression. Patients with NAFLD present a 30 % increase in hepatic TG derived from DNL while 60 % came from the circulating FFA (NEFA) and less than 10 % came from diet (Matsuzaka and Shimano 2011; Donnelly et al. 2005). Many studies have clearly associated the high levels of sugar consumption with activation of DNL and NAFLD development (Moore et al. 2014). Contribution of hepatic DNL to TG secretion was also significantly higher in NAFLD patients than in healthy patients while the inverse was shown for the contribution of NEFA re-esterification (Diraison et al. 2003). Lambert and collaborators recently demonstrated that subjects with high hepatic fat content present higher DNL than those with hepatic low fat content. However, this was only moderately correlated with the hepatic TG content (Lambert et al. 2014).

The DNL pathway is highly regulated in response to body needs. Nutrients and hormones are the major regulators of this pathway. The expressions of the key lipogenic genes expression are mainly regulated at the transcriptional level by LXR (liver X receptor), RXR (retinoid X receptor), SREBP (Sterol regulatory element binding protein) and ChREBP (carbohydrate responsive element binding protein) transcription factors (Strable and Ntambi 2010).

Many studies based on "OMIC" approaches as well as studies using human subject or engineered animals revealed that more than the rate of hepatic DNL is the type of stored FA associated with NAFLD progression, suggesting the implication of the specific genes involved in the DNL pathway. Patients diagnosed with NAFLD have more hepatic endogenous C12:0 to C22:0 lipids (monounsaturated and polyunsaturated) than control patients. In NASH patients, the C18:0/C16:0 ratio is increased and strongly associated with steatosis. This suggests an important role for ELOVL6 in the progression of fatty liver diseases. In parallel, transcriptional expression levels of SCD-1, ELOVL6, SREBP-1C, FAS, and PPARγ were markedly enhanced in NASH group compared to simple steatosis but less associated with the different histological manifestations scores of NAFLD (Yamada et al. 2015).

Elovl6$^{-/-}$ mice are protected from atherogenic HFD-induced liver injuries, inflammation, oxidative stress, fibrosis and NASH progression, while transgenic mice overexpressing human ELOVL6 showed an opposite phenotype (Matsuzaka et al. 2012). In the same study, examination of NASH-human biopsy samples showed a correlation between ELOVL6 expression and liver injury. The authors suggest that ELOVL6 mediates its effect on NASH by promoting the palmitate-induced activation of the NLR family pyrin domain-containing 3 inflammasome. At the opposite, another Elovl6$^{-/-}$ mice model accumulated significantly more hepatic TG than control mice under fat-free/high carbohydrate diet and HFD. In these conditions, the Elovl6$^{-/-}$ mice are not protected against obesity, fatty liver and IR (Moon et al. 2014). The authors also showed that abolishing Elovl6 expression in *ob/ob* mice does not change their phenotype. However, these mice produced more C18:1n7 than controls, suggesting the presence of an ELOVL6 alternative pathway to generate stearic acid. Therefore, the precise role for ELOVL6 in NAFLD and NASH development remains to be determined.

Other lipogenic genes can predispose to NAFLD, namely ACC, FAS and SCD-1 as well as the associated transcription factors SREBP-1c, LXR and ChREBP. Recently, a study showed that hepatic human biopsies obtained from patients with NAFLD showed a significant decreased expression of sirtuins (SIRT) proteins compared to control patients. This is associated with a higher expression of ACC, FAS and SREBP-1c (Wu et al. 2014). Interestingly it was previously demonstrated that SIRT proteins have an anti-NAFLD potential by reducing the expression of lipogenic genes (Yamazaki et al. 2009).

Deletion of ACC1 in mice is lethal while ACC2$^{-/-}$ mice are leaner mice and resistant to HFD-induced obesity and IR. These mice present a lower level of hepatic mitochondrial malonyl-CoA leading to activation of CPT1 and subsequently to β-oxidation (Abu-Elheiga et al. 2003; Postic and Girard 2008). In fact, NASH is characterized by an elevated concentration of malonyl-CoA probably due to an increased activity of ACC. In turn, malonyl-CoA inhibits the carnitine palmitoyl-transferase-1 (CPT-1) leading to a decrease of long chain fatty acids import into the mitochondria, and consequently a reduction in β-oxidation. Consequently, TG accumulate in the liver aggravating the steatosis (Rolo et al. 2012; Vega et al. 2000). Liver specific ACC1 KO mice (LACC1 KO) do not present any health problem under normal diet. However, the decreased contents of hepatic malonyl-CoA and TG are associated with a lower lipogenic yield (twofold) (Mao et al. 2006). Unexpectedly, these mice fed with HFD showed an up-regulation of both FAS and PPARγ but this is still associated with a decrease in DNL and hepatic TG (Mao et al. 2006). In another LACC1 KO mice model, Kohjima and collaborators showed that hepatic deletion of ACC1 activates both expression and activity of the ACC2. This compensatory phenomenon was associated with a null impact on DNL capacity compared to control animals. The authors conclude that silencing both isoforms of ACC is necessary to clearly establish a role for this enzyme in NAFLD (Postic and Girard 2008).

Given its central role in the DNL process, FAS is an ideal target to prevent the NAFLD progression but also the progression of a large spectrum of other metabolic diseases such as obesity, diabetes, hepatocarcinoma, and cardiovascular complications (Berlanga et al. 2014). Studies performed on human NAFLD biopsies (Yamada et al. 2015; Kohjima et al. 2007), on human primary hepatocytes exposed to palmitic acid as well as on murine models of steatohepatitis (Dorn et al. 2010) revealed a marked elevation in FAS expression associated with NAFLD or NAFLD-like conditions. Unexpectedly, liver-specific FASKO mice developed fatty liver and hypoglycemia under a zero-fat diet and this was reversed by addition of dietary fat. A similar observation was made under fasting. This was corrected by addition of PPARα agonist suggesting that FAS is responsible for the synthesis of PPARα ligands (Chakravarthy et al. 2005). Indeed, silencing FAS expression results in a decrease of de novo synthesized LCFA known to be the activating-ligands of PPARα (Chakravarthy et al. 2005; Postic and Girard 2008). Therefore, through the activation of PPARα, FAS can induce β-oxydation decreasing TG accumulation in the liver. However, silencing FAS gene expression could also lead to accumulation of malonyl-CoA, inhibiting CPT1 activity and consequently decreasing β-oxidation. This latter pathway seems to be predominant as most of the published data confirms the relevance of modulating FAS expression to prevent NAFLD progression.

Several reports also demonstrated a marked elevation in hepatic LXR expression associated with NAFLD (Higuchi et al. 2008; Berlanga et al. 2014; Ducheix et al. 2013). Immunostaining of hepatic human biopsies revealed that the rate of positive LXRα expression was 30 % in healthy people, 50 % in NAFLD patients, and 97 % in NASH. LXRα expression was also positively associated with SREBP-1c expression as well as with inflammation, hepatic fibrosis progression and elevated expression of genes involved in fatty acid uptake (Ahn et al. 2014). In agreement with this study, Higuchi and collaborators showed that in hepatic biopsies of NAFLD patients, overexpression of LXR is correlated with higher expression of SREBP-1c (Higuchi et al. 2008). Treatment of diet-induced NAFLD mice with SR9238, a selective LXRα/β inhibitor, abrogated hepatic DNL, lowered inflammation progression and hepatic lipids accumulation. In these conditions, no hepatic damage was observed (Griffett et al. 2013). In another study Sim and collaborators showed that treating mice fed with a HFD with the LXRα antagonist MDGA (meso-dihydroguaiaretic) decreases hepatic lipid accumulation. This is directly correlated with a marked drop in lipogenic genes expression (Sim et al. 2014).

Silencing SREBP-1c in *ob/ob* mice in both liver and adipose tissues is associated with a mark reduction in fatty liver progression, while obesity and IR remained unchanged. This correlated with a reduction in hepatic TG content and a lower expression of lipogenic genes. This suggests an implication of SREBP-1c in fatty liver (and maybe in NAFLD) but not in obesity (Yahagi et al. 2002). However, the knockout in both liver and adipose tissue may influence the data and a specific hepatic knockout may clarify the implication of SREBP-1c in obesity and IR progression. Interestingly, it was recently showed that Dec1 (differentiated embryo chondrocyte expressed gene 1), a negative regulator of SREBP-1c expression, is under expressed in NAFLD while its overexpression led to a significant drop in SREBP-1c and other lipogenic genes such as FAS and ACC decreasing accumulation of hepatic TG (Shen et al. 2014). Another study showed that comparing healthy, simple steatosis and NASH patients does not revealed any significant difference in fatty acid uptake, hepatic lipid oxidation and VLDL secretion. However, NASH patients show a lower hepatic SREBP-1c and lipogenic genes expressions (Nagaya et al. 2010). This suggests that down-regulation of SREBP-1c and lipogenic enzymes are characteristics of a fibrosis state probably mediated by a higher level of TNFα. Therefore, expressions of lipogenic enzymes must be specific for each step of the NAFLD progression spectrum. Recently, a potential role of SREBP-1a in NAFLD was also suggested as SREBP-1a knock-down induced a significant drop in SREBP-1c expression associated with a radical decrease in the expression level of all lipogenic genes (Bitter et al. 2015).

While LXRα, SREBP-1c and lipogenic genes were upregulated in NAFLD-diagnosed patients, ChREBP expression is not modify (Higuchi et al. 2008). However patients with NASH show higher ChREBP expression when steatosis was upper than 50 % and lowered if IR was present (Benhamed et al. 2012). In the same study, it was showed that mice fed with HFD showed elevated expression of ChREBP that can dissociate hepatic steatosis to the IR ameliorating lipid metabolism and glucose metabolism. However, liver-specific inhibition of ChREBP in *ob/ob* mice led to decrease plasma TG and NEFA levels improving steatohepatitis. This is associated with elevated lipogenic rate and enhanced insulin sensitivity in liver, skeletal muscles and white adipose tissue (Dentin et al. 2006).

Taken together, the studies reveal that elevated DNL is associated with increased hepatic TG and NAFLD development. This is probably the result of increased SREBP-1c expression. However, the progression of the disease leading to NASH seems to be associated with a lower DNL probably through the activation of TNFα secretion inhibiting SREBP-1c. However, even if a general role of the DNL is established, a role of each individual enzyme remains to be precisely defined due to presence of compensatory mechanisms.

4 Role of Delta 5 (Δ5) and Delta 6 (Δ6) Desaturases in NAFLD

Livers of obese patients with NAFLD show depletion in the concentration of long-chain poly-unsaturated fatty acids (LCPUFA) especially the n-3 and n-6 series of TG (Araya et al. 2004, 2010). In particular, a strong decrease of arachidonic acid (AA; 20:4, n-6), eicosapentaenoic acid (EPA; 20:5, n-3) and docosahexaenoic acid (DHA; 22:6, n-3) was observed in hepatic steatosis and steatohepatitis (Araya et al. 2004). Depletion of such fatty acids may be responsible for the progression of the disease leading to cirrhosis (Gormaz et al. 2010). Modification in the availability of LCPUFA has also been associated with numerous other metabolic pathologies often associated with NAFLD such as cardio-vascular diseases, obesity, type II diabetes and metabolic syndrome (Gormaz et al. 2010). LCPUFA are formed by series of desaturations and elongations. **Δ**5 and **Δ**6 desaturases are the key desaturases in this process catalyzing the synthesis of n-3 and n-6 LCPUFA (Vessby et al. 2002). The 18:2(*n*-6) and 18:3(*n*-3) are desaturated by the **Δ**6 desaturase to form the 18:3(*n*-6) and 18:4(*n*-3) while the **Δ**5 desaturase will form the AA, EPA and DHA. In this process, the Δ6 desaturase is the rate limiting enzyme (Lenihan-Geels et al. 2013).

If the role of SCD1 in the development of NASH has been clearly established, the roles of the **Δ**5 and **Δ**6 desaturases appear more complex. In a study published in 2005, Tovar and collaborators showed that feeding hyperinsulinemic obese Zucker *fa/fa* rats with Soy protein decreases liver steatosis and lipotoxicity. This is associated with a potent inhibition of both **Δ**5 and **Δ**6 mRNA levels (40 % and 69 % respectively) (Tovar et al. 2005). However liver of mice fed with methionine and choline deficient diet (MCD), a nutritional model of steatohepatitis displays a significant increase in **Δ**5 and **Δ**6 desaturases mRNA (Larter et al. 2008). In these studies, the type of hepatic LCPUFA was not evaluated. Recently, a study performed in liver biopsies of NASH patients showed a clear increased in both **Δ**5 and **Δ**6 desaturases mRNA expression. This was associated with an impaired desaturation of ω3 and ω6 fatty acids, with an increase in the ω6:ω3 ratio and a reduction in the ω3 desaturation index (Lopez-Vicario et al. 2014). Several other studies performed in obese NAFLD patients also revealed a decrease in **Δ**5 and **Δ**6 desaturases activities and an increase in the ω6:ω3 ratio (Araya et al. 2004, 2010; Kotronen et al. 2009). Interestingly, it was recently showed that the decrease in Δ6 desaturase expression is associated with an up-regulation of miR-140, and miR2885, two miRNA previously associated with hepatic disorders and NAFLD (Fatima et al. 2014).

The molecular mechanisms of LCPUFA depletion on NAFLD development can be explained by the effect of these fatty acids on the transcription factors involved in lipogenesis and β-oxydation (Pettinelli et al. 2009). LCPUFA inhibits the expression of both ChREBP (Dentin et al. 2005) and SREBP-1c (Hannah et al. 2001), two transcription factors known to activate lipogenic genes expression in response to sugar and insulin respectively. At the opposite, LCPUFA activate PPARα promoting β-oxydation (Nakamura and Nara 2004).

At the difference of the previous published reports, a study performed in 1100 non-obese elderly men showed that an increased in Δ6 desaturase index is positively associated with the plasmatic concentration of alanine aminotransferase (ALT), a marker of liver injury. Even if in this study, the authors did not directly measured the presence of lipid in the liver, they observed a positive correlation between plasmatic fatty acid composition known to be associated with fat liver, and Δ6 desaturase activity (Petersson et al. 2010). The authors explained the differences between their observations and those made by the other groups by the fact that their study was performed in non-obese elderly men (BMI: 26.3 ± 3.4) compared to the other studies performed with severely obese subject (BMI around 50) (Araya et al. 2004, 2010; Kotronen et al. 2009). According to the authors, the decrease in Δ6 desaturase activity observed by the other studies may be a consequence of long term obesity and hyperinsulinemia. They also argue that the low number of patients in the other studies (10 in average *vs.* 1100 subjects in the present study) may also influence the results observed by other groups. Finally, the authors mentioned that their study was performed on Swedish patients while most of the other studies were performed on patients from Chile (Araya et al. 2004, 2010). According to them, the differences observed may be the result of different genetic background and dietary habits. However, a study performed with Finn patients that have probably similar life style habits than patients from Sweden, also showed a decrease in the ω6:ω3 ratio (Kotronen et al. 2009).

Despite the differences observed in the latest described study, a consensus seems to emerge at least in obese patients, on an increase in the ω6:ω3 ratio associated with the development of NAFLD and the metabolic disorders. This prompts the researchers to propose a dietary ω3 supplementation to reduce the hepatic TG content and the associated liver injury (Capanni et al. 2006; Spadaro et al. 2008). However, a recently published trial performed on 103 obese patients treated with EPA and DHA for 15–18 months showed a decrease in fat liver content but no effect on liver fibrosis biomarker scores. At the opposite, a study performed in children shows that DHA supplementation improved liver steatosis and insulin sensitivity (Nobili et al. 2013) while lack of ω3 consumption increase lobular inflammation (St-Jules et al. 2013).

Taken together, in obese patients with NAFLD, LCPUFA are depleted and the ω6:ω3 ratio elevated. This results in hepatic AA accumulation, activation of DNL and inhibition of β-oxydation. This increase in the ω6:ω3 ratio probably further aggravates the hepatic pathology increasing the inflammation and the fatty acid deposit. This could be ameliorates by ω3 supplementation however the time and dose for the treatment may be adjusted. Difference in patients should also be taken in consideration.

5 Role of the Stearoyl CoA Desaturase 1 in NAFLD

Numerous studies have associated an elevated stearoyl-CoA desaturase 1 (SCD1) activity with the pathophysiology of fatty liver disease in both mice and humans. SCD1, also known as fatty acid desaturase or Δ9-desaturase, is a microsomal enzyme involved in the biosynthesis of monounsaturated fatty acids (MUFA), primarily oleate (C18:1) and palmitoleate (C16:1). These MUFA are the major substrates for the synthesis of complex lipids such as DAG, phospholipids, TG, wax esters, and cholesterol esters (Sampath and Ntambi 2006; Nakamura and Nara 2004; Mauvoisin and Mounier 2011). SCD1 is now considered as one of the major enzymes in the control of lipid metabolism (Flowers and Ntambi 2008).

Mice with naturally occurring SCD1 null mutations and those with global deletion ($SCD1^{-/-}$) display a hyper metabolic phenotype that protects the animal from obesity, IR and hepatic steatosis under high-carbohydrate diet or HFD (Flowers and Ntambi 2008; Miyazaki et al. 2001, 2004; Ntambi et al. 2002). The $SCD1^{-/-}$ mice also display reduced hepatic TG and cholesterol esters. This is associated with up-regulation of the carnitine palmitoyltransferase 1 (CPT1), the rate limiting enzyme of lipid β-oxydation while the genes encoding for the enzymes implicated in lipid synthesis (FAS and glycerol phosphate acyl-CoA transferase (*GPAT*)) are reduced (Ntambi et al. 2002). Liver specific SCD1 knockout mice (LKO mice) are also protected from carbohydrate but not from HFD-induced hepatic steatosis (Miyazaki et al. 2007). Elevated hepatic SCD1 expression appears therefore associated with carbohydrate-induced hepatic steatosis while an extra-hepatic expression seems necessary to promote HFD-induced steatosis. However, mice intraperitoneally injected with SCD1 targeted antisense oligonucleotide (ASO) are protected from development of NAFLD under HFD (Brown et al. 2008). The discrepancy between the two studies can be explained by the fact that ASO-injected mice showed inhibition of SCD1 expression both in liver and adipose tissue while in LKO mice SCD1 is only inhibited in liver (Flowers and Ntambi 2008). In addition, LKO mice display reduced nuclear content of ChREBP and SREBP-1c, two key transcription factors regulating the expression of lipogenic genes in response to glucose and insulin respectively. The exact mechanism by which SCD1 affects the maturation or the translocation of these two transcriptional factors remains however to be identified. A possible explanation could be that decreased in oleate concentration in LKO mice is responsible for the inhibition of ChREBP and SREBP-1c expressions as oleate supplementation normalized their levels (Miyazaki et al. 2007).

In human studies, the role of SCD1 in hepatic fat metabolism has been mainly evaluated by the measure of the both hepatic (Kotronen et al. 2009; Silbernagel et al. 2012) and plasmatic (Stefan et al. 2008; Petersson et al. 2010; Lee et al. 2015) 18:1 n-9/18:0 ratio referred as the desaturase index. The hepatic desaturase index in total lipids from individuals with or without NAFLD was found to be positively correlated with the percentage of hepatic fat (Kotronen et al. 2009). In this latter study, the authors state that fatty liver is characterized by an increase in hepatic SCD1 activity. In contrast to this data, another study showed that the desaturase index in

VLDL-TG was negatively correlated with the hepatic fat content particularly in obese patients while no apparent correlation was observed in lean patients (Stefan et al. 2008). The authors suggested that the liver specific down regulation of SCD1 may impair VLDL assembly and subsequently may reduce the capacity for the liver to clear intra-hepatic TG. The authors stated that the differences between their observations and those made in mice could be explained by the fact that $SCD1^{-/-}$ mice present a lean and not an obese phenotype. In addition, the desaturation index in VLDL-TG can reflect both hepatic and fat SCD1 desaturation indexes while most of the studies performed in mice have directly measured the hepatic SCD1 index. At the opposite, Peters and collaborators showed that in 50 non-obese Caucasians, the SCD1 index correlates with the hepatic TG content while no correlation was observed with the PL fraction. Interestingly, in the same study, the authors did not find any correlation with the hepatic SCD1 mRNA level showing the importance of measuring specific FA fraction to evaluate SCD1 activity (Peter et al. 2011). In another study also performed in non-obese healthy patients, the basal SCD1 activity index does not appeared to correlate with the hepatic fat content (Silbernagel et al. 2012). The authors also showed that addition of a sugar-enriched lipogenic diet increases the hepatic fat content that is negatively correlated with the hepatic SCD1 activity (Silbernagel et al. 2012). The authors suggest that SCD1 may protect from the adverse effects of a lipogenic enriched sugar diet. Interestingly, the authors proposed the hypothesis that low hepatic SCD1 activity observed in sugar-enriched lipogenic diet, can be the consequence of long-chain saturated fatty acids accumulation such as palmitate, leading to ER stress (Cao et al. 2012). This in turn inhibits VLDL secretion leading to hepatic TG accumulation and steatosis (Caviglia et al. 2011). Adding to the complexity, a recent study showed that in obese patients, the 16:1 n-7/16:0 desaturation index correlates with the hepatic fat content but not with the 18:1 n-9/18:0 index (Lee et al. 2015). Finally, Li and collaborators investigated the role of hepatic SCD1 in the prevention of hepatic inflammation, apoptosis and fibrosis in a model of steatosis using both in vitro (murine and human hepatocytes) and in vivo models (Li et al. 2009). Their findings suggest that up-regulation of hepatic SCD1 is a crucial adaptive mechanism in the prevention of liver damage and hepatitis by decreasing the amount of SFA in hepatocytes.

Taken together, all this evidences point for a central role of SCD1 in the pathogenesis of NAFLD. SCD1 activity is indeed necessary for the generation of hepatic TG, the main lipids stored in LD, generating the initial step of steatosis. However the association between the hepatic level of SCD1 activity and the TG content appears to depend on several conditions such as the presence or not of obesity (low or high FA flux), the site of SCD1 expression (liver *vs.* adipose tissue), the type of fatty acids examined (C16 *vs.* C18) and finally the type of diet used in the studies. At the opposite, a consensus emerged that high SCD1 activity may protect liver from lipotoxicity by decreasing the pool of hepatic SFA and increasing of the pool of MUFA.

6 Conclusion

NAFLD is characterized by the accumulation of TG in hepatic LD. The disease can range from a simple steatosis to a NASH characterized by the apparition of hepatic injury, inflammation, and eventually fibrosis. To date evidences point for a central role of hepatic DNL in the establishment of NAFLD especially in the case of energy excess as observed in obesity. In these conditions, a large part of hepatic TG comes from DNL. This is associated with an elevated expression of the lipogenic enzymes and the associated transcription factor SREBP-1c. However, when the disease progress, TNFα concentration raises leading to increased inflammation. Consequently, TNFα down-regulates SREBP-1c expression leading to transcriptional inhibition of the lipogenic genes. FAS is also responsible for the synthesis of LCFA. These fatty acids are the direct activators of PPARα activating β-oxidation while high SCD1 activity decreases the pool of LCFA such as palmitate, a fatty acid inducing inflammation. Therefore, apparition of NASH associated with elevated TNFα will inhibit FAS and SCD1 activities aggravating the disease by decreasing β-oxidation and increasing palmitate concentration. Taken together, activation of DNL appears implicated in the establishment of NAFLD while DNL inhibition will be more associated with the progression of the disease and the apparition of NASH. Finally, the roles of Δ5 and Δ6 desaturases appear more controversial. However, a consensus seems to be established that in obese patients, the ω6:ω3 ratio raises associated with an elevated concentration of AA, a ω6 FA implicated in the development of inflammation. Consequently, the NAFLD progresses in NASH and eventually in cirrhosis. In conclusion, treatment of NAFLD could be achieved by inhibition of DNL while for NASH, a decrease in the ω6:ω3 ratio could be a better target that can be achieved by ω3 supplementation.

References

Abu-Elheiga L, Oh W, Kordari P, Wakil SJ (2003) Acetyl-CoA carboxylase 2 mutant mice are protected against obesity and diabetes induced by high-fat/high-carbohydrate diets. Proc Natl Acad Sci U S A 100:10207–10212

Ahn SB, Jang K, Jun DW, Lee BH, Shin KJ (2014) Expression of liver X receptor correlates with intrahepatic inflammation and fibrosis in patients with nonalcoholic fatty liver disease. Dig Dis Sci 59:2975–2982

Ameer F, Scandiuzzi L, Hasnain S, Kalbacher H, Zaidi N (2014) De novo lipogenesis in health and disease. Metabolism 63:895–902

Angulo P (2002) Nonalcoholic fatty liver disease. N Engl J Med 346:1221–1231

Araya J, Rodrigo R, Videla LA, Thielemann L, Orellana M, Pettinelli P, Poniachik J (2004) Increase in long-chain polyunsaturated fatty acid n-6/n-3 ratio in relation to hepatic steatosis in patients with non-alcoholic fatty liver disease. Clin Sci (Lond) 106:635–643

Araya J, Rodrigo R, Pettinelli P, Araya AV, Poniachik J, Videla LA (2010) Decreased liver fatty acid delta-6 and delta-5 desaturase activity in obese patients. Obesity (Silver Spring) 18:1460–1463

Beller M, Bulankina AV, Hsiao HH, Urlaub H, Jackle H, Kuhnlein RP (2010) PERILIPIN-dependent control of lipid droplet structure and fat storage in Drosophila. Cell Metab 12:521–532

Benhamed F, Denechaud PD, Lemoine M, Robichon C, Moldes M, Bertrand-Michel J, Ratziu V, Serfaty L, Housset C, Capeau J, Girard J, Guillou H, Postic C (2012) The lipogenic transcription factor ChREBP dissociates hepatic steatosis from insulin resistance in mice and humans. J Clin Invest 122:2176–2194

Berlanga A, Guiu-Jurado E, Porras JA, Auguet T (2014) Molecular pathways in non-alcoholic fatty liver disease. Clin Exp Gastroenterol 7:221–239

Bitter A, Nussler AK, Thasler WE, Klein K, Zanger UM, Schwab M, Burk O (2015) Human sterol regulatory element-binding protein 1a contributes significantly to hepatic lipogenic gene expression. Cell Physiol Biochem 35:803–815

Boden G, Song W, Duan X, Cheung P, Kresge K, Barrero C, Merali S (2011) Infusion of glucose and lipids at physiological rates causes acute endoplasmic reticulum stress in rat liver. Obesity (Silver Spring) 19:1366–1373

Brasaemle DL, Wolins NE (2012) Packaging of fat: an evolving model of lipid droplet assembly and expansion. J Biol Chem 287:2273–2279

Brown JM, Chung S, Sawyer JK, Degirolamo C, Alger HM, Nguyen T, Zhu X, Duong MN, Wibley AL, Shah R, Davis MA, Kelley K, Wilson MD, Kent C, Parks JS, Rudel LL (2008) Inhibition of stearoyl-coenzyme A desaturase 1 dissociates insulin resistance and obesity from atherosclerosis. Circulation 118:1467–1475

Buhman KK, Chen HC, Farese RV Jr (2001) The enzymes of neutral lipid synthesis. J Biol Chem 276:40369–40372

Cao J, Dai DL, Yao L, Yu HH, Ning B, Zhang Q, Chen J, Cheng WH, Shen W, Yang ZX (2012) Saturated fatty acid induction of endoplasmic reticulum stress and apoptosis in human liver cells via the PERK/ATF4/CHOP signaling pathway. Mol Cell Biochem 364:115–129

Capanni M, Calella F, Biagini MR, Genise S, Raimondi L, Bedogni G, Svegliati-Baroni G, Sofi F, Milani S, Abbate R, Surrenti C, Casini A (2006) Prolonged n-3 polyunsaturated fatty acid supplementation ameliorates hepatic steatosis in patients with non-alcoholic fatty liver disease: a pilot study. Aliment Pharmacol Ther 23:1143–1151

Caviglia JM, Gayet C, Ota T, Hernandez-Ono A, Conlon DM, Jiang H, Fisher EA, Ginsberg HN (2011) Different fatty acids inhibit apoB100 secretion by different pathways: unique roles for ER stress, ceramide, and autophagy. J Lipid Res 52:1636–1651

Chakravarthy MV, Pan Z, Zhu Y, Tordjman K, Schneider JG, Coleman T, Turk J, Semenkovich CF (2005) "New" hepatic fat activates PPARalpha to maintain glucose, lipid, and cholesterol homeostasis. Cell Metab 1:309–322

Chavez JA, Summers SA (2012) A ceramide-centric view of insulin resistance. Cell Metab 15:585–594

Choi CS, Savage DB, Kulkarni A, Yu XX, Liu ZX, Morino K, Kim S, Distefano A, Samuel VT, Neschen S, Zhang D, Wang A, Zhang XM, Kahn M, Cline GW, Pandey SK, Geisler JG, Bhanot S, Monia BP, Shulman GI (2007) Suppression of diacylglycerol acyltransferase-2 (DGAT2), but not DGAT1, with antisense oligonucleotides reverses diet-induced hepatic steatosis and insulin resistance. J Biol Chem 282:22678–22688

Cohen JC, Horton JD, Hobbs HH (2011) Human fatty liver disease: old questions and new insights. Science 332:1519–1523

Coleman RA, Mashek DG (2011) Mammalian triacylglycerol metabolism: synthesis, lipolysis, and signaling. Chem Rev 111:6359–6386

Crespo J, Cayon A, Fernandez-Gil P, Hernandez-Guerra M, Mayorga M, Dominguez-Diez A, Fernandez-Escalante JC, Pons-Romero F (2001) Gene expression of tumor necrosis factor alpha and TNF-receptors, p55 and p75, in nonalcoholic steatohepatitis patients. Hepatology 34:1158–1163

Day CP, James OF (1998) Steatohepatitis: a tale of two "hits"? Gastroenterology 114:842–845

Dentin R, Benhamed F, Pegorier JP, Foufelle F, Viollet B, Vaulont S, Girard J, Postic C (2005) Polyunsaturated fatty acids suppress glycolytic and lipogenic genes through the inhibition of ChREBP nuclear protein translocation. J Clin Invest 115:2843–2854

Dentin R, Benhamed F, Hainault I, Fauveau V, Foufelle F, Dyck JR, Girard J, Postic C (2006) Liver-specific inhibition of ChREBP improves hepatic steatosis and insulin resistance in ob/ob mice. Diabetes 55:2159–2170

Diraison F, Moulin P, Beylot M (2003) Contribution of hepatic de novo lipogenesis and reesterification of plasma non esterified fatty acids to plasma triglyceride synthesis during non-alcoholic fatty liver disease. Diabetes Metab 29:478–485

Donnelly KL, Smith CI, Schwarzenberg SJ, Jessurun J, Boldt MD, Parks EJ (2005) Sources of fatty acids stored in liver and secreted via lipoproteins in patients with nonalcoholic fatty liver disease. J Clin Invest 115:1343–1351

Dorn C, Riener MO, Kirovski G, Saugspier M, Steib K, Weiss TS, Gabele E, Kristiansen G, Hartmann A, Hellerbrand C (2010) Expression of fatty acid synthase in nonalcoholic fatty liver disease. Int J Clin Exp Pathol 3:505–514

Ducheix S, Montagner A, Theodorou V, Ferrier L, Guillou H (2013) The liver X receptor: a master regulator of the gut-liver axis and a target for non alcoholic fatty liver disease. Biochem Pharmacol 86:96–105

Endo M, Masaki T, Seike M, Yoshimatsu H (2007) TNF-alpha induces hepatic steatosis in mice by enhancing gene expression of sterol regulatory element binding protein-1c (SREBP-1c). Exp Biol Med (Maywood) 232:614–621

Fabbrini E, Sullivan S, Klein S (2010) Obesity and nonalcoholic fatty liver disease: biochemical, metabolic, and clinical implications. Hepatology 51:679–689

Farese RV Jr, Walther TC (2009) Lipid droplets finally get a little R-E-S-P-E-C.-T. Cell 139:855–860

Fatima A, Waters S, O'Boyle P, Seoighe C, Morris DG (2014) Alterations in hepatic miRNA expression during negative energy balance in postpartum dairy cattle. BMC Genomics 15:28

Feng B, Yao PM, Li Y, Devlin CM, Zhang D, Harding HP, Sweeney M, Rong JX, Kuriakose G, Fisher EA, Marks AR, Ron D, Tabas I (2003) The endoplasmic reticulum is the site of cholesterol-induced cytotoxicity in macrophages. Nat Cell Biol 5:781–792

Flowers MT, Ntambi JM (2008) Role of stearoyl-coenzyme A desaturase in regulating lipid metabolism. Curr Opin Lipidol 19:248–256

Frangioudakis G, Burchfield JG, Narasimhan S, Cooney GJ, Leitges M, Biden TJ, Schmitz-Peiffer C (2009) Diverse roles for protein kinase C delta and protein kinase C epsilon in the generation of high-fat-diet-induced glucose intolerance in mice: regulation of lipogenesis by protein kinase C delta. Diabetologia 52:2616–2620

Fromenty B, Robin MA, Igoudjil A, Mansouri A, Pessayre D (2004) The ins and outs of mitochondrial dysfunction in NASH. Diabetes Metab 30:121–138

Fu S, Watkins SM, Hotamisligil GS (2012) The role of endoplasmic reticulum in hepatic lipid homeostasis and stress signaling. Cell Metab 15:623–634

Fujii H, Ikura Y, Arimoto J, Sugioka K, Iezzoni JC, Park SH, Naruko T, Itabe H, Kawada N, Caldwell SH, Ueda M (2009) Expression of perilipin and adipophilin in nonalcoholic fatty liver disease; relevance to oxidative injury and hepatocyte ballooning. J Atheroscler Thromb 16:893–901

Galbo T, Perry RJ, Jurczak MJ, Camporez JP, Alves TC, Kahn M, Guigni BA, Serr J, Zhang D, Bhanot S, Samuel VT, Shulman GI (2013) Saturated and unsaturated fat induce hepatic insulin resistance independently of TLR-4 signaling and ceramide synthesis in vivo. Proc Natl Acad Sci U S A 110:12780–12785

Gormaz JG, Rodrigo R, Videla LA, Beems M (2010) Biosynthesis and bioavailability of long-chain polyunsaturated fatty acids in non-alcoholic fatty liver disease. Prog Lipid Res 49:407–419

Griffett K, Solt LA, El-Gendy Bel D, Kamenecka TM, Burris TP (2013) A liver-selective LXR inverse agonist that suppresses hepatic steatosis. ACS Chem Biol 8:559–567

Guo Y, Cordes KR, Farese RV Jr, Walther TC (2009) Lipid droplets at a glance. J Cell Sci 122:749–752

Hannah VC, Ou J, Luong A, Goldstein JL, Brown MS (2001) Unsaturated fatty acids down-regulate srebp isoforms 1a and 1c by two mechanisms in HEK-293 cells. J Biol Chem 276:4365–4372

Higuchi N, Kato M, Shundo Y, Tajiri H, Tanaka M, Yamashita N, Kohjima M, Kotoh K, Nakamuta M, Takayanagi R, Enjoji M (2008) Liver X receptor in cooperation with SREBP-1c is a major lipid synthesis regulator in nonalcoholic fatty liver disease. Hepatol Res 38:1122–1129

Hooper AJ, Adams LA, Burnett JR (2011) Genetic determinants of hepatic steatosis in man. J Lipid Res 52:593–617

Inoue M, Ohtake T, Motomura W, Takahashi N, Hosoki Y, Miyoshi S, Suzuki Y, Saito H, Kohgo Y, Okumura T (2005) Increased expression of PPARgamma in high fat diet-induced liver steatosis in mice. Biochem Biophys Res Commun 336:215–222

Jacquier N, Choudhary V, Mari M, Toulmay A, Reggiori F, Schneiter R (2011) Lipid droplets are functionally connected to the endoplasmic reticulum in Saccharomyces cerevisiae. J Cell Sci 124:2424–2437

Jacquier N, Mishra S, Choudhary V, Schneiter R (2013) Expression of oleosin and perilipins in yeast promotes formation of lipid droplets from the endoplasmic reticulum. J Cell Sci 126:5198–5209

Kantartzis K, Machicao F, Machann J, Schick F, Fritsche A, Haring HU, Stefan N (2009) The DGAT2 gene is a candidate for the dissociation between fatty liver and insulin resistance in humans. Clin Sci (Lond) 116:531–537

Kanuri G, Bergheim I (2013) In vitro and in vivo models of non-alcoholic fatty liver disease (NAFLD). Int J Mol Sci 14:11963–11980

Kaplowitz N, Ji C (2006) Unfolding new mechanisms of alcoholic liver disease in the endoplasmic reticulum. J Gastroenterol Hepatol 21(Suppl 3):S7–S9

Kersten S (2001) Mechanisms of nutritional and hormonal regulation of lipogenesis. EMBO Rep 2:282–286

Kessler SM, Simon Y, Gemperlein K, Gianmoena K, Cadenas C, Zimmer V, Pokorny J, Barghash A, Helms V, VAN Rooijen N, Bohle RM, Lammert F, Hengstler JG, Mueller R, Haybaeck J, Kiemer AK (2014) Fatty acid elongation in non-alcoholic steatohepatitis and hepatocellular carcinoma. Int J Mol Sci 15:5762–5773

Kohjima M, Enjoji M, Higuchi N, Kato M, Kotoh K, Yoshimoto T, Fujino T, Yada M, Yada R, Harada N, Takayanagi R, Nakamuta M (2007) Re-evaluation of fatty acid metabolism-related gene expression in nonalcoholic fatty liver disease. Int J Mol Med 20:351–358

Kohlwein SD, Veenhuis M, van der Klei IJ (2013) Lipid droplets and peroxisomes: key players in cellular lipid homeostasis or a matter of fat—store 'em up or burn 'em down. Genetics 193:1–50

Kotronen A, Seppanen-Laakso T, Westerbacka J, Kiviluoto T, Arola J, Ruskeepaa AL, Oresic M, Yki-Jarvinen H (2009) Hepatic stearoyl-CoA desaturase (SCD)-1 activity and diacylglycerol but not ceramide concentrations are increased in the nonalcoholic human fatty liver. Diabetes 58:203–208

Lambert JE, Ramos-Roman MA, Browning JD, Parks EJ (2014) Increased de novo lipogenesis is a distinct characteristic of individuals with nonalcoholic fatty liver disease. Gastroenterology 146:726–735

Larter CZ, Yeh MM, Williams J, Bell-Anderson KS, Farrell GC (2008) MCD-induced steatohepatitis is associated with hepatic adiponectin resistance and adipogenic transformation of hepatocytes. J Hepatol 49:407–416

Lee JJ, Lambert JE, Hovhannisyan Y, Ramos-Roman MA, Trombold JR, Wagner DA, Parks EJ (2015) Palmitoleic acid is elevated in fatty liver disease and reflects hepatic lipogenesis. Am J Clin Nutr 101:34–43

Lenihan-Geels G, Bishop KS, Ferguson LR (2013) Alternative sources of omega-3 fats: can we find a sustainable substitute for fish? Nutrients 5:1301–1315

Li Z, Diehl AM (2003) Innate immunity in the liver. Curr Opin Gastroenterol 19:565–571
Li ZZ, Berk M, McIntyre TM, Feldstein AE (2009) Hepatic lipid partitioning and liver damage in nonalcoholic fatty liver disease: role of stearoyl-CoA desaturase. J Biol Chem 284:5637–5644
Listenberger LL, Han X, Lewis SE, Cases S, Farese RV Jr, Ory DS, Schaffer JE (2003) Triglyceride accumulation protects against fatty acid-induced lipotoxicity. Proc Natl Acad Sci U S A 100:3077–3082
Lodhi IJ, Wei X, Semenkovich CF (2011) Lipoexpediency: de novo lipogenesis as a metabolic signal transmitter. Trends Endocrinol Metab 22:1–8
Long AP, Manneschmidt AK, Verbrugge B, Dortch MR, Minkin SC, Prater KE, Biggerstaff JP, Dunlap JR, Dalhaimer P (2012) Lipid droplet de novo formation and fission are linked to the cell cycle in fission yeast. Traffic 13:705–714
Lopez-Vicario C, Gonzalez-Periz A, Rius B, Moran-Salvador E, Garcia-Alonso V, Lozano JJ, Bataller R, Cofan M, Kang JX, Arroyo V, Claria J, Titos E (2014) Molecular interplay between Delta5/Delta6 desaturases and long-chain fatty acids in the pathogenesis of non-alcoholic steatohepatitis. Gut 63:344–355
Mao J, Demayo FJ, Li H, Abu-Elheiga L, Gu Z, Shaikenov TE, Kordari P, Chirala SS, Heird WC, Wakil SJ (2006) Liver-specific deletion of acetyl-CoA carboxylase 1 reduces hepatic triglyceride accumulation without affecting glucose homeostasis. Proc Natl Acad Sci U S A 103:8552–8557
Martin S, Parton RG (2006) Lipid droplets: a unified view of a dynamic organelle. Nat Rev Mol Cell Biol 7:373–378
Matsusue K, Kusakabe T, Noguchi T, Takiguchi S, Suzuki T, Yamano S, Gonzalez FJ (2008) Hepatic steatosis in leptin-deficient mice is promoted by the PPARgamma target gene Fsp27. Cell Metab 7:302–311
Matsuzaka T, Shimano H (2011) Molecular mechanisms involved in hepatic steatosis and insulin resistance. J Diabetes Investig 2:170–175
Matsuzaka T, Atsumi A, Matsumori R, Nie T, Shinozaki H, Suzuki-Kemuriyama N, Kuba M, Nakagawa Y, Ishii K, Shimada M, Kobayashi K, Yatoh S, Takahashi A, Takekoshi K, Sone H, Yahagi N, Suzuki H, Murata S, Nakamuta M, Yamada N, Shimano H (2012) Elovl6 promotes nonalcoholic steatohepatitis. Hepatology 56:2199–2208
Mauvoisin D, Mounier C (2011) Hormonal and nutritional regulation of SCD1 gene expression. Biochimie 93:78–86
Milic S, Stimac D (2012) Nonalcoholic fatty liver disease/steatohepatitis: epidemiology, pathogenesis, clinical presentation and treatment. Dig Dis 30:158–162
Miyazaki M, Kim YC, Ntambi JM (2001) A lipogenic diet in mice with a disruption of the stearoyl-CoA desaturase 1 gene reveals a stringent requirement of endogenous monounsaturated fatty acids for triglyceride synthesis. J Lipid Res 42:1018–1024
Miyazaki M, Dobrzyn A, Man WC, Chu K, Sampath H, Kim HJ, Ntambi JM (2004) Stearoyl-CoA desaturase 1 gene expression is necessary for fructose-mediated induction of lipogenic gene expression by sterol regulatory element-binding protein-1c-dependent and -independent mechanisms. J Biol Chem 279:25164–25171
Miyazaki M, Flowers MT, Sampath H, Chu K, Otzelberger C, Liu X, Ntambi JM (2007) Hepatic stearoyl-CoA desaturase-1 deficiency protects mice from carbohydrate-induced adiposity and hepatic steatosis. Cell Metab 6:484–496
Moon YA, Shah NA, Mohapatra S, Warrington JA, Horton JD (2001) Identification of a mammalian long chain fatty acyl elongase regulated by sterol regulatory element-binding proteins. J Biol Chem 276:45358–45366
Moon YA, Ochoa CR, Mitsche MA, Hammer RE, Horton JD (2014) Deletion of ELOVL6 blocks the synthesis of oleic acid but does not prevent the development of fatty liver or insulin resistance. J Lipid Res 55:2597–2605
Moore JB, Gunn PJ, Fielding BA (2014) The role of dietary sugars and de novo lipogenesis in non-alcoholic fatty liver disease. Nutrients 6:5679–5703

Motomura W, Inoue M, Ohtake T, Takahashi N, Nagamine M, Tanno S, Kohgo Y, Okumura T (2006) Up-regulation of ADRP in fatty liver in human and liver steatosis in mice fed with high fat diet. Biochem Biophys Res Commun 340:1111–1118

Musso G, Gambino R, Cassader M, Pagano G (2011) Meta-analysis: natural history of non-alcoholic fatty liver disease (NAFLD) and diagnostic accuracy of non-invasive tests for liver disease severity. Ann Med 43:617–649

Nagaya T, Tanaka N, Suzuki T, Sano K, Horiuchi A, Komatsu M, Nakajima T, Nishizawa T, Joshita S, Umemura T, Ichijo T, Matsumoto A, Yoshizawa K, Nakayama J, Tanaka E, Aoyama T (2010) Down-regulation of SREBP-1c is associated with the development of burned-out NASH. J Hepatol 53:724–731

Nakamura MT, Nara TY (2004) Structure, function, and dietary regulation of delta6, delta5, and delta9 desaturases. Annu Rev Nutr 24:345–376

Neuschwander-Tetri BA, Caldwell SH (2003) Nonalcoholic steatohepatitis: summary of an AASLD Single Topic Conference. Hepatology 37:1202–1219

Nobili V, Alisi A, Della Corte C, Rise P, Galli C, Agostoni C, Bedogni G (2013) Docosahexaenoic acid for the treatment of fatty liver: randomised controlled trial in children. Nutr Metab Cardiovasc Dis 23:1066–1070

Ntambi JM, Miyazaki M, Stoehr JP, Lan H, Kendziorski CM, Yandell BS, Song Y, Cohen P, Friedman JM, Attie AD (2002) Loss of stearoyl-CoA desaturase-1 function protects mice against adiposity. Proc Natl Acad Sci U S A 99:11482–11486

Nye CK, Hanson RW, Kalhan SC (2008) Glyceroneogenesis is the dominant pathway for triglyceride glycerol synthesis in vivo in the rat. J Biol Chem 283:27565–27574

Odegaard JI, Ricardo-Gonzalez RR, Red Eagle A, Vats D, Morel CR, Goforth MH, Subramanian V, Mukundan L, Ferrante AW, Chawla A (2008) Alternative M2 activation of Kupffer cells by PPARdelta ameliorates obesity-induced insulin resistance. Cell Metab 7:496–507

Peter A, Cegan A, Wagner S, Elcnerova M, Konigsrainer A, Konigsrainer I, Haring HU, Schleicher ED, Stefan N (2011) Relationships between hepatic stearoyl-CoA desaturase-1 activity and mRNA expression with liver fat content in humans. Am J Physiol Endocrinol Metab 300:E321–E326

Petersson H, Arnlov J, Zethelius B, Riserus U (2010) Serum fatty acid composition and insulin resistance are independently associated with liver fat markers in elderly men. Diabetes Res Clin Pract 87:379–384

Pettinelli P, del Pozo T, Araya J, Rodrigo R, Araya AV, Smok G, Csendes A, Gutierrez L, Rojas J, Korn O, Maluenda F, Diaz JC, Rencoret G, Braghetto I, Castillo J, Poniachik J, Videla LA (2009) Enhancement in liver SREBP-1c/PPAR-alpha ratio and steatosis in obese patients: correlations with insulin resistance and n-3 long-chain polyunsaturated fatty acid depletion. Biochim Biophys Acta 1792:1080–1086

Postic C, Girard J (2008) Contribution of de novo fatty acid synthesis to hepatic steatosis and insulin resistance: lessons from genetically engineered mice. J Clin Invest 118:829–838

Puri P, Mirshahi F, Cheung O, Natarajan R, Maher JW, Kellum JM, Sanyal AJ (2008) Activation and dysregulation of the unfolded protein response in nonalcoholic fatty liver disease. Gastroenterology 134:568–576

Robertson G, Leclercq I, Farrell GC (2001) Nonalcoholic steatosis and steatohepatitis. II. Cytochrome P-450 enzymes and oxidative stress. Am J Physiol Gastrointest Liver Physiol 281:G1135–G1139

Rolo AP, Teodoro JS, Palmeira CM (2012) Role of oxidative stress in the pathogenesis of nonalcoholic steatohepatitis. Free Radic Biol Med 52:59–69

Rui L (2014) Energy metabolism in the liver. Compr Physiol 4:177–197

Sampath H, Ntambi JM (2006) Stearoyl-coenzyme A desaturase 1, sterol regulatory element binding protein-1c and peroxisome proliferator-activated receptor-alpha: independent and interactive roles in the regulation of lipid metabolism. Curr Opin Clin Nutr Metab Care 9:84–88

Samuel VT, Liu ZX, Qu X, Elder BD, Bilz S, Befroy D, Romanelli AJ, Shulman GI (2004) Mechanism of hepatic insulin resistance in non-alcoholic fatty liver disease. J Biol Chem 279:32345–32353

Samuel VT, Liu ZX, Wang A, Beddow SA, Geisler JG, Kahn M, Zhang XM, Monia BP, Bhanot S, Shulman GI (2007) Inhibition of protein kinase Cepsilon prevents hepatic insulin resistance in nonalcoholic fatty liver disease. J Clin Invest 117:739–745

Samuel VT, Petersen KF, Shulman GI (2010) Lipid-induced insulin resistance: unravelling the mechanism. Lancet 375:2267–2277

Shen L, Cui A, Xue Y, Cui Y, Dong X, Gao Y, Yang H, Fang F, Chang Y (2014) Hepatic differentiated embryo-chondrocyte-expressed gene 1 (Dec1) inhibits sterol regulatory element-binding protein-1c (Srebp-1c) expression and alleviates fatty liver phenotype. J Biol Chem 289:23332–23342

Silbernagel G, Kovarova M, Cegan A, Machann J, Schick F, Lehmann R, Haring HU, Stefan N, Schleicher E, Fritsche A, Peter A (2012) High hepatic SCD1 activity is associated with low liver fat content in healthy subjects under a lipogenic diet. J Clin Endocrinol Metab 97:E2288–E2292

Sim WC, Park S, Lee KY, Je YT, Yin HQ, Choi YJ, Sung SH, Park SJ, Park HJ, Shin KJ, Lee BH (2014) LXR-alpha antagonist meso-dihydroguaiaretic acid attenuates high-fat diet-induced nonalcoholic fatty liver. Biochem Pharmacol 90:414–424

Spadaro L, Magliocco O, Spampinato D, Piro S, Oliveri C, Alagona C, Papa G, Rabuazzo AM, Purrello F (2008) Effects of n-3 polyunsaturated fatty acids in subjects with nonalcoholic fatty liver disease. Dig Liver Dis 40:194–199

Speliotes EK, Yerges-Armstrong LM, Wu J, Hernaez R, Kim LJ, Palmer CD, Gudnason V, Eiriksdottir G, Garcia ME, Launer LJ, Nalls MA, Clark JM, Mitchell BD, Shuldiner AR, Butler JL, Tomas M, Hoffmann U, Hwang SJ, Massaro JM, O'Donnell CJ, Sahani DV, Salomaa V, Schadt EE, Schwartz SM, Siscovick DS, Voight BF, Carr JJ, Feitosa MF, Harris TB, Fox CS, Smith AV, Kao WH, Hirschhorn JN, Borecki IB (2011) Genome-wide association analysis identifies variants associated with nonalcoholic fatty liver disease that have distinct effects on metabolic traits. PLoS Genet 7:e1001324

Stefan N, Haring HU (2011) The metabolically benign and malignant fatty liver. Diabetes 60:2011–2017

Stefan N, Peter A, Cegan A, Staiger H, Machann J, Schick F, Claussen CD, Fritsche A, Haring HU, Schleicher E (2008) Low hepatic stearoyl-CoA desaturase 1 activity is associated with fatty liver and insulin resistance in obese humans. Diabetologia 51:648–656

St-Jules DE, Watters CA, Brunt EM, Wilkens LR, Novotny R, Belt P, Lavine JE (2013) Estimation of fish and omega-3 fatty acid intake in pediatric nonalcoholic fatty liver disease. J Pediatr Gastroenterol Nutr 57:627–633

Strable MS, Ntambi JM (2010) Genetic control of de novo lipogenesis: role in diet-induced obesity. Crit Rev Biochem Mol Biol 45:199–214

Straub BK, Stoeffel P, Heid H, Zimbelmann R, Schirmacher P (2008) Differential pattern of lipid droplet-associated proteins and de novo perilipin expression in hepatocyte steatogenesis. Hepatology 47:1936–1946

Teli MR, James OF, Burt AD, Bennett MK, Day CP (1995) The natural history of nonalcoholic fatty liver: a follow-up study. Hepatology 22:1714–1719

Tovar AR, Torre-Villalvazo I, Ochoa M, Elias AL, Ortiz V, Aguilar-Salinas CA, Torres N (2005) Soy protein reduces hepatic lipotoxicity in hyperinsulinemic obese Zucker fa/fa rats. J Lipid Res 46:1823–1832

Vega RB, Huss JM, Kelly DP (2000) The coactivator PGC-1 cooperates with peroxisome proliferator-activated receptor alpha in transcriptional control of nuclear genes encoding mitochondrial fatty acid oxidation enzymes. Mol Cell Biol 20:1868–1876

Vessby B, Gustafsson IB, Tengblad S, Boberg M, Andersson A (2002) Desaturation and elongation of fatty acids and insulin action. Ann N Y Acad Sci 967:183–195

Wei Y, Wang D, Topczewski F, Pagliassotti MJ (2006) Saturated fatty acids induce endoplasmic reticulum stress and apoptosis independently of ceramide in liver cells. Am J Physiol Endocrinol Metab 291:E275–E281

Wilfling F, Haas JT, Walther TC, Farese RV Jr (2014) Lipid droplet biogenesis. Curr Opin Cell Biol 29:39–45

Wu T, Liu YH, Fu YC, Liu XM, Zhou XH (2014) Direct evidence of sirtuin downregulation in the liver of non-alcoholic fatty liver disease patients. Ann Clin Lab Sci 44:410–418

Yahagi N, Shimano H, Hasty AH, Matsuzaka T, Ide T, Yoshikawa T, Amemiya-Kudo M, Tomita S, Okazaki H, Tamura Y, Iizuka Y, Ohashi K, Osuga J, Harada K, Gotoda T, Nagai R, Ishibashi S, Yamada N (2002) Absence of sterol regulatory element-binding protein-1 (SREBP-1) ameliorates fatty livers but not obesity or insulin resistance in Lep(ob)/Lep(ob) mice. J Biol Chem 277:19353–19357

Yamada M, Wolfe D, Han G, French SW, Ross MG, Desai M (2011) Early onset of fatty liver in growth-restricted rat fetuses and newborns. Congenit Anom (Kyoto) 51:167–173

Yamada K, Mizukoshi E, Sunagozaka H, Arai K, Yamashita T, Takeshita Y, Misu H, Takamura T, Kitamura S, Zen Y, Nakanuma Y, Honda M, Kaneko S (2015) Characteristics of hepatic fatty acid compositions in patients with nonalcoholic steatohepatitis. Liver Int 35:582–590

Yamazaki Y, Usui I, Kanatani Y, Matsuya Y, Tsuneyama K, Fujisaka S, Bukhari A, Suzuki H, Senda S, Imanishi S, Hirata K, Ishiki M, Hayashi R, Urakaze M, Nemoto H, Kobayashi M, Tobe K (2009) Treatment with SRT1720, a SIRT1 activator, ameliorates fatty liver with reduced expression of lipogenic enzymes in MSG mice. Am J Physiol Endocrinol Metab 297:E1179–E1186

Yki-Jarvinen H (2014) Non-alcoholic fatty liver disease as a cause and a consequence of metabolic syndrome. Lancet Diabetes Endocrinol 2:901–910

Chapter 8
Lipid Droplet Proteins and Hepatic Lipid Metabolism

Yumi Imai, Michelle B. Trevino, and Rexford S. Ahima

Abstract Lipid droplet (LD) is an intracellular organelle that stores neutral lipids including triglycerides, cholesterol ester, and retinyl ester. LD is not a passive depot of excess lipids but is a metabolically active organelle that interacts with endoplasmic reticulum, mitochondria and other cellular compartment to regulate fate of intracellular lipids. As the increase in size and number of LD is a key feature of hepatosteatosis, information regarding LD biogenesis and mobilization is highly relevant to hepatic lipid metabolism. Lipid droplet proteins represented by perilipins (PLIN) reside on the surface of LD and contribute to stabilization and degradation of LD through their interactions with neutral lipids and lipid metabolic enzymes. For example PLIN1, a prototypical PLIN enriched in adipocytes, prevents lipolysis at un-stimulated state but supports robust increase in lipolysis upon cAMP dependent protein kinase (PKA) activation by regulating access of lipases to LD. The presence of 4 PLINs (PLIN1, PLIN2, PLIN3, and PLIN5) with distinct characteristics is reported in normal liver and that affected by hepatosteatosis. Here, we discuss the current knowledge regarding expression and functions of each PLIN in the liver at physiological and pathological status.

Keywords Perilipin 1 • Perilipin 2 • Perilipin 3 • Perilipin 5 • Non alcoholic fatty liver disease • Triglycerides

---

Y. Imai, M.D. (✉) • R.S. Ahima, M.D., Ph.D.
Division of Endocrinology, Diabetes and Metabolism, Perelman School of Medicine, University of Pennsylvania, 3400 Civic Center Blvd, Building 421, Philadelphia, PA 19104-5160, USA
e-mail: imaiy@evms.edu; ahima@mail.med.upenn.edu

M.B. Trevino
Department of Internal Medicine, The Strelitz Diabetes Center, Eastern Virginia Medical School, 700 West Olney Road, LH2156, Norfolk, VA 23507-1696, USA

J.M. Ntambi (ed.), *Hepatic De Novo Lipogenesis and Regulation of Metabolism*,
DOI 10.1007/978-3-319-25065-6_8

1 Introduction

As a key organ for the whole body energy and nutrient homeostasis, the liver performs a wide range of anabolic and catabolic functions in lipid metabolism including lipogenesis, fatty acid (FA) oxidation, ketogenesis, and lipoprotein secretion (Fig. 8.1). Moreover, multiple pathologies involving the liver manifest as an increase in liver lipids known as hepatosteatosis. The most common cause of abnormal liver function tests is non-alcoholic fatty liver disease (NAFLD) (Perry et al. 2014). In addition, alcoholic liver disease, hepatitis C, Wilson disease, starvation, drug induced liver injury, and liver ischemia all present with hepatosteatosis, indicating that derangement in lipid metabolism is a common indicator of dysfunctional liver (Pawella et al. 2014). Therefore, knowledge about the regulation of hepatic lipid metabolism in normal and pathological conditions is critically important.

Lipid droplets (LDs) are organelles that store neutral lipids in cells including hepatocytes (Fig. 8.2). A LD is composed of a lipid core mainly of triglycerides (TG) and cholesterol ester encased by a phospholipid monolayer studded with proteins that regulate both the formation and mobilization of a lipid core (Greenberg et al. 2011; Khor et al. 2013). Studies have shown that LDs are not just a passive end result of excessive lipid accumulation in cells. Rather, LDs are actively engaged in the regulation of intracellular lipid metabolism affecting the cellular metabolic activity, viability, and other functions (Greenberg et al. 2011). Indeed, proteomic studies and electron microscopy have shown that LDs are closely associated with other organelles that play important roles in cellular metabolism such as the endoplasmic reticulum (ER), mitochondria, and peroxisome (Arrese et al. 2014). Thus, understanding structure and functions of LDs in healthy and diseased liver will increase our understanding of lipid metabolism and provide insights into the development of hepatosteatosis in various diseases. In this review, we will discuss the regulation of LD biogenesis, alterations in LD responses under physiological and pathological conditions, and regulation of lipid metabolism by LD and LD associated proteins.

2 Regulation of Lipid Droplet Formation

2.1 *The Formation of Lipid Droplets Is Integral to Cellular Lipid Metabolism*

The primary source of FA for intracellular TG in the liver is FA derived from the circulation (Donnelly et al. 2005), while chylomicron remnant and de novo FA synthesis contribute to a lesser extent (Mashek 2013). The liver possesses high capacity to extract FA from circulation removing 20–30 % of circulatory FA during the first-pass. As in other cells, FA enters hepatocytes by "flip-flop" from the outer to inner layer of the plasma membrane (Schaffer 2002), but the majority of hepatic FA

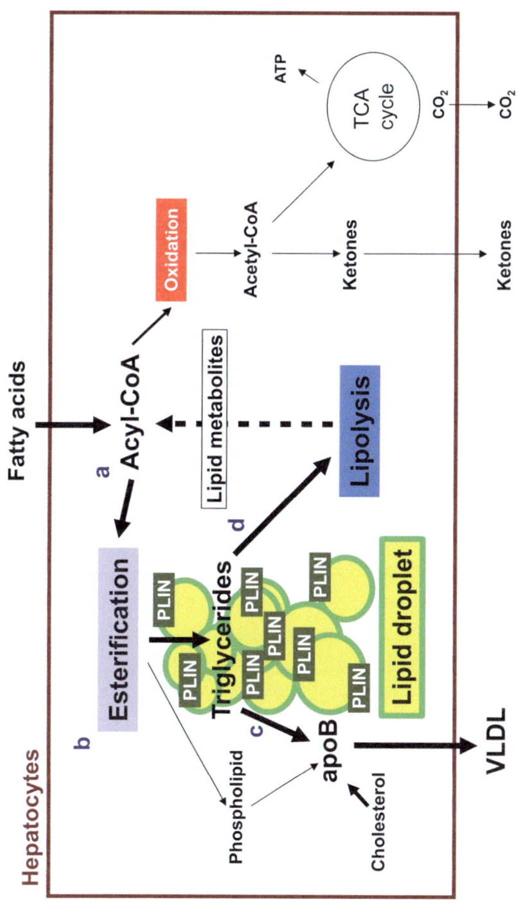

Fig. 8.1 The lipid droplet plays an active role in lipid metabolism in hepatocytes. (**a**) Hepatocytes efficiently take up fatty acids from the circulation and convert them to acyl-CoA, a metabolically active form of fatty acids. (**b**) Acyl-CoA can be oxidized to produce ketones and ATP. Ketones secreted into the circulation supports energy needs during fasting, and ATP provides intracellular energy needs. Acyl-CoA can be esterified to produce triglycerides that are stored in lipid droplets coated with perilipins (PLIN). (**c**) Triglycerides formed in hepatocytes are precursors for VLDL. (**d**) Triglycerides also provide acyl-CoA and other lipid metabolites, e.g., glycerol, diacylglycerides, and monoacylglyceride, through lipolysis

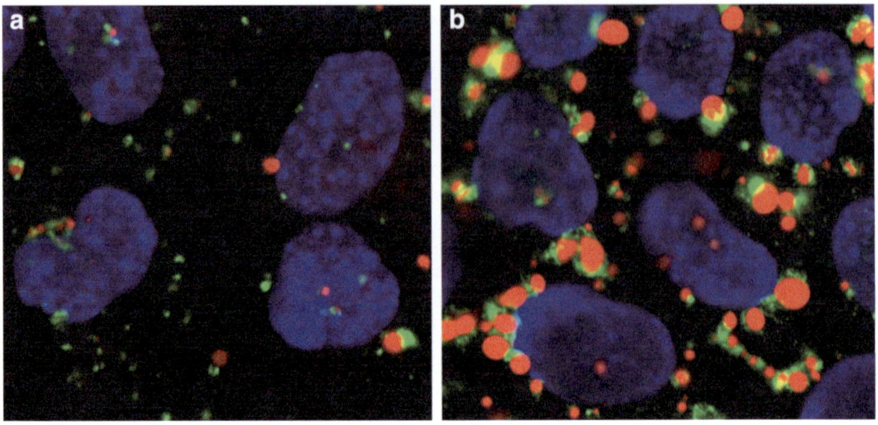

Fig. 8.2 PLIN2 coats lipid droplets in AML12 murine hepatocytes. AML12 cells are cultured in the absence (**a**) or the presence of 0.1 mM oleic acid (**b**) overnight, fixed with formalin, and stained with PLIN2 antibody (*green color*). Neutral lipids are stained with oil red O (*red color*), and the nuclei are stained with 4′,6-diamidino-2-phenylindole (DAPI; *blue color*)

uptake is actively mediated by membrane protein facilitators including CD36 and fatty acid transporter proteins (FATP) (Mashek 2013). Importantly, FATP functions as acyl-CoA synthases to produce metabolically active FA that serves as substrates for TG synthesis and supports a feed forward pathway that increases TG synthesis efficiently when circulatory FA is elevated. The swift esterification of FA and storage into LD likely serves several functions for cells. In addition to the storage of excessive FA for a future caloric needs, the esterification of FA also limits intracellular levels of FA that can be cytotoxic (Holzer et al. 2011; Garbarino et al. 2009). The incorporation of lipids into LD also allows for the organized usage of lipids by regulating the access of lipases and facilitating the transfer of FA to mitochondria for beta-oxidation (Wang et al. 2011a). The increase in hepatic LD content during fasting exemplifies concurrence of FA utilization and esterification. Therefore, LD is a metabolically highly active organelle that is tightly integrated into cellular lipid metabolism.

2.2 *Biogenesis of Lipid Droplets in Endoplasmic Reticulum*

Neutral lipid synthesisinvolves a series of reactions. The primary pathway for TG synthesis in the liver is mediated by glycerol-3-phosphate acyltransferase (GPAT), 1-acylglycerol-3-phosphate-O-acyltransferase (AGPAT), lipin, and acyl CoA:diacylglycerol acyltransferase (DGAT). Although its precise role is not fully understood, the MGAT pathway in the liver is also known to contribute to diacylglycerol (DG), a precursor of TG (Hall et al. 2014). Importantly, each step has multiple isoenzymes with distinct preferences for FA chains and saturations resulting in the production of complex and heterogeneous lipid species (Pol et al. 2014). It is widely accepted that the major site of TG synthesis is the endoplasmic reticulum (ER) where these synthetic enzymes are found. As extensively covered in a recent review (Pol et al. 2014), small depositions of TG, globules, between two leaflets of the phospholipid bilayer of the ER membrane are considered to be the initial step of LD formation. The small TG globule is mobile within the phospholipid bilayer and can depart from synthetic enzymes that reside in the ER. Further expansion of the neutral lipid core requires a remodeling of the phospholipid monolayer layer to create a curvature over the neutral lipid core (Thiam et al. 2013). The enrichment of the phospholipid monolayer with lysophosphatidic acid and lysophosphatidic choline is considered to aid in the creation of a round curved phospholipid membrane due to their conical shape (Thiam et al. 2013). Membrane phospholipid over LD is also enriched with phosphatidyl choline (PC) considered to be critically important for preventing LD coalescence. DG is not only the intermediate product of TG synthesis but also may aid the formation of the LD membrane curvature and anchoring of proteins such as perilipin 3 (PLIN3), protein kinase C, and CTP:phosphocholine cytidylyltransferase (CCT) at the surface of an emerging LD (Pol et al. 2014). As the neutral lipid core grows under a convex phospholipid monolayer, the distinct characteristics of the monolayer membrane and hydrophobic core likely promotes the association of proteins that recognize the unique environment of LD, such as perilipins, which aid to stabilize the LD.

2.3 Lipid Droplet Expansion

It has been proposed that LD continues to maintain a connection with the ER through a tiny tubular bridge and never becomes completely detached from ER (Pol et al. 2014; Wilfling et al. 2014). However, LD expansion is also achieved by synthesizing TG on the surface of LD to increase LD diameter that allows for the volume of neutral lipid core to increase relative to the LD surface area (Wilfling et al. 2013). Enzymes required for TG synthesis, e.g., GPAT4, AGPAT3, lipin1g and DGAT2, are found in the LD. A comparison of protein structures indicates that isoenzymes with positive amino acids (Ingelmo-Torres et al. 2009) and those with bended hairpin structure (Stevanovic and Thiele 2013; Wilfling et al. 2013) preferentially reside in LD, while those without positive amino acid (Pol et al. 2014) or membrane-spanning helices are retained in the ER, thus creating differential distributions of isoenzymes for TG synthesis within LD and ER compartments. For example, DGAT2 has a bended hairpin structure and resides in the LD, while DGAT1 has membrane-spanning helices and is retained inside the ER (Thiam et al. 2013). The fusion of LD expands its size and promotes TG storage. FSP27 (fat specific protein 27; Cidec) is known to promote LD fusion in adipocytes. The knockdown of Rab8a, a protein with affinity to FSP27 and potential regulator of FSP27, reduced LD size and TG accumulation in leptin deficient *ob/ob* mice indicating a potential contribution of LD fusion in the expansion of hepatic LD (Wu et al. 2014).

Another study in mice also demonstrated that reduced FSP27 levels in the liver attenuated TG accumulation in response to fasting or a high fat diet (Langhi and Baldan 2015).

3 Perilipins: Structures and Functions

3.1 PLIN Family of Proteins

The PLIN family of LD proteins is the best known LD-associated protein shown to actively regulate LD formation and mobilization through interactions with enzymes mediating lipid metabolism. The PLIN family consists of five proteins (PLIN1–PLIN5) that share structural features conferring preferential distribution on the LD surface, but differ in their tissue distribution, subcellular localization, LD binding properties, and interactions with other proteins (Fig. 8.3). Perilipin 1 (PLIN1) (Greenberg et al., 1991), perilipin 2 (PLIN2) (Jiang et al. 1992) and PLIN3 (Diaz and Pfeffer 1998; Than et al. 1998) make up an trio of PLIN proteins termed the "PAT family" based on their original names, i.e., perilipin, adipocyte differentiation-related protein, and tail-interacting protein of 47 kDa (kDa). The later discovery of perilipin 4 (PLIN4, S3-12) (Scherer et al. 1998; Wolins et al. 2003) and perilipin 5 (PLIN5, lipid storage droplet protein 5, myocardial LD protein, OXPAT) (Yamaguchi et al. 2006; Dalen et al. 2007; Wolins et al. 2006) completed the current list of family members, and a new name "Perilipin (PLIN) family" was adopted in 2010 (Kimmel et al. 2010).

Defining structural characteristics of PLIN proteins include a conserved PAT domain (Lu et al. 2001) and the 11-mer repeating helical organization (Bussell and

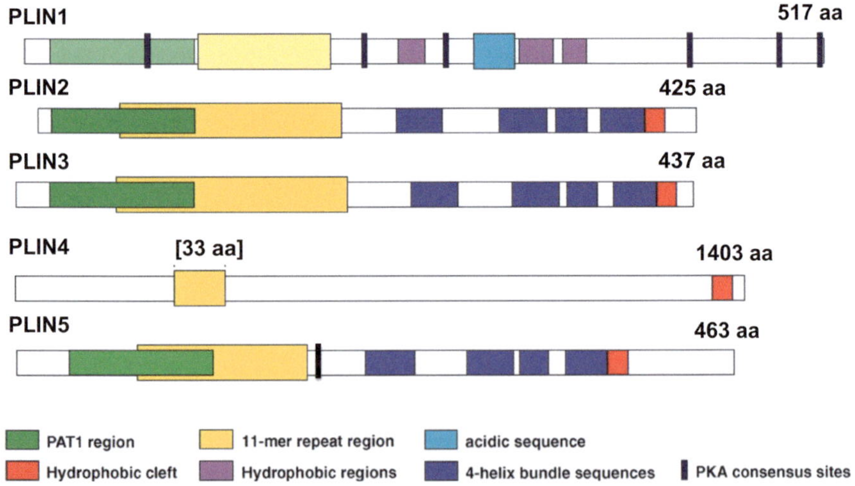

Fig. 8.3 Structures of PLIN family of lipid droplet proteins

Eliezer 2003). The ~100 residues located within the extreme amino-terminal end are most homologous among the PLIN proteins (ranging from 40 % between PLIN1 and PLIN2 and 65 % between PLIN3 and PLIN2) and named the PAT-1 domain; the PAT-2 domain resides toward the carboxyl-terminal portion and is less conserved (20 % identity between PLIN1 and PLIN2 and 40 % between PLIN3 and PLIN2) (Londos et al. 2005). An 11-mer repeating helical organization is also shared among all PLIN members but its location and sequence differ. The functions of these domains regarding the targeting and regulation of LD size are still not fully understood. Although it was originally presumed the PAT domains were required for LD targeting, further studies indicated that these domains were not essential. Apolipoproteins that aid the formation of lipoproteins consisting of lipid core and phospholipid monolayer are structurally close to the PLIN family and also contain an 11-mer motif (Bussell and Eliezer 2003). However, apolipoproteins do not possess a PAT domain (Kimmel et al. 2010).

Functionally, the PLIN family shares a LD binding ability in response to lipid loading when LDs are formed, enlarged and mature. However, the PLIN proteins can be classified into either a constitutive or exchangeable LD-associated group based on their stability when they are not bound to the LD. PLIN1 and PLIN2 are constitutively bound to the LD, and the unbound proteins are degraded through ubiquitin-dependent pathways (Xu et al. 2005; Brasaemle et al. 1997). In contrast, PLIN3, PLIN4, and PLIN5 are exchangeable LD proteins (ePAT) stable both within the cytosol and when bound to the LD (Wolins et al. 2001, 2003; Yamaguchi et al. 2006). Each PLIN shows a preference for different sizes of LD as shown in adipocytes where PLIN3 and PLIN4 are found in smaller LD, while PLIN1 and PLIN2 preferentially bind large LD (Wolins et al. 2005). Possible explanations for the LD size preference of PLINs may lie in protein structures creating better stability within certain membrane curvature (corresponding to LD size) or their interactions with the phospholipid composition of the LD monolayer membrane (Walther and Farese 2009; Thiam et al. 2013). Moreover, the composition of the LD affects PLIN binding: PLIN1 and PLIN5 tend to bind TG-rich LD, PLIN2 and PLIN3 bind both TG and cholesterol ester enriched LDs, and PLIN4 show preference to cholesterol ester enriched LD (Hsieh et al. 2012). Lastly, the tissue distribution of PLINs varies: PLIN1 is abundant in white adipocytes and steroidogenic cells, PLIN2 and PLIN3 are ubiquitous, PLIN4 is present in white adipocytes, and PLIN5 is highly expressed in oxidative tissues, e.g., heart, liver, and skeletal muscle. In the liver, PLIN2, PLIN3, and PLIN5 are the most abundant PLINs (Okumura 2011). To date, PLIN4 expression has not been convincingly demonstrated in normal or diseased liver in humans or animals.

3.2 Regulation of Lipolysis in Adipocytes by PLIN1

PLIN1 was the first PLIN family discovered through studies addressing the regulation of lipolysis in adipocytes (Egan et al. 1990; Greenberg et al. 1991). There are four splice variant transcripts of PLIN1 translated into four PLIN1 protein isoforms named PLIN1a, PLIN1b, PLIN1c and PLIN1d, of which PLIN1a and PLIN1b are the major forms (Greenberg et al. 1993). Lipolysis in adipocytes dynamically responds to whole body energy state; starvation markedly increases lipolysis in adipocytes to provide FA as an energy source for the rest of the body, while the postprandial state shuts down lipolysis in adipocytes. It is well known that lipolysis in adipocytes is regulated through activation of cAMP dependent protein kinase A (PKA) related pathways (Honnor et al. 1985; Londos et al. 1985). PLIN1 plays a dual role in adipocytes supporting the formation of large LD by preventing lipolysis in the non-stimulated state, and promoting mobilization of LD through activation of lipolysis in a PKA dependent manner. Although PLIN1 does not possess lipase activity, it affects the degree and rate of lipolysis through PLIN1-lipase interactions (Zechner et al. 2012). In adipocytes, adipocyte triglyceride lipase (ATGL) mediates the hydrolysis of TG to DG, and hormone sensitive lipase (HSL) hydrolyzes DG to monoacylglycerol (MG). Multiple PKA phosphorylation sites of PLIN1 mediate its association with major lipases, directly with HSL and indirectly with ATGL through interaction of PLIN1 and a co-activator of ATGL, comparative gene identification 58 (CGI-58, aka ABHD5) (Zechner et al. 2012). Under basal conditions, PLIN1 sequestrates CGI-58 preventing its access to ATGL, thereby preventing TG hydrolysis (Granneman et al. 2009a; Shen et al. 2009; Wang et al. 2009). Under stimulated conditions when cellular cAMP levels are increased due to beta adrenergic stimulation, phosphorylation of PLIN1 at PKA consensus sequences promotes lipolysis in the following ways: PLIN1 releases CGI-58 to promotes its binding to ATGL to initiate lipolysis, and PKA phosphorylates both PLIN1 and HSL that promotes the recruitment of HSL to LD (Zechner et al. 2012). In line with this model, adipocytes isolated from PLIN1 whole body knockout mouse show increased basal lipolysis but a blunted response to catecholamine-stimulated lipolysis (Tansey et al. 2001; Martinez-Botas et al. 2000). In humans, genetic variations of the *PLIN1* gene locus have been discovered (Qi et al. 2004b, 2005) revealing increased lipolysis and greater weight loss (from visceral fat) in response to dieting in individuals with certain *PLIN1* polymorphisms (Mottagui-Tabar et al. 2003; Qi et al. 2004a; Jang et al. 2006; Corella et al. 2005). Interestingly, the activity of ATGL is also regulated by an inhibitory protein G0/G1 switch gene (G0S2) (Jenkins et al. 2004; Lass et al. 2006), but its association with PLIN1 has not been documented so far.

3.3 Regulation of Lipolysis in Non-adipose Tissues

Lipolysis in non-adipose tissue provides FA as substrate for beta-oxidation to support cellular energy needs. In addition, lipolysis in non-adipocytes generates lipid metabolites that modulate gene expression through PPARs activation (Mashek 2013; Zechner et al. 2012). Various studies have demonstrated that the interaction between ATGL and CGI-58 is a key step for the degradation of neutral lipid core in hepatocyte LD (Kimmel and Sztalryd 2014; Wu et al. 2011; Ong et al. 2011; Schweiger et al. 2009). Human mutations of CGI-58 are associated with Chanarin–Dorfman syndrome, a neutral lipid storage disease characterized by ichthyosis and hepatosteatosis (Schweiger et al. 2009). ATGL accounts for less than 50 % of total lipase activity during fasting in mice based on residual lipase activity in ATGL KO mice (Reid et al. 2008). However, the down regulation of hepatic ATGL is sufficient to induce steatosis indicating that hepatic ATGL plays an important role in lipolysis (Wu et al. 2011; Ong et al. 2011). Although PLIN1 is a key spatial organizer of lipolysis through its interaction with CGI-58 and HSL in adipocytes, its expression is very low in non-adipocytes including hepatocytes. Among PLINs highly expressed in the liver, i.e., PLIN2, 3, and 5, PLIN5 has the strongest evidence for its interactions with ATGL and CGI-58 to regulate lipolysis (Mason and Watt 2015; Kimmel and Sztalryd 2014). Fluorescence resonance energy transfer (FRET) and other methods show a close proximity of PLIN5 and ATGL and CGI-58 when overexpressed in cultured cells (Wang et al. 2011a; Granneman et al. 2011; 2009b). Moreover, PLIN5 is co-localized with ATGL and CGI-58 at the LD surface in a mouse model of acute hepatic steatosis PLIN5 (Wang et al. 2011a; Granneman et al. 2009b, 2011). However, the exact mode of interaction between PLIN5, ATGL and CGI-58 has not been determined (Mason and Watt 2015; Kimmel and Sztalryd 2014). Interestingly, PLIN5 binds to either ATGL or CGI-58, not both at the same time when tested under an un-stimulated condition (Granneman et al. 2011). PLIN5 contains one PKA consensus site (S155) that can be phosphorylated by PKA, as PKA activation increased the incorporation of [^{32}P] to PLIN5 (Wang et al. 2011a). Like for PLIN1 in adipocytes, overexpression of PLIN5 in AML12 hepatocyte cell line (Wang et al. 2011a) and MIN6 beta-cell line (Trevino, 2015) increased lipolysis in a PKA dependent manner. More recently, the mutation of S155 to alanine to prevent phosphorylation by PKA was shown to prevent the increase in lipolysis in response to PKA activation in COS7 cells, further supporting the notion that PKA dependent phosphorylation of PLIN5 increases lipolysis as does the phosphorylation of PLIN1 (Pollak et al. 2015). Thus, the current model is that PLIN5 binds ATGL or CGI58 to prevent these proteins from interacting with each other and with LD, thus preventing basal (un-stimulated) lipolysis (Granneman et al. 2009b, 2011; Wang et al. 2014; Kimmel and Sztalryd 2014). Once phosphorylated, conformational changes in PLIN5 favor lipolysis through poorly defined mechanisms (Mason and Watt 2015; Kimmel and Sztalryd 2014). Further studies are needed to determine whether PLIN5 phosphorylation increases lipolysis by affecting its interactions with ATGL and CGI-58. PLIN5 possesses only one PKA target sites while PLIN1 may be phosphorylated at six serine residues (Fig. 8.3). Therefore, there are substantial conformational differences upon PKA activation, which may confer cell type specific patterns of lipolysis. PKA activation in PLIN1 expressing adipocytes leads to >50 fold stimulation of lipolysis, the levels that are sufficient to raise circulatory FA (Viswanadha and Londos 2008). In contrast, the increase in lipolysis with PKA activation under PLIN5 overexpression is significant but less dramatic at around two-fold that could be sufficient to produce metabolites for intracellular signaling (Trevino et al. 2015; Wang et al. 2011b).

4 Perilipins in the Liver

4.1 PLIN1

As mentioned earlier, PLIN1 expression is restricted to adipocytes and steroidogenic cells. PLIN1 expression has not been convincingly demonstrated in the normal livers of humans or mice (Chang et al. 2006; Wolins et al. 2006; Straub et al. 2008). However, increased PLIN1 levels have been detected in human steatotic liver and may play an important role in large LD formation. Microarray analysis comparing human liver with low and high fat contents revealed that *PLIN1* is among genes whose expression positively correlate with liver fat contents (Greco et al. 2008). A positive correlation between steatosis and PLIN1 was also demonstrated at protein levels (Western blot, immunohistochemistry) (Straub et al. 2008). PLIN1 immunostaining at the rim of LD has been observed in human steatotic liver in wide range of diseases including NASH, alcoholic fatty liver disease, hepatitis C, drugs, and Wilson disease (Fujii et al. 2009; Pawella et al. 2014). While PLIN1 in human liver seems to be limited to chronic hepatosteatosis especially macrovesicular steatosis, PLIN1 is not associated with acute fatty liver resulting from ischemic injury, parental nutrition, or acute drug injuries (Pawella et al. 2014).

In mice, the expression of PLIN1 in the liver is less common and is not seen in fatty liver due to high fat diet or leptin deficiency (Okumura 2011). However, C57BL/6 mice fed a Lieber–DeCarli ethanol diet containing 45 % kcal fat diet and ethanol (LD-E) showed PLIN1 associated with large LD, while control mice fed an isocaloric 45 % kcal fat diet lacked large LD (Orlicky et al. 2011). This pronounced change in LD size was observed at slightly higher levels of hepatic TG in LD-E fed mice compared with control mice (Orlicky et al. 2011). Collectively, PLIN1 expression in the liver may promote the emergence of large LD based on aforementioned characteristics of hepatosteatosis with PLIN1 expression in human (Pawella et al. 2014) and mouse (Orlicky et al. 2011). In support, the induction of PLIN1 in cultured cells (stable transfection of PLIN1 in 293 cells (Orlicky et al. 2011) or by chemical treatment of HuH7 cells (Pawella et al. 2014) demonstrated that PLIN1 expression correlates with large LD compared with the expression of PLIN2 and PLIN3 indicating that structural features of PLIN1 may contribute to macrosteatotic LD (Orlicky et al. 2011). However, the up-regulation of PLIN1 was associated with smaller LD despite the increase in liver TG contents when CD36 deficient *ob/ob* mice were compared with *ob/ob* mice indicating that PLIN1 expression may not be sufficient to determine LD size (Nassir et al. 2013). Currently, little is known about how PLIN1 is regulated in the liver is in vivo. Considering that PLIN1 expression in the liver requires chronic hepatic steatosis in humans and multiple stressors (alcohol + high fat diet, leptin deficiency and CD36 deficiency) in mice, the induction of PLIN1 expression in the liver likely requires a combination of factors. It is possible that PLIN1 regulates large LDs, which increase TG storage in hepatocytes as in adipocytes. Also, PLIN1 is phosphorylated in human hepatocytes suggesting a potential role in regulation of lipolysis in response to PKA activation (Pawella et al. 2014).

4.2 PLIN2

4.2.1 Expression of PLIN2 in the Liver

Among the five members of PLINs, PLIN2 is the most abundantly expressed in multiple cells including hepatocytes, myocytes, and pancreatic beta cells (Trevino et al. 2015; Greenberg et al. 2011). In addition to being the principal PLIN in hepatocytes, PLIN2 is highly expressed in hepatic stellate cells where it coats LDs that store retinyl ester (Motomura et al. 2006; Yu et al. 2003). The overexpression of PLIN2 increases LD formation while its down-regulation reduces LD formation in multiple types of cells supporting the role of PLIN2 as the major LD protein in non-adipocytes (Greenberg et al. 2011).

Human polymorphism in PLIN2 is associated with lower serum TG levels (Magne et al. 2013). PLIN2 S251 substituted with proline (S251P) is predicted to disrupt alpha-helix domain within PLIN2 (Magne et al. 2013). When overexpressed in human embryonic kidney 293 cells, S251P mutant decreases lipolysis and increases intracellular accumulation of TG with a shift in LD size toward lower diameters (Magne et al. 2013). However, it is currently unknown whether PLIN2 S251P alters hepatic lipid metabolism.

The expression of PLIN2 is highly responsive to an increase in FA loading and it is regulated by peroxisome proliferator-activated receptors (PPARs) (Okumura 2011). PLIN2 is increased during the formation of LD in response to ethanol in WIF-B cells, a polarized hepatocyte cell line (McVicker et al. 2012). In vivo, PLIN2 expression is almost always increased when TG accumulates in the liver of both humans and animals (Motomura et al. 2006; Straub et al. 2008; Pawella et al. 2014). In humans, PLIN2 protein expression closely correlates liver TG contents (Straub et al. 2008) and is increased in livers affected by NASH, alcoholic fatty liver disease, hepatitis C, drugs, Wilson disease, acute steatosis from parental nutrition, and acute complications in transplanted liver (Pawella et al. 2014). While PLIN1 is commonly seen in large LD of human liver affected by NASH, PLIN2 coats LDs of variable sizes and is especially prominent in ballooned hepatocytes (Fujii et al. 2009). The same study of human liver with NASH also noted a strong correlation between the staining of PLIN2 and oxidized LDL levels within ballooned hepatocytes (Fujii et al. 2009). In mouse models of hepatosteatosis, it is rare to find cases in which PLIN2 expression is not increased. PLIN2 is increased in fatty liver of mice in response to fasting (Dalen et al. 2006), PPAR-alpha agonist (Dalen et al. 2006), high fat diet (Motomura et al. 2006), leptin-deficiency (Motomura et al. 2006), ER stress (Yamamoto et al. 2010), ethanol (Carr et al. 2013, 2014; Mak et al. 2008; Orlicky et al. 2011), partial hepatectomy (Kohjima et al. 2013), and Glycine N-methyltransferase (GNMT) deficiency (Martínez-Uña et al. 2013). The latter increases hepatic TG synthesis by increasing the conversion of phosphatidylethanolamine to phosphatidylcholine (Martínez-Uña et al. 2013).

4.2.2 Regulation of Hepatic Lipid Metabolism by PLIN2

The importance of PLIN2 in the formation of liver LD is supported by reduction in hepatosteatosis in PLIN2 deficiency. Whole body PLIN2 knockout mice created by deletion of exon 2–3 did not show developmental abnormalities and weighed similar to wild type control mice (Chang et al. 2006). However, these mice showed reduced liver TG on regular rodent chow (Chang et al. 2006) and were protected against hepatosteatosis in response to high fat diet (Chang et al. 2006), leptin deficiency (Chang et al. 2010), post-partial hepatectomy (Kohjima et al. 2013), and ethanol (Carr et al. 2014). Interestingly, exon 2-3 targeted PLIN2 deficient mice in *ob/ob* mice showed 25 % reduction in hepatic TG contents, but the LDs were larger than those seen in *ob/ob* or wild type mice (Chang et al. 2010). Down-regulation of PLIN2 by antisense oligonucleotide (ASO) that primarily targets liver decreased fatty liver in diet-induced and leptin deficient *ob/ob* mice (Imai et al. 2007). Recently, another model of PLIN2 deficient mice was created by deleting exon 5, and these mice were also protected from diet-induced hepatosteatosis (McManaman et al. 2013). However, the improvement in fatty liver and resistance to diet-induced obesity in this model may be secondary to a decrease in food intake and increase in beige fat (McManaman et al. 2013).

A proposed mechanism by which PLIN2 increases TG accumulation is the reduction of lipolysis through prevention of ATGL association with LD (Listenberger et al. 2007). Interestingly, the reduction of PLIN2 by ASO in the liver with established steatosis decreased the expression of wide range of lipogenic genes, e.g., fatty acid synthase, acetyl-CoA carboxylase, and diacylglycerol O-acyltransferase 2 (Imai et al. 2007) implying that PLIN2 level itself exert transcriptional regulation over lipid metabolism possibly through alterations in lipid metabolite levels (Zechner et al. 2012). PLIN2 expression also increases glucose uptake (Magnusson et al. 2006; Sun et al. 2012) and could enhance the switching of energy substrates to lipogenesis in the liver (Sun et al. 2012), again supporting the idea that PLIN2 may actively shift cellular metabolism toward lipid storage in addition to preventing lipolysis.

VLDL secretion is one of the unique functions of hepatocytes and this is highly relevant to LD biology in hepatocytes as a significant portion of TG in VLDL is derived from TG formed within hepatocytes (Fig. 8.1). Studies in McA-RH7777 cells and primary rat hepatocytes showed that PLIN2 reduced the transfer of TG to VLDL causing a retention of TG in hepatocytes (Magnusson et al. 2006). In support of this observation, whole body PLIN2-deficient mice created by exon 2-3 deletion showed an increase in VLDL secretion (Chang et al. 2010; Kohjima et al. 2013). However, it is important to note that VLDL secretion is affected by multiple factors in hepatocyte including TG content, apolipoproteins, microsomal triacylglycerol transfer protein (MTP), and other processes regulating VLDL assembly. When PLIN2 was down-regulated by ASO in livers of diet-induced and leptin deficient obese mice, hepatic TG contents were significantly reduced and both VLDL secretion and serum TG levels were reduced as well (Imai et al. 2007). A competition between cell death-inducing DFF45-like effector B (CIDEB) and PLIN2 in hepatic VLDL regulation has been reported (Li et al. 2012b). CIDEB is found in ER, LD, and Golgi apparatus where it increases lipidation and maturation of VLDL. CIDEB-deficient mice develop hepatosteatosis with reduction in VLDL secretion, which is relieved by additional deletion of PLIN2 that increases TG export in the form of VLDL (Li et al. 2012b).

While the critical contribution of PLIN2 in LD formation in the liver is well supported, the results in mouse models have implicated variable mechanisms. Thus, further studies in models where PLIN2 is specifically modulated in hepatocytes are necessary. It also is noteworthy that PLIN3 and PLIN5 may compensate for the loss of PLIN2 in PLIN2 deficient mice (Chang et al. 2010; McManaman et al. 2013).

4.2.3 Roles of Hepatic PLIN2 beyond the Regulation of Lipid Metabolism

As hepatosteatosis is closely associated with insulin resistance (Perry et al. 2014), glucose homeostasis has been evaluated in several models of PLIN2 deficiency. Improved insulin tolerance and insulin sensitivity in hyperinsulinemic euglycemic clamp have been reported when hepatosteatosis is improved in exon 2-3 deleted PLIN2 knockout mice (*ob/ob* mice and ethanol fed mice), as well as in PLIN2 ASO treated mice (diet-induced obese mice and *ob/ob* mice) (Carr et al. 2014; Chang et al. 2010; Imai et al. 2007; Varela et al. 2008). However, the adenovirus mediated overexpression of PLIN2 in the liver also improved glucose tolerance at least within the timeframe studied, likely by increasing hepatic glucose utilization through an increase in lipogenesis (Sun et al. 2012). Unbiased microarray analyses of livers of PLIN2 ASO treated diet-induced obese mice showed an up-regulation of alpha feto-protein (*Afp*) and genes associated with remodeling of extracellular matrix (Imai et al. 2012). Thus, the down-regulation of PLIN2 may decrease hepatosteatosis and increase susceptibility to injury. In a model of partial hepatectomy, the proliferation of hepatocytes was delayed in whole body PLIN2 deficient mice indicating that an increase in LD is required to support a mitotic response that necessitates the production of phospholipids and other lipid metabolites for cell structures (Kohjima et al. 2013). It also needs to be noted that quiescent hepatic stellate cells possess LD filled with retinyl esters and coated with PLIN2 (Friedman 2008; Straub et al. 2008). Lee et al. reported in their *ex vivo* model that the down-regulation of PLIN2 in LX-2, immortalized human hepatic stellate cells, induced features of stellate cell activation (Lee et al. 2010). Thus, PLIN2 in hepatic stellate cells plays a role in the maintenance of quiescence and its down-regulation may lead to steatohepatitis. This aspect could have been under appreciated in mouse models of fatty liver that are resistant to the development of steatohepatitis and fibrosis.

4.3 *PLIN3*

At the time of discovery, PLIN3 was believed to function as a cytoplasmic sorting factor to traffic mannose 6-phosphate receptors from endosomes to Golgi apparatus (Diaz and Pfeffer 1998). Then, a later study revealed that PLIN3 is a LD protein since it is recruited to LD surface and its knockdown reduced LD biogenesis (Bulankina et al. 2009; Wolins et al. 2001; Miura et al. 2002). When compared with other PLINs, PLIN3 is most homologous to PLIN2; approximately 80 % of PLIN3 amino-terminal end is homologous to PLIN2 (Wolins et al. 2001). Like PLIN2, PLIN3 are widely expressed in a number of tissues including macrophages, neutrophils, enterocytes and retinal epithelial cells (Buers et al. 2009; Lee et al. 2009; Bulankina et al. 2009; Tsuiki et al. 2007), and highly expressed in murine and human liver (Bickel et al. 2009; Straub et al. 2008; Dalen et al. 2006). To date, the factors that regulate PLIN3 levels in the liver remain unclear. As previously noted, PLIN2 expression in the liver is closely tied with changes in both extra- and intracellular lipids and/or lipid metabolites partly through the activation of PPAR-alpha (Okumura 2011). However, PPAR-alpha agonists had no effect on PLIN3 expression in mouse liver, rat hepatoma cell lines or human hepatoma cell lines (Dalen et al. 2006). In contrast to PLIN2 whose expression levels are correlated with tissue TG content, PLIN3 levels are not correlated with the amount of LDs or degree of steatosis in human liver (Straub et al. 2008). One potential explanation is that PLIN3 is stable in the cytosol unbound to LD, while PLIN2 is unstable when unbound to LD. The mode of lipid exposure also appears to influence PLIN3 levels in the liver. Human primary hepatocytes cultured in media supplemented with oleic acid, cholesterol, LDL and ethanol for less than 10 days did not alter protein levels of PLIN3 nor PLIN5 (Pawella et al. 2014). However, long-term incubation (up to 40 days) of human hepatoma cell lines in media supplemented with oleic acid and compounds that are known to promote adipogenesis resulted in the upregulation of PLIN proteins in sequential order: PLIN3, PLIN5, PIN2 and PLIN1 (Pawella et al. 2014). The expression levels in the liver under pathological conditions are also different between PLIN3 and PLIN2. Histologic analyses of PLIN3 in human livers affected by various disease including NAFLD, NASH, cirrhosis, hepatitis B and hepatitis C infected livers showed relatively limited PLIN3 staining when PLIN2 is increased compared with non diseased human liver (Straub et al. 2008). However, PLIN3 was expressed in steatotic liver specimens of patients with hepatocellular carcinoma (HCC) and HCC cell lines (Straub et al. 2008). Although its implication is still unknown, PLIN3 is one of genes up-regulated in microarray analysis of liver from Sprague-Dawley rats that developed chemically induced liver fibrosis (AbdulHameed et al. 2014). There clearly is a need for further studies to determine how hepatic PLIN3 expression is regulated.

PLIN2 and PLIN3 have distinct cellular distributions in the liver. In normal human liver, PLIN3 is mainly localized on retinyl ester containing LDs within hepatic stellate cells and on LDs of very small size in hepatocytes, whereas PLIN2 displays a more ubiquitous staining (Straub et al. 2008). Under conditions of acute steatosis in human liver, such as liver transplantation, ischemia-reperfusion injury, and acute drug-induced steatosis, there is an increase in small LD coated with PLIN3 (Pawella et al. 2014). The preferential association of PLIN3 to small LD in the liver is analogues to emergence of PLIN3 coated small LD in cells exposed to stimuli to promote adipogenesis (Pawella et al. 2014; Skinner et al. 2009), and implies that PLIN3 may play a functional role in coating nascent LDs in the liver as well. Additionally, PLIN3's ability to remain in the cytosol as an unbound protein may provide a pool of LD proteins to compensate for PLIN2 if needed. PLIN3 shuttles from a cytosolic pool to the LD to maintain TG laden LD in PLIN2 deficient adipocytes, THP-1 macrophages and hepatocytes (Sztalryd et al. 2006; Bell et al. 2008; Buers et al. 2009). In exon 2-3 deleted PLIN2 knockout *ob/ob* mice, PLIN3 became more associated with the LDs as well (Chang et al. 2010).

Although immunohistological analyses did not demonstrate an increase of PLIN3 in human livers affected by NAFLD (Pawella et al. 2014), PLIN3 protein was increased in steatotic liver from male C57BL/6J mice on high fat diet (Carr et al. 2012). To address whether PLIN3 contributes to hepatosteatosis and insulin resistance in response to high fat diet, 70 % reduction of PLIN3 mRNA levels in the liver was achieved by ASO treatment of high fat fed mice (Carr et al. 2012). Notably, PLIN3 ASO treatment did not alter PLIN2 mRNA or protein levels (Carr et al. 2012). In this study, PLIN3 ASO treatment significantly reduced hepatic TG contents, decreased hepatic TG secretion, and improved insulin sensitivity in liver, adipose tissue, and muscle measured by hyperinsulinemic-euglycemic clamp. The study indicates that PLIN3 actively participates to hepatosteatosis and associated metabolic abnormalities in a mouse model of NAFLD.

An additional role of PLIN3 in liver pathology is implicated in the study of hepatitis C infection. Interestingly, hepatitis C virus infection is tightly integrated into lipid metabolism of infected hepatocytes. Hepatitis C virus is targeted to LD in DGAT1 dependent manner (Herker et al. 2010), reduces lipolysis, and stabilizes LD (Harris et al. 2011). In LD, hepatitis C virus takes advantage of lipid as a nutritional source and evades the immune response. It was shown that PLIN3 interacted with hepatitis C virus and aided its targeting to LD (Ploen et al. 2013).

5 PLIN5

PLIN5 is strongly upregulated in the liver during fasting (Wolins et al. 2006; Dalen et al. 2007; Yamaguchi et al. 2006; Wang et al. 2014; Mason and Watt 2015). In contrast, mixed reports exist whether PLIN5 is increased in fatty liver associated with human or mouse obesity (Wolins et al. 2006; Dalen et al. 2007; Yamaguchi et al. 2006; Wang et al. 2014; Mason and Watt 2015). FA and PPARs regulate the expression of PLIN5 (Grasselli et al. 2010; Dalen et al. 2007; Yamaguchi et al. 2006; Wolins et al. 2006; Bindesboll et al. 2013). Regular chow supplemented with a PPAR-alpha agonist increased *Plin5* in the liver and heart in C57BL6 mice, while *Plin5* was undetectable in the livers of *Ppar-alpha* null mice fed ad libitum (Dalen et al. 2007). Two independent groups demonstrated that hepatic PLIN5 expression did not respond to fasting and insulin deficiency in *Ppar-alpha* null mouse (Wolins et al. 2006; Yamaguchi et al. 2006). However, fasting increased hepatic PLIN5 expression in *Ppar-alpha* null mice to levels seen in control mice in another study indicating that the up-regulation of *Plin5* in the liver upon fasting is independent of PPAR-alpha (Dalen et al. 2007). Furthermore, FA did not increase PLIN5 expression in mouse primary hepatocytes or human pancreatic islets (Trevino et al. 2015; Asimakopoulou et al. 2014). In contrast to the liver, skeletal muscle *Plin5* is regulated by either PPAR-alpha (Wolins et al. 2006) or PPAR-delta (Bindesboll et al. 2013), and PPAR-gamma was shown to affect *Plin5* levels in human white adipose tissue (Wolins et al. 2006). Recently, atorvastatin, an HMG-CoA reductase inhibitor that reduces cholesterol synthesis, was shown to suppress PLIN5 levels in mouse liver and hepatocytes through Sterol regulatory element-binding protein 2 (SREBP2) (Langhi et al. 2014).

Several studies addressed roles of PLIN5 in the regulation of hepatic lipid metabolism and cellular function using hepatocytes in culture. The overexpression of PLIN5 in AML12 cells and primary hepatocytes increased TG laden LD that was attributed to increasing in FA uptake and decreasing in basal lipolysis, changes similar to those observed in the overexpression of PLIN2 (Dalen et al. 2007; Wolins et al. 2006; Li et al. 2012a; Wang et al. 2011a). As expected, knockdown of PLIN5 in AML12 cells decreased TG laden LD, increased parameters of lipolysis, and modestly increased numbers of mitochondria and FA oxidation (Li et al. 2012a). The up-regulation of mitochondria and FA oxidation in PLIN5 deficient AML12 cells is at least partly through PPAR-alpha activation (Li et al. 2012a), which is likely driven by the increase in lipolysis based on previous studies that indicated the regulation of PPAR-alpha activity by lipolysis derived metabolites (Zechner et al. 2012; Haemmerle et al. 2011). However, it should be noted that these studies were done in hepatocytes and hepatoma cells lines cultured without modulations of PKA activity. As mentioned in the preceding section, PLIN5 stimulates lipolysis in response to PKA, and PLIN5 may have dual functions in the regulation of lipid metabolism in hepatocytes similar to PLIN1 in adipocytes that prevents lipolysis at the base line and promotes lipolysis upon PKA activation (Mason and Watt 2015).

Currently, there are three independent reports of whole-body *Plin5* knockout mice (Wang et al. 2014; Mason et al. 2014; Kuramoto et al. 2012). As summarized in a recent review (Mason and Watt 2015), all models show that the loss of PLIN5 has little effects on growth, weight, glucose homeostasis, adiposity, or serum lipids levels both on regular chow and high fat diet. The major phenotype of the whole body PLIN5 deficient mice is the lack of cardiac lipid droplet upon prolonged fasting and increased FA oxidation in cardiomyocytes resulting in oxidative stress in heart of aged mice (Kuramoto et al. 2012). Hepatic phenotypes of PLIN5 deficient mice seem to vary among the three mouse models reported making the data interpretation difficult at this point (Mason and Watt 2015). VLDL secretion is decreased in primary hepatocytes of PLIN5 deficient mice in one study (Wang et al. 2014), but the circulating VLDL-TG did not differ between PLIN5 deficient mice and wild type mice (Wang et al. 2014; Kuramoto et al. 2012). PLIN5 deficiency reduced (Wang et al. 2014), did not change, or increased hepatic TG contents after fasting (Mason et al. 2014; Kuramoto et al. 2012). As PLIN2 is also highly upregulated during fasting in the liver (Trevino et al. 2015), the extent of compensation by PLIN2 may modify TG accumulation among different lines of PLIN5 deficient mice. Genetic background and environmental factors such as gut microbiota may also be responsible for divergent phenotypes of PLIN5 deficient mice.

Although not always present (Mason and Watt 2015), PLIN5 expression is reported to be increased in some NAFLD affected livers in human and mouse (Granneman et al. 2011; Wang et al. 2014). Wang et al. focused their study of PLIN5 deficient mice on the liver after high fat diet and noted that hepatic TG along with enzymes catalyzing FA synthesis and *Plin2* were decreased in PLIN5 deficient mice compared with wild-type controls (Wang et al. 2014). Insulin sensitivity under normal or high fat diet was unchanged (Wang et al. 2014). However, PLIN5 deficiency led to a number of changes reflective of liver injury, ER stress and inflammation after high fat diet. Significant increases in serum aminotransferase and bilirubin levels along with major changes in liver lobular architecture including an increase in binucleate cells and intrahepatic cholestasis were noted with PLIN5 deficiency. Similarly, markers for lipid peroxidation, inflammation and ER stress were increased in the livers of PLIN5 deficient mice (Wang et al. 2014). Collectively, PLIN5 may protect hepatocytes from lipotoxic stress under overnutrition. It will be important to determine whether PLIN5 deficiency accelerates liver injury in other models such as methionine choline deficient diet that causes steatohepatitis and is known to induce PLIN5 in a lipocailin two dependent manner (Asimakopoulou et al. 2014).

6 Concluding Remarks

Since the excellent review on the topic of hepatic LD was published in 2011 (Okumura 2011), there has been significant progress in the field of LD and PLINs. We have better information regarding the characteristics of different PLINs. It has been firmly established that PLIN2 is the major hepatic PLIN that is tightly correlated with the extent of hepatosteatosis from various causes, and is critical for the maintenance of hepatosteatosis. PLIN1 is aberrantly expressed in certain liver diseases with macrosteatosis. PLIN5 is markedly increased in hepatosteatosis associated with energy deprivation and may play a key role in the regulation of lipolysis, while its role in obesity associated hepatosteatosis is less clear. PLIN3 expression in the liver is less responsive to nutritional statue of the liver but may serve as a pool of a recruitable LD protein due to its distribution in both cytosol and LD surface.

The contribution of LD in liver diseases such as NAFLD and NASH is complex. In general, the formation of LD coated with PLINs in the liver is associated with worsening of hepatosteatosis in terms of TG accumulation. However, it simultaneously may prevent the production of cytotoxic lipid metabolites. There are increasing numbers of cases where hepatosteatosis is dissociated from insulin sensitivity when PLINs and other lipid metabolic factors are altered (Farese et al. 2012). Thus, future studies should discern how hepatic PLINs specifically affect lipid and glucose metabolism in normal and pathological states, and what lipid species mediate the metabolic effects of PLINS.

References

Abdulhameed MD, Tawa GJ, Kumar K, Ippolito DL, Lewis JA, Stallings JD, Wallqvist A (2014) Systems level analysis and identification of pathways and networks associated with liver fibrosis. PLoS One 9, e112193

Arrese EL, Saudale FZ, Soulages JL (2014) Lipid droplets as signaling platforms linking metabolic and cellular functions. Lipid Insights 7:7–16

Asimakopoulou A, Borkham-Kamphorst E, Henning M, Yagmur E, Gassler N, Liedtke C, Berger T, Mak TW, Weiskirchen R (2014) Lipocalin-2 (LCN2) regulates PLIN5 expression and intracellular lipid droplet formation in the liver. Biochim Biophys Acta 1842:1513–1524

Bell M, Wang H, Chen H, McLenithan JC, Gong DW, Yang RZ, Yu D, Fried SK, Quon MJ, Londos C, Sztalryd C (2008) Consequences of lipid droplet coat protein downregulation in liver cells: abnormal lipid droplet metabolism and induction of insulin resistance. Diabetes 57:2037–2045

Bickel PE, Tansey JT, Welte MA (2009) PAT proteins, an ancient family of lipid droplet proteins that regulate cellular lipid stores. Biochim Biophys Acta 1791:419–440

Bindesboll C, Berg O, Arntsen B, Nebb HI, Dalen KT (2013) Fatty acids regulate perilipin5 in muscle by activating PPARdelta. J Lipid Res 54:1949–1963

Brasaemle DL, Barber T, Kimmel AR, Londos C (1997) Post-translational regulation of perilipin expression. Stabilization by stored intracellular neutral lipids. J Biol Chem 272:9378–9387

Buers I, Robenek H, Lorkowski S, Nitschke Y, Severs NJ, Hofnagel O (2009) TIP47, a lipid cargo protein involved in macrophage triglyceride metabolism. Arterioscler Thromb Vasc Biol 29:767–773

Bulankina AV, Deggerich A, Wenzel D, Mutenda K, Wittmann JG, Rudolph MG, Burger KN, Honing S (2009) TIP47 functions in the biogenesis of lipid droplets. J Cell Biol 185:641–655

Bussell R Jr, Eliezer D (2003) A structural and functional role for 11-mer repeats in alpha-synuclein and other exchangeable lipid binding proteins. J Mol Biol 329:763–778

Carr RM, Patel RT, Rao V, Dhir R, Graham MJ, Crooke RM, Ahima RS (2012) Reduction of TIP47 improves hepatic steatosis and glucose homeostasis in mice. Am J Physiol Regul Integr Comp Physiol 302:R996–R1003

Carr RM, Dhir R, Yin X, Agarwal B, Ahima RS (2013) Temporal effects of ethanol consumption on energy homeostasis, hepatic steatosis, and insulin sensitivity in mice. Alcohol Clin Exp Res 37:1091–1099

Carr RM, Peralta G, Yin X, Ahima RS (2014) Absence of perilipin 2 prevents hepatic steatosis, glucose intolerance and ceramide accumulation in alcohol-fed mice. PLoS One 9, e97118

Chang BH, Li L, Paul A, Taniguchi S, Nannegari V, Heird WC, Chan L (2006) Protection against fatty liver but normal adipogenesis in mice lacking adipose differentiation-related protein. Mol Cell Biol 26:1063–1076

Chang BH, Li L, Saha P, Chan L (2010) Absence of adipose differentiation related protein upregulates hepatic VLDL secretion, relieves hepatosteatosis, and improves whole body insulin resistance in leptin-deficient mice. J Lipid Res 51:2132–2142

Corella D, Qi L, Sorli JV, Godoy D, Portoles O, Coltell O, Greenberg AS, Ordovas JM (2005) Obese subjects carrying the 11482G>A polymorphism at the perilipin locus are resistant to weight loss after dietary energy restriction. J Clin Endocrinol Metab 90:5121–5126

Dalen KT, Ulven SM, Arntsen BM, Solaas K, Nebb HI (2006) PPARalpha activators and fasting induce the expression of adipose differentiation-related protein in liver. J Lipid Res 47:931–943

Dalen KT, Dahl T, Holter E, Arntsen B, Londos C, Sztalryd C, Nebb HI (2007) LSDP5 is a PAT protein specifically expressed in fatty acid oxidizing tissues. Biochim Biophys Acta 1771:210–227

Diaz E, Pfeffer SR (1998) TIP47: a cargo selection device for mannose 6-phosphate receptor trafficking. Cell 93:433–443

Donnelly KL, Smith CI, Schwarzenberg SJ, Jessurun J, Boldt MD, Parks EJ (2005) Sources of fatty acids stored in liver and secreted via lipoproteins in patients with nonalcoholic fatty liver disease. J Clin Invest 115:1343–1351

Egan JJ, Greenberg AS, Chang MK, Londos C (1990) Control of endogenous phosphorylation of the major cAMP-dependent protein kinase substrate in adipocytes by insulin and beta-adrenergic stimulation. J Biol Chem 265:18769–18775

Farese RV Jr, Zechner R, Newgard CB, Walther TC (2012) The problem of establishing relationships between hepatic steatosis and hepatic insulin resistance. Cell Metab 15:570–573

Friedman SL (2008) Hepatic stellate cells: protean, multifunctional, and enigmatic cells of the liver. Physiol Rev 88:125–172

Fujii H, Ikura Y, Arimoto J, Sugioka K, Iezzoni JC, Park SH, Naruko T, Itabe H, Kawada N, Caldwell SH, Ueda M (2009) Expression of perilipin and adipophilin in nonalcoholic fatty liver disease; relevance to oxidative injury and hepatocyte ballooning. J Atheroscler Thromb 16:893–901

Garbarino J, Padamsee M, Wilcox L, Oelkers PM, D'Ambrosio D, Ruggles KV, Ramsey N, Jabado O, Turkish A, Sturley SL (2009) Sterol and diacylglycerol acyltransferase deficiency triggers fatty acid-mediated cell death. J Biol Chem 284:30994–31005

Granneman JG, Moore HP, Krishnamoorthy R, Rathod M (2009a) Perilipin controls lipolysis by regulating the interactions of AB-hydrolase containing 5 (Abhd5) and adipose triglyceride lipase (Atgl). J Biol Chem 284:34538–34544

Granneman JG, Moore HP, Mottillo EP, Zhu Z (2009b) Functional interactions between Mldp (LSDP5) and Abhd5 in the control of intracellular lipid accumulation. J Biol Chem 284:3049–3057

Granneman JG, Moore HP, Mottillo EP, Zhu Z, Zhou L (2011) Interactions of perilipin-5 (Plin5) with adipose triglyceride lipase. J Biol Chem 286:5126–5135

Grasselli E, Voci A, Pesce C, Canesi L, Fugassa E, Gallo G, Vergani L (2010) PAT protein mRNA expression in primary rat hepatocytes: Effects of exposure to fatty acids. Int J Mol Med 25:505–512

Greco D, Kotronen A, Westerbacka J, Puig O, Arkkila P, Kiviluoto T, Laitinen S, Kolak M, Fisher RM, Hamsten A, Auvinen P, Yki-Jarvinen H (2008) Gene expression in human NAFLD. Am J Physiol Gastrointest Liver Physiol 294:G1281–G1287

Greenberg AS, Egan JJ, Wek SA, Garty NB, Blanchette-Mackie EJ, Londos C (1991) Perilipin, a major hormonally regulated adipocyte-specific phosphoprotein associated with the periphery of lipid storage droplets. J Biol Chem 266:11341–11346

Greenberg AS, Egan JJ, Wek SA, Moos MC Jr, Londos C, Kimmel AR (1993) Isolation of cDNAs for perilipins A and B: sequence and expression of lipid droplet-associated proteins of adipocytes. Proc Natl Acad Sci U S A 90:12035–12039

Greenberg AS, Coleman RA, Kraemer FB, McManaman JL, OBIN MS, Puri V, Yan QW, Miyoshi H, Mashek DG (2011) The role of lipid droplets in metabolic disease in rodents and humans. J Clin Invest 121:2102–2110

Haemmerle G, Moustafa T, Woelkart G, Büttner S, Schmidt A, van de Weijer T, Hesselink M, Jaeger D, Kienesberger PC, Zierler K, Schreiber R, Eichmann T, Kolb D, Kotzbeck P, Schweiger M, Kumari M, Eder S, Schoiswohl G, Wongsiriroj N, Pollak NM, Radner FP, Preiss-Landl K, Kolbe T, Rülicke T, Pieske B, Trauner M, Lass A, Zimmermann R, Hoefler G, Cinti S, Kershaw EE, Schrauwen P, Madeo F, Mayer B, Zechner R (2011) ATGL-mediated fat catabolism regulates cardiac mitochondrial function via PPAR-alpha and PGC-1. Nat Med 17:1076–1085

Hall AM, Soufi N, Chambers KT, Chen Z, Schweitzer GG, McCommis KS, Erion DM, Graham MJ, Su X, Finck BN (2014) Abrogating monoacylglycerol acyltransferase activity in liver improves glucose tolerance and hepatic insulin signaling in obese mice. Diabetes 63:2284–2296

Harris C, Herker E, Farese RV Jr, Ott M (2011) Hepatitis C virus core protein decreases lipid droplet turnover: a mechanism for core-induced steatosis. J Biol Chem 286:42615–42625

Herker E, Harris C, Hernandez C, Carpentier A, Kaehlcke K, Rosenberg AR, Farese RV Jr, Ott M (2010) Efficient hepatitis C virus particle formation requires diacylglycerol acyltransferase-1. Nat Med 16:1295–1298

Holzer RG, Park EJ, Li N, Tran H, Chen M, Choi C, Solinas G, Karin M (2011) Saturated fatty acids induce c-Src clustering within membrane subdomains, leading to JNK activation. Cell 147:173–184

Honnor RC, Dhillon GS, Londos C (1985) cAMP-dependent protein kinase and lipolysis in rat adipocytes. II. Definition of steady-state relationship with lipolytic and antilipolytic modulators. J Biol Chem 260:15130–15138

Hsieh K, Lee YK, Londos C, Raaka BM, Dalen KT, Kimmel AR (2012) Perilipin family members preferentially sequester to either triacylglycerol-specific or cholesteryl-ester-specific intracellular lipid storage droplets. J Cell Sci 125:4067–4076

Imai Y, Varela GM, Jackson MB, Graham MJ, Crooke RM, Ahima RS (2007) Reduction of hepatosteatosis and lipid levels by an adipose differentiation-related protein antisense oligonucleotide. Gastroenterology 132:1947–1954

Imai Y, Boyle S, Varela GM, Caron E, Yin X, Dhir R, Dhir R, Graham MJ, Ahima RS (2012) Effects of perilipin 2 antisense oligonucleotide treatment on hepatic lipid metabolism and gene expression. Physiol Genomics 44:1125–1131

Ingelmo-Torres M, González-Moreno E, Kassan A, Hanzal-Bayer M, Tebar F, Herms A, Grewal T, Hancock JF, Enrich C, Bosch M, Gross SP, Parton RG, Pol A (2009) Hydrophobic and basic domains target proteins to lipid droplets. Traffic 10:1785–1801

Jang Y, Kim OY, Lee JH, Koh SJ, Chae JS, Kim JY, Park S, Cho H, Lee JE, Ordovas JM (2006) Genetic variation at the perilipin locus is associated with changes in serum free fatty acids and abdominal fat following mild weight loss. Int J Obes (Lond) 30:1601–1608

Jenkins CM, Mancuso DJ, Yan W, Sims HF, Gibson B, Gross RW (2004) Identification, cloning, expression, and purification of three novel human calcium-independent phospholipase A2 family members possessing triacylglycerol lipase and acylglycerol transacylase activities. J Biol Chem 279:48968–48975

Jiang HP, Harris SE, Serrero G (1992) Molecular cloning of a differentiation-related mRNA in the adipogenic cell line 1246. Cell Growth Differ 3:21–30

KHOR VK, Shen WJ, Kraemer FB (2013) Lipid droplet metabolism. Curr Opin Clin Nutr Metab Care 16:632–637

Kimmel AR, Sztalryd C (2014) Perilipin 5, a lipid droplet protein adapted to mitochondrial energy utilization. Curr Opin Lipidol 25:110–117

Kimmel AR, Brasaemle DL, McAndrews-Hill M, Sztalryd C, Londos C (2010) Adoption of PERILIPIN as a unifying nomenclature for the mammalian PAT-family of intracellular lipid storage droplet proteins. J Lipid Res 51:468–471

Kohjima M, Tsai TH, Tackett BC, Thevananther S, Li L, Chang BH, Chan L (2013) Delayed liver regeneration after partial hepatectomy in adipose differentiation related protein-null mice. J Hepatol 59:1246–1254

Kuramoto K, Okamura T, Yamaguchi T, Nakamura TY, Wakabayashi S, Morinaga H, Nomura M, Yanase T, Otsu K, Usuda N, Matsumura S, Inoue K, Fushiki T, Kojima Y, Hashimoto T, Sakai F, Hirose F, Osumi T (2012) Perilipin 5, a lipid droplet-binding protein, protects heart from oxidative burden by sequestering fatty acid from excessive oxidation. J Biol Chem 287:23852–23863

Langhi C, Baldan A (2015) CIDEC/Fsp27 is regulated by PPARalpha and plays a critical role in fasting- and diet-induced hepatosteatosis. Hepatology 61(4):1227–1238

Langhi C, Marquart TJ, Allen RM, Baldan A (2014) Perilipin-5 is regulated by statins and controls triglyceride contents in the hepatocyte. J Hepatol 61:358–365

Lass A, Zimmermann R, Haemmerle G, Riederer M, Schoiswohl G, Schweiger M, Kienesberger P, Strauss JG, Gorkiewicz G, Zechner R (2006) Adipose triglyceride lipase-mediated lipolysis of cellular fat stores is activated by CGI-58 and defective in Chanarin-Dorfman Syndrome. Cell Metab 3:309–319

Lee B, Zhu J, Wolins NE, Cheng JX, Buhman KK (2009) Differential association of adipophilin and TIP47 proteins with cytoplasmic lipid droplets in mouse enterocytes during dietary fat absorption. Biochim Biophys Acta 1791:1173–1180

Lee TF, Mak KM, Rackovsky O, Lin YL, Kwong AJ, Loke JC, Friedman SL (2010) Downregulation of hepatic stellate cell activation by retinol and palmitate mediated by adipose differentiation-related protein (ADRP). J Cell Physiol 223:648–657

Li H, Song Y, Zhang LJ, Gu Y, Li FF, Pan SY, Jiang LN, Liu F, Ye J, Li Q (2012a) LSDP5 enhances triglyceride storage in hepatocytes by influencing lipolysis and fatty acid beta-oxidation of lipid droplets. PLoS One 7, e36712

Li X, Ye J, Zhou L, Gu W, Fisher EA, Li P (2012b) Opposing roles of cell death-inducing DFF45-like effector B and perilipin 2 in controlling hepatic VLDL lipidation. J Lipid Res 53:1877–1889

Listenberger LL, Ostermeyer-FAY AG, Goldberg EB, Brown WJ, Brown DA (2007) Adipocyte differentiation-related protein reduces the lipid droplet association of adipose triglyceride lipase and slows triacylglycerol turnover. J Lipid Res 48:2751–2761

Londos C, Honnor RC, Dhillon GS (1985) cAMP-dependent protein kinase and lipolysis in rat adipocytes. III. Multiple modes of insulin regulation of lipolysis and regulation of insulin responses by adenylate cyclase regulators. J Biol Chem 260:15139–15145

Londos C, Sztalryd C, Tansey JT, Kimmel AR (2005) Role of PAT proteins in lipid metabolism. Biochimie 87:45–49

Lu X, Gruia-Gray J, Copeland NG, Gilbert DJ, Jenkins NA, LONDOS C, Kimmel AR (2001) The murine perilipin gene: the lipid droplet-associated perilipins derive from tissue-specific, mRNA splice variants and define a gene family of ancient origin. Mamm Genome 12:741–749

Magne J, Aminoff A, Perman Sundelin J, Mannila MN, Gustafsson P, Hultenby K, Wernerson A, Bauer G, Listenberger L, Neville MJ, Karpe F, Boren J, Ehrenborg E (2013) The minor allele of the missense polymorphism Ser251Pro in perilipin 2 (PLIN2) disrupts an alpha-helix, affects lipolysis, and is associated with reduced plasma triglyceride concentration in humans. FASEB J 27:3090–3099

Magnusson B, ASP L, Bostrom P, Ruiz M, Stillemark-Billton P, Linden D, Boren J, Olofsson SO (2006) Adipocyte differentiation-related protein promotes fatty acid storage in cytosolic triglycerides and inhibits secretion of very low-density lipoproteins. Arterioscler Thromb Vasc Biol 26:1566–1571

Mak KM, Ren C, Ponomarenko A, Cao Q, Lieber CS (2008) Adipose differentiation-related protein is a reliable lipid droplet marker in alcoholic fatty liver of rats. Alcohol Clin Exp Res 32:683–689

Martinez-Botas J, Anderson JB, Tessier D, Lapillonne A, Chang BH, Quast MJ, Gorenstein D, Chen KH, Chan L (2000) Absence of perilipin results in leanness and reverses obesity in Lepr(db/db) mice. Nat Genet 26:474–479

Martínez-Uña M, Varela-Rey M, Cano A, Fernández-Ares L, Beraza N, Aurrekoetxea I, Martínez-Arranz I, García-Rodríguez JL, Buqué X, Mestre D, Luka Z, Wagner C, Alonso C, Finnell RH, Lu SC, Martínez-Chantar ML, Aspichueta P, Mato JM (2013) Excess S-adenosylmethionine reroutes phosphatidylethanolamine towards phosphatidylcholine and triglyceride synthesis. Hepatology 58:1296–1305

Mashek DG (2013) Hepatic fatty acid trafficking: multiple forks in the road. Adv Nutr 4:697–710

Mason RR, Watt MJ (2015) Unraveling the roles of PLIN5: linking cell biology to physiology. Trends Endocrinol Metab 26:144–152

Mason RR, Mokhtar R, Matzaris M, Selathurai A, KOWALSKI GM, Mokbel N, Meikle PJ, Bruce CR, Watt MJ (2014) PLIN5 deletion remodels intracellular lipid composition and causes insulin resistance in muscle. Mol Metab 3:652–663

McManaman JL, Bales ES, Orlicky DJ, Jackman M, Maclean PS, Cain S, Crunk AE, Mansur A, Graham CE, Bowman TA, Greenberg AS (2013) Perilipin-2-null mice are protected against diet-induced obesity, adipose inflammation, and fatty liver disease. J Lipid Res 54:1346–1359

McVicker BL, Rasineni K, Tuma DJ, McNiven MA, Casey CA (2012) Lipid droplet accumulation and impaired fat efflux in polarized hepatic cells: consequences of ethanol metabolism. Int J Hepatol 2012:978136

Miura S, Gan JW, Br Zostowski J, PARISI MJ, SCHULTZ CJ, LONDOS C, OLIVER B, KIMMEL AR (2002) Functional conservation for lipid storage droplet association among Perilipin, ADRP, and TIP47 (PAT)-related proteins in mammals, Drosophila, and Dictyostelium. J Biol Chem 277:32253–32257

Motomura W, Inoue M, Ohtake T, Takahashi N, Nagamine M, Tanno S, Kohgo Y, Okumura T (2006) Up-regulation of ADRP in fatty liver in human and liver steatosis in mice fed with high fat diet. Biochem Biophys Res Commun 340:1111–1118

Mottagui-Tabar S, Ryden M, Lofgren P, Faulds G, Hoffstedt J, Brookes AJ, Andersson I, Arner P (2003) Evidence for an important role of perilipin in the regulation of human adipocyte lipolysis. Diabetologia 46:789–797

Nassir F, Adewole OL, Brunt EM, Abumrad NA (2013) CD36 deletion reduces VLDL secretion, modulates liver prostaglandins, and exacerbates hepatic steatosis in ob/ob mice. J Lipid Res 54:2988–2997

Okumura T (2011) Role of lipid droplet proteins in liver steatosis. J Physiol Biochem 67:629–636

Ong KT, Mashek MT, Bu SY, Greenberg AS, Mashek DG (2011) Adipose triglyceride lipase is a major hepatic lipase that regulates triacylglycerol turnover and fatty acid signaling and partitioning. Hepatology 53:116–126

Orlicky DJ, Roede JR, Bales E, Greenwood C, Greenberg A, Petersen D, McManaman JL (2011) Chronic ethanol consumption in mice alters hepatocyte lipid droplet properties. Alcohol Clin Exp Res 35:1020–1033

Pawella LM, Hashani M, Eiteneuer E, Renner M, Bartenschlager R, Schirmacher P, Straub BK (2014) Perilipin discerns chronic from acute hepatocellular steatosis. J Hepatol 60:633–642

Perry RJ, Samuel VT, Petersen KF, Shulman GI (2014) The role of hepatic lipids in hepatic insulin resistance and type 2 diabetes. Nature 510:84–91

Ploen D, Hafirassou ML, Himmelsbach K, Schille SA, Biniossek ML, Baumert TF, Schuster C, Hildt E (2013) TIP47 is associated with the hepatitis C virus and its interaction with Rab9 is required for release of viral particles. Eur J Cell Biol 92:374–382

Pol A, Gross SP, Parton RG (2014) Review: biogenesis of the multifunctional lipid droplet: lipids, proteins, and sites. J Cell Biol 204:635–646

Pollak NM, Jaeger D, Kolleritsch S, Zimmermann R, Zechner R, Lass A, Haemmerle G (2015) The interplay of protein kinase A and perilipin 5 regulates cardiac lipolysis. J Biol Chem 290(3):1295–1306

Qi L, Corella D, Sorli JV, Portoles O, Shen H, Coltell O, Godoy D, Greenberg AS, Ordovas JM (2004a) Genetic variation at the perilipin (PLIN) locus is associated with obesity-related phenotypes in White women. Clin Genet 66:299–310

Qi L, Shen H, Larson I, Schaefer EJ, Greenberg AS, Tregouet DA, Corella D, Ordovas JM (2004b) Gender-specific association of a perilipin gene haplotype with obesity risk in a white population. Obes Res 12:1758–1765

Qi L, Tai ES, Tan CE, Shen H, Chew SK, Greenberg AS, Corella D, Ordovas JM (2005) Intragenic linkage disequilibrium structure of the human perilipin gene (PLIN) and haplotype association with increased obesity risk in a multiethnic Asian population. J Mol Med (Berl) 83:448–456

Reid BN, Ables GP, Otlivanchik OA, Schoiswohl G, Zechner R, Blaner WS, Goldberg IJ, Schwabe RF, Chua SC Jr, Huang LS (2008) Hepatic overexpression of hormone-sensitive lipase and adipose triglyceride lipase promotes fatty acid oxidation, stimulates direct release of free fatty acids, and ameliorates steatosis. J Biol Chem 283:13087–13099

Schaffer JE (2002) Fatty acid transport: the roads taken. Am J Physiol Endocrinol Metab 282:E239–E246

Scherer PE, Bickel PE, Kotler M, Lodish HF (1998) Cloning of cell-specific secreted and surface proteins by subtractive antibody screening. Nat Biotechnol 16:581–586

Schweiger M, Lass A, Zimmermann R, Eichmann TO, Zechner R (2009) Neutral lipid storage disease: genetic disorders caused by mutations in adipose triglyceride lipase/PNPLA2 or CGI-58/ABHD5. Am J Physiol Endocrinol Metab 297:E289–E296

Shen WJ, Patel S, Miyoshi H, Greenberg AS, Kraemer FB (2009) Functional interaction of hormone-sensitive lipase and perilipin in lipolysis. J Lipid Res 50:2306–2313

Skinner JR, Shew TM, Schwartz DM, Tzekov A, Lepus CM, Abumrad NA, Wolins NE (2009) Diacylglycerol enrichment of endoplasmic reticulum or lipid droplets recruits perilipin 3/TIP47 during lipid storage and mobilization. J Biol Chem 284:30941–30948

Stevanovic A, Thiele C (2013) Monotopic topology is required for lipid droplet targeting of ancient ubiquitous protein 1. J Lipid Res 54:503–513

Straub BK, Stoeffel P, Heid H, Zimbelmann R, Schirmacher P (2008) Differential pattern of lipid droplet-associated proteins and de novo perilipin expression in hepatocyte steatogenesis. Hepatology 47:1936–1946

Sun Z, Miller RA, Patel RT, Chen J, Dhir R, Wang H, Zhang D, Graham MJ, Unterman TG, Shulman GI, Sztalryd C, Bennett MJ, AHIMA RS, Birnbaum MJ, Lazar MA (2012) Hepatic Hdac3 promotes gluconeogenesis by repressing lipid synthesis and sequestration. Nat Med 18:934–942

Sztalryd C, Bell M, Lu X, Mertz P, Hickenbottom S, Chang BH, Chan L, Kimmel AR, Londos C (2006) Functional compensation for adipose differentiation-related protein (ADFP) by Tip47 in an ADFP null embryonic cell line. J Biol Chem 281:34341–34348

Tansey JT, Sztalryd C, Gruia-Gray J, Roush DL, Zee JV, Gavrilova O, Reitman ML, Deng CX, Li C, Kimmel AR, Londos C (2001) Perilipin ablation results in a lean mouse with aberrant adipocyte lipolysis, enhanced leptin production, and resistance to diet-induced obesity. Proc Natl Acad Sci U S A 98:6494–6499

Than NG, Sumegi B, Than GN, Kispal G, Bohn H (1998) Cloning and sequence analysis of cDNAs encoding human placental tissue protein 17 (PP17) variants. Eur J Biochem 258:752–757

Thiam AR, Farese RV Jr, Walther TC (2013) The biophysics and cell biology of lipid droplets. Nat Rev Mol Cell Biol 14:775–786

Trevino M, Machida Y, Hallinger D, Garcia E, Christensen A, Dutta S, Peake D, Ikeda Y, Imai Y (2015) Perilipin 5 regulates islet lipid metabolism and insulin secretion in a cyclic AMP dependent manner: implication of its role in the postprandial insulin secretion. Diabetes 64(4):1299–1310

Tsuiki E, Fujita A, Ohsaki Y, CHENG J, Irie T, Yoshikawa K, Senoo H, Mishima K, Kitaoka T, Fujimoto T (2007) All-trans-retinol generated by rhodopsin photobleaching induces rapid recruitment of TIP47 to lipid droplets in the retinal pigment epithelium. Invest Ophthalmol Vis Sci 48:2858–2867

Varela GM, Antwi DA, Dhir R, Yin X, Singhal NS, Graham MJ, Crooke RM, Ahima RS (2008) Inhibition of ADRP prevents diet-induced insulin resistance. Am J Physiol Gastrointest Liver Physiol 295:G621–G628

Viswanadha S, Londos C (2008) Determination of lipolysis in isolated primary adipocytes. Methods Mol Biol 456:299–306

Walther TC, Farese RV Jr (2009) The life of lipid droplets. Biochim Biophys Acta 1791:459–466

Wang H, Hu L, Dalen K, Dorward H, Marcinkiewicz A, Russell D, Gong D, Londos C, Yamaguchi T, Holm C, Rizzo MA, Brasaemle D, Sztalryd C (2009) Activation of hormone-sensitive lipase requires two steps, protein phosphorylation and binding to the PAT-1 domain of lipid droplet coat proteins. J Biol Chem 284:32116–32125

Wang H, Bell M, Sreenivasan U, Hu H, Liu J, Dalen K, Londos C, Yamaguchi T, Rizzo MA, Coleman R, Gong D, Brasaemle D, Sztalryd C (2011a) Unique regulation of adipose triglyceride lipase (ATGL) by perilipin 5, a lipid droplet-associated protein. J Biol Chem 286:15707–15715

Wang H, Sreenivasan U, Hu H, Saladino A, Polster BM, Lund LM, Gong DW, Stanley WC, Sztalryd C (2011b) Perilipin 5, a lipid droplet-associated protein, provides physical and metabolic linkage to mitochondria. J Lipid Res 52:2159–2168

Wang C, Zhao Y, Gao X, Li L, Yuan Y, Liu F, Zhang L, Wu J, Hu P, Zhang X, Gu Y, Xu Y, Li Z, Zhang H, Ye J (2014) Perilipin 5 improves hepatic lipotoxicity by inhibiting lipolysis. Hepatology 61(3):870–882

Wilfling F, Wang H, Haas JT, Krahmer N, Gould TJ, Uchida A, Cheng JX, Graham M, Christiano R, Fröhlich F, Liu X, Buhman KK, Coleman RA, Bewersdorf J, Farese RV Jr, Walther TC (2013) Triacylglycerol synthesis enzymes mediate lipid droplet growth by relocalizing from the ER to lipid droplets. Dev Cell 24:384–399

Wilfling F, Thiam AR, Olarte MJ, Wang J, Beck R, Gould TJ, Allgeyer ES, Pincet F, Bewersdorf J, Farese RV Jr, Walther TC (2014) Arf1/COPI machinery acts directly on lipid droplets and enables their connection to the ER for protein targeting. Elife 3, e01607

Wolins NE, Rubin B, Brasaemle DL (2001) TIP47 associates with lipid droplets. J Biol Chem 276:5101–5108

Wolins NE, Skinner JR, Schoenfish MJ, Tzekov A, Bensch KG, Bickel PE (2003) Adipocyte protein S3-12 coats nascent lipid droplets. J Biol Chem 278:37713–37721

Wolins NE, Quaynor BK, Skinner JR, Schoenfish MJ, Tzekov A, Bickel PE (2005) S3-12, Adipophilin, and TIP47 package lipid in adipocytes. J Biol Chem 280:19146–19155

Wolins NE, Quaynor BK, Skinner JR, Tzekov A, Croce MA, Gropler MC, Varma V, Yao-Borengasser A, Rasouli N, Kern PA, Finck BN, Bickel PE (2006) OXPAT/PAT-1 is a PPAR-induced lipid droplet protein that promotes fatty acid utilization. Diabetes 55:3418–3428

Wu JW, Wang SP, Alvarez F, Casavant S, Gauthier N, Abed L, Soni KG, Yang G, Mitchell GA (2011) Deficiency of liver adipose triglyceride lipase in mice causes progressive hepatic steatosis. Hepatology 54:122–132

Wu L, Xu D, Zhou L, Xie B, Yu L, Yang H, Huang L, Ye J, Deng H, Yuan YA, Chen S, Li P (2014) Rab8a-AS160-MSS4 regulatory circuit controls lipid droplet fusion and growth. Dev Cell 30:378–393

Xu G, Sztalryd C, Lu X, Tansey JT, Gan J, Dorward H, Kimmel AR, Londos C (2005) Post-translational regulation of adipose differentiation-related protein by the ubiquitin/proteasome pathway. J Biol Chem 280:42841–42847

Yamaguchi T, Matsushita S, Motojima K, Hirose F, Osumi T (2006) MLDP, a novel PAT family protein localized to lipid droplets and enriched in the heart, is regulated by peroxisome proliferator-activated receptor alpha. J Biol Chem 281:14232–14240

Yamamoto K, Takahara K, Oyadomari S, Okada T, Sato T, Harada A, Mori K (2010) Induction of liver steatosis and lipid droplet formation in ATF6alpha-knockout mice burdened with pharmacological endoplasmic reticulum stress. Mol Biol Cell 21:2975–2986

Yu S, Matsusue K, Kashireddy P, Cao WQ, Yeldandi V, Yeldandi AV, Rao MS, Gonzalez FJ, Reddy JK (2003) Adipocyte-specific gene expression and adipogenic steatosis in the mouse liver due to peroxisome proliferator-activated receptor gamma1 (PPARgamma1) overexpression. J Biol Chem 278:498–505

Zechner R, Zimmermann R, Eichmann TO, Kohlwein SD, Haemmerle G, Lass A, Madeo F (2012) Fat signals—lipases and lipolysis in lipid metabolism and signaling. Cell Metab 15:279–291

Chapter 9
Influence of Gut Microbiota on Hepatic Lipogenesis and Disease Pathogenesis

Vishal Singh, Beng San Yeoh, James M. Ntambi, and Matam Vijay-Kumar

Abstract Liver, a major metabolic organ in the human body, plays a key role in the macronutrient metabolism. Accordingly, the liver is intertwined with an array of biochemical pathways that are central in keeping energy homeostasis. Loss of this delicate balance, however, may lead to a spectrum of liver diseases such as NAFLD, NASH, fibrosis, cirrhosis and HCC. The etiology of liver disorders is often multifactorial and greatly influenced by host genetics, diet and various environmental factors. Among the environmental factors, the gut microbiota has emerged as one of the key player that can profoundly influence host metabolism. Akin to a double-edged sword, the gut microbiota and their metabolites exert both beneficial and deleterious extra-intestinal effects to the host. As a vascular firewall that receives 70 % of blood enriched with gut-derived factors via portal vein, the liver is thus aptly placed at the crossroad between host health and disease. In this chapter, we focus on the influence of gut microbiota and their metabolites in hepatic lipogenesis and disease pathogenesis.

Keywords Gut bacteria • Dietary fiber • Short chain fatty acids • Endotoxemia • Insulin resistance • SCD-1 • NASH

1 Introduction

Liver, the largest internal organ in the human body, literally means '*life*' in the old Germanic and Romance languages (Riva et al. 2011). As the name suggests, the liver occupies a central position in energy, metabiotic, endobiotic and xenobiotic metabolism essential to life. It can be viewed as an altruistic organ due to its

V. Singh • B.S. Yeoh • M. Vijay-Kumar, Ph.D. (✉)
Department of Nutritional Sciences, The Pennsylvania State University, University Park, PA 16802, USA

J.M. Ntambi
Department of Biochemistry, University of Wisconsin-Madison, Madison, WI, USA

Department of Nutritional Sciences, University of Wisconsin-Madison, Madison, WI, USA
e-mail: mvk13@psu.edu

J.M. Ntambi (ed.), *Hepatic De Novo Lipogenesis and Regulation of Metabolism*,
DOI 10.1007/978-3-319-25065-6_9

continuous supply of nutrients during starvation. For instance, the liver synthesizes ketone bodies for other organs but does not use them because of the lack of the enzyme thiophorase/β-ketoacyl-CoA transferase (Laffel 1999). During inflammation, the liver has an additional responsibility of secreting numerous acute phase proteins [*e.g.*, secretory IL-1 receptor antagonist (sIL-1Ra)] to counter the progression of acute inflammation.

The liver is strategically placed so that absorbed intestinal contents must first pass through it via the hepatic portal vein before entering systemic circulation. As a vascular 'firewall,' the liver is well-equipped with an array of enzymes to detoxify gut bacterial-derived metabolites including ammonia, ethanol, acetaldehyde, phenols and benzodiazepines (Abu-Shanab and Quigley 2010). Additionally, this organ houses an abundant population of Kupffer cells (resident macrophages) that mediate rapid clearance of bacteria and prevent their translocation into systemic circulation, especially during intestinal inflammation or injury (Balmer et al. 2014). Nevertheless, the liver and gut commensals have mutually co-evolved so that their combined activities ultimately dictate the host metabolism. For example, gut microbiota generate several metabolites including short-chain fatty acids (SCFAs), which influence host metabolism including hepatic lipogenesis. In this chapter, we mostly focus on the interplay between the gut microbiota and the liver and how that influences hepatic lipogenesis and disease pathogenesis.

2 Gut Microbiota: An Organ Within an Organ

The total number of microbes that inhabit our body is collectively termed the 'microbiota,' and it has been proven to play a versatile function in our energy homeostasis. Microbes start colonizing the human gut immediately after birth and grow to number in the trillions, which outnumbers our body cells. Microbiota are remarkably and unexpectedly diverse, and many of them specialize in fermenting the soluble fiber in natural foods, such as grains, legumes and fruits (2015; Dorrestein et al. 2014). Intriguingly, the collective genome of the microbiota known as 'microbiome' is estimated to be 150 times larger than the human genome, which prompted many researchers to regard the microbiota as our 'second genome' (Grice and Segre 2012). The human gut microbiota is mainly composed of the phyla Bacteroidetes and Firmicutes, whereas Proteobacteria, Verrucomicrobia, Actinobacteria, Fusobacteria and Cyanobacteria are present in minor proportions (Fig. 9.1) (Eckburg et al. 2005; Sekirov et al. 2010). Several factors can affect the gut microbial composition, yet surprisingly the composition of the human microbiota remains fairly stable at the phylum level (Sekirov et al. 2010).

The amount and type of dietary fiber are among the major factors that influence the gut microbial composition (Fernandes et al. 2014; Musso et al. 2010). Consumption of a low-fiber Western obesogenic diet significantly alters the microbial composition of the murine gut (Turnbaugh et al. 2008). Interestingly, dietary manipulations that limited weight gain were able to reverse the effects of diet-

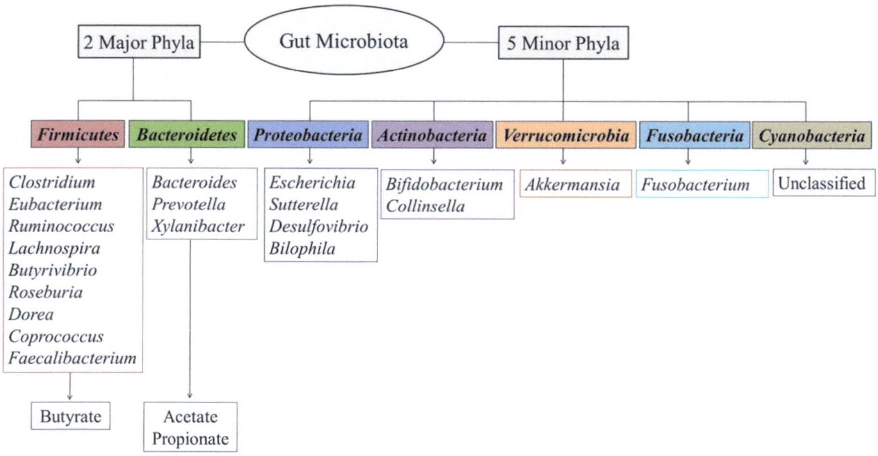

Fig. 9.1 Microbiota composition in the human gut. Human gut microbiota is mainly composed of the phyla Bacteroidetes and Firmicutes, whereas Proteobacteria, Verrucomicrobia, Actinobacteria, Fusobacteria and Cyanobacteria are present in minor proportions. A non-exhaustive list of bacteria genera from each phyla represents bacteria populations that are commonly found in the human gut. Short-chain fatty acid butyrate are mainly produced by members of the Firmicutes phyla whereas acetate and propionate are mainly produced by members of the Bacteroidetes phyla

induced obesity on the microbiota (Sekirov et al. 2010). Feeding mice with a Western-style high-fat diet increases the ratio of Firmicutes/Bacteroidetes, a phenomenon termed as the "shift in obese-type microbiota" by several researchers (Conterno et al. 2011). Microbiota can influence food intake (Vijay-Kumar et al. 2010), dietary nutrient harvest (Semova et al. 2012), regulate intestinal absorption (Carmody and Turnbaugh 2012; Semova et al. 2012), gastrointestinal motility (Anitha et al. 2012) and contribute in development of metabolic syndrome (Vijay-Kumar et al. 2010). Potential mechanisms by which the microbiota influence host metabolism include increased calorie extraction from ingested food (Ley et al. 2006; Turnbaugh et al. 2006), metabolic endotoxemia that drives insulin resistance (Cani et al. 2008), and generation of toxic metabolites (Wang et al. 2011). Studies have demonstrated that the conventionally-raised mice have significantly more body fat than their germ-free counterparts (Backhed et al. 2004, 2005; 2007). More importantly, germ-free mice are resistant to diet-induced obesity which suggest that the gut microbiota is an important environmental factor that affects energy harvest from the diet and its storage in the host (Backhed et al. 2004).

Most of the knowledge we gained on microbiota are due to the generation of germ-free mice and microbiotal transplantation studies. Further development of high-throughput sequencing also contributed substantially in this endeavor. Bäckhed et al. found that germ-free mice that received microbiota transplantation display suppressed production of fasting-induced adipocyte factor (FIAF), a potent inhibitor of lipoprotein lipase (Backhed et al. 2007). Microbiota-mediated suppression of FIAF was found to increase the induction of de novo hepatic lipogenesis, followed by increasing triglycerides accumulation in the liver and adipose tissues. The same group also discovered that germ-free mice resistance to high-fat diet-induced obesity is attributable to their constitutive activation of the enzyme adenosine monophosphate-activated protein kinase (AMPK) in the liver and muscle tissues (Backhed et al. 2007). AMPK primarily serves as a rheostat for a energy status sensor and activates ATP-generating catabolic pathways, enhances fatty acid oxidation in mitochondria, and reduces hepatic glycogen synthase activity. Interestingly, conventionalized germ-free mice display suppressed AMPK activity in the liver (in similar fashion to FIAF) (Backhed et al. 2007). All together, these studies implicated the active role of gut microbiota in modulating host energy extraction, and glucose and lipid metabolism. However, the microbiotal-associated factor(s) that influence FIAF and AMPK need to be further investigated.

Many studies over the past decades have begun to shed light on various contributions of the gut microbiota to liver metabolism, but what we currently know may be just the tip of the iceberg. As a major source for hormones and secondary metabolites, the gut microbiota is aptly dubbed the 'virtual endocrine organ'. Indeed, the gut microbiota are prominent producers of serotonin, tryptophan, indoles, dopamine, norepinephrine, histamine, and gamma-aminobutyric acid, and also are capable of inducing the release of host gut hormones such as incretin GLP-1 and peptide YY (Lyte 2013; Madsbad 2014; Samuel et al. 2008; Wall et al. 2014). Further, the gut microbiota constantly provide the host with many important nutrients such as thiamine, riboflavin, biotin, vitamin K, vitamin B6 and vitamin B12 (Said 2011). More importantly, we have come to appreciate that the host metabolic activities are, to an extent, regulated by bacterial-derived metabolites; SCFAs, secondary bile acids,

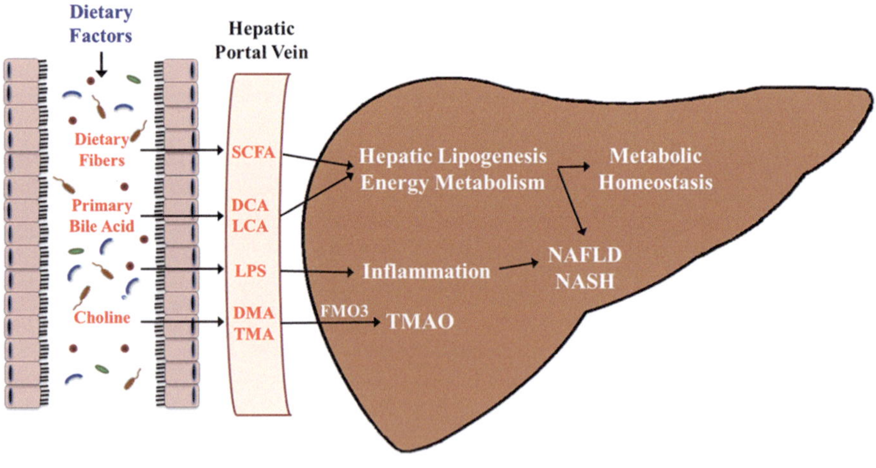

Fig. 9.2 Distinct gut microbiota-derived metabolites reach the liver and mediate effects on health and disease. *SCFA* short-chain fatty acids (acetate, propionate, butyrate), *DCA* deoxycholate, *LCA* lithocholate, *LPS* lipopolysaccharide, *DMA* dimethylamine, *TMA* trimethylamine, *FMO3* flavin monooxygenase 3, *TMAO* trimethylamine N-oxide, *NAFLD* non-alcoholic fatty liver disease, *NASH* non-alcoholic steatohepatitis

choline-derived metabolites and microbe-associated molecular patterns (MAMPs) are among those that are strongly linked to liver metabolism (Fig. 9.2).

3 SCFAs and Its Effects on Hepatic Lipogenesis

The gut microbiota functions as an environmental factor that regulates fat storage, and thus metabolic syndrome (Backhed et al. 2004, 2007). A plethora of research studies suggest that the gut microbiota functions as a metabolically active organ that ferments dietary fibers and generates several metabolites, which then can be absorbed and metabolized by the human body (Backhed et al. 2005; Shoaie et al. 2013). Many bacterial species have been implicated in the metabolism of dietary soluble fiber to SCFAs (*e.g.*, acetate, propionate and butyrate), the principal end products of dietary fiber fermentation by the anaerobic intestinal microbiota (Sekirov et al. 2010). Similar to gut microbial composition, SCFAs production by gut bacteria also are influenced by different factors, including host diet composition (Macfarlane and Macfarlane 2003). In both mice and humans, the three phyla: Bacteroidetes (gram-negative), Firmicutes (gram-positive) and Actinobacteria (gram-positive) are predominantly present in the intestine. The Bacteroidetes mainly produces acetate and propionate, whereas the Firmicutes produce butyrate as its primary metabolic end product (den Besten et al. 2013b; Macfarlane and Macfarlane 2003). Various population survey data show that fecal SCFAs content is in the order of acetate > propionate ≥ butyrate in a molar ratio of approximately 60:20:20, respectively (Topping and Clifton 2001; Wong et al. 2006). In humans, the SCFAs produced account for 5–10 % of total dietary energy (Fernandes et al. 2014; Royall et al. 1990) and known to have a substantial influence on host lipid metabolism. The microbiota can increase fat storage in adipose tissue and cause significant changes in secondary lipid metabolites (Backhed et al. 2004; Rabot et al. 2010; Velagapudi et al. 2010). The SCFAs not only serve as a major fuel (i.e., butyrate) for the colonocytes, but also are absorbed via enterohepatic circulation and contribute to hepatic lipogenesis (Flint et al. 2008; Kim et al. 2013; McNeil 1984; Samuel et al. 2008; Wolever et al. 1989). The major de novo synthesized lipotoxic saturated fatty acids (SFA) in the liver are palmitic acid (C16:0) and stearic acid (C18:0), which are converted into the less lipotoxic monounsaturated fatty acids (MUFA) palmitoleate (C16:1 *n*7) and oleate (C18:1 *n*9), a steatogenic agent. The addition of a double bond is catalyzed by hepatic microsomal, lipogenic rate limiting enzyme delta 9-desaturase or stearoyl CoA desaturase-1 (SCD1), a key regulatory enzyme in the homeostasis of SFA and MUFA (Hodson and Fielding 2013). Despite oleate being a major MUFA from the diet, the expression of SCD1 is highly regulated by developmental, dietary, hormonal and environmental factors including microbiota.

4 Metabolic Functions of SCFAs

Acetate, propionate, and butyrate are the major SCFAs that arise from cecal and colonic fermentation of dietary fiber, which has important signaling functions in the gut through G-protein-coupled receptors (GPCRs) GPR41 (also known as free fatty acid 3) and GPR43 (also known as free fatty acid 2). Acetate, propionate and butyrate are known to protect against diet-induced obesity (Lin et al. 2012). Interestingly, concomitant dietary administration of propionate and butyrate completely block the high-fat diet-induced weight gain, whereas acetate led to a moderate suppression of excess weight gain (Lin et al. 2012). More importantly, both propionate and butyrate acutely stimulate gut hormones, namely glucagon-like peptide-1 (GLP-1), peptide YY (PYY) and amylin (rank order of stimulation butyrate > propionate > acetate), thereby exerting anorexic effect in mice. Studies from SCFA receptor GPR41- and GPR43-deficient mice revealed that interaction of these receptors with SCFA has a profound effect on host immune and metabolic functions (Maslowski et al. 2009). For example, GPR43 deficiency induces obese phenotype in mice on regular chow diet (Kimura et al. 2013), whereas mice overexpressing GPR43 remain lean even when fed a high-fat diet (Kimura et al. 2013). Mice lacking GPR41 also display the obese phenotype, which is accompanied with low-energy expenditure (Bellahcene et al. 2013). Both GPR41 and GPR43 mediate the immunomodulatory effect of SCFAs by regulating leukocyte recruitment and activating effector T cells in the intestines (Kim et al. 2013; Maslowski et al. 2009).

SCFAs are substrates for energy metabolism in the host (den Besten et al. 2013b). Average daily diet in Western societies yields approximately 300–600 mmol SCFA/day, which is equivalent to 10 % of the daily caloric requirements in humans (Bergman 1990). SCFAs are actively metabolized by the colonocytes in lumen; and also serve as substrate for de novo lipogenesis in colonic epithelial cells (Zambell et al. 2003). The fraction of SCFAs that is not consumed by the colonocytes is transported across the basolateral membrane to portal vein circulation and then to the liver, possibly via $SCFA^{-}$ -HCO_3^{-} antiport and the cation-SCFA anion symport (Bloemen et al. 2009; den Besten et al. 2013b). Using stable isotopes of SCFA in mice, den Besten et al., demonstrated that colonic SCFAs reaches the liver through the hepatic portal vein and participate in de novo lipogenesis. Additionally, their study suggests that acetate and butyrate are lipogenic, whereas propionate is gluconeogenic (den Besten et al. 2013a). Interestingly, rectal infusion of acetate plus propionate in humans resulted in significant elevation of serum cholesterol and triglycerides (Wolever et al. 1989). Metabolic function and relative contribution of individual SCFAs in hepatic lipogenesis is described below.

Acetate: Acetate is one of the most abundant SCFAs generated during bacterial fermentation of dietary fibers in the distal parts of the intestine. Acetate produced in the gut has different metabolic fates: (1) locally utilized as substrate for de novo lipogenesis by colonic epithelial cells (Zambell et al. 2003); (2) converted to butyrate by gut bacteria (den Besten et al. 2013a); (3) oxidized via the tricarboxylic acid

(TCA) cycle in the liver; (4) used as a substrate for synthesis of cholesterol, ketone bodies and long-chain fatty acids (den Besten et al. 2013a). Another major source of acetate is the liver itself; under starved condition, adipose tissue-derived fatty acids are oxidized by the liver and produce ketone bodies and acetate as end product (Yamashita et al. 2001). Acetate, either produced in the liver or reached from the gut, is provided to the extrahepatic tissue via blood, where it serves as fuel or is utilize in various metabolic processes, including the TCA cycle, fatty acid synthesis and cholesterol synthesis (Ballard 1972; Skutches et al. 1979; Yamashita et al. 2007); and it plays a key metabolic role for peripheral tissues (Shoaie et al. 2013; Yamashita et al. 2007). Oppositely, in vitro stimulation of bovine hepatocytes with acetate has led to increased lipid oxidation and decreased lipid synthesis via AMP-activated protein kinase α activation (Li et al. 2013b).

Apart from its use in lipogenic processes, acetate performs various functions via its receptors GPR41 and GPR43. A recent study demonstrated that increased local formation by pancreatic islets of acetate, an endogenous agonist of GPR41 and GPR43 might contribute to the insufficient capacity of beta cells to respond to hyperglycemia in obese and type 2 diabetic mice (Tang et al. 2015). In contrast, acetate infusion improves obesity and glucose tolerance in type 2 diabetic rat (Yamashita et al. 2007). Acetate also modulate hepatic inflammatory response as in acute alcoholic hepatitis; Ethanol-derived acetate reduces histone deacetylase activity and enhances inflammatory response (Kendrick et al. 2010).

Propionate: Propionate is another major microbial fermentation metabolite in the human gut with putative health effects that extend beyond the gut epithelium (Hosseini et al. 2011). In contrast to acetate, propionate is thought to lower lipogenesis and serum cholesterol levels in both hepatic and non-hepatic tissues (Hosseini et al. 2011). Accordingly, dietary propionate reduces fasting blood glucose and hepatic cholesterol in obese rats (Berggren et al. 1996; Boillot et al. 1995). However, studies performed using radiolabelled propionate demonstrates that propionate is the preferred substrate for hepatic gluconeogenesis (den Besten et al. 2013a). Propionate is first converted into propionyl-CoA by propionate:CoA ligase, and subsequently propionyl-CoA is converted to succinyl-CoA via propionyl-CoA carboxylase, methylmalonyl-CoA epimerase, and methylmalonyl-CoA mutase. Ultimately, succinyl-CoA enters the TCA cycle and is converted to oxaloacetate, the precursor of gluconeogenesis (Bloemen et al. 2010; den Besten et al. 2013a). However, limited data are available to demonstrate the extent to which propionate is partaking in gluconeogenesis and energy metabolism in humans. Concentrations of propionate in portal blood and hepatic venous blood suggest that around 30 % of propionate is taken up by the liver (Cummings et al. 1987). In another study, it is estimated that humans use 50 % of the propionate as a substrate for hepatic gluconeogenesis (den Besten et al. 2013a).

Butyrate: Similar to acetate, butyrate is another major substrate for de novo lipogenesis in colonocytes (Zambell et al. 2003). Butyrate level in the gut, at least in part, comes from gut bacterial dependent conversion of acetate (den Besten et al. 2013a). The major part of butyrate is used as fuel for colonocytes and also helps in maintaining intestinal epithelial cells. The remainder is mostly oxidized by hepatocytes, which prevents its accumulation to a toxic level (Bloemen et al. 2010; den Besten et al. 2013a).

5 Extra-hepatic Effect of SCFAs

Despite their contribution in hepatic lipogenesis, SCFAs have been shown to exert multiple effects on mammalian energy metabolism (den Besten et al. 2013b; Sekirov et al. 2010). For example, SCFAs protect from high-fat diet-induced obesity via a peroxisome proliferator-activated receptor gamma (PPAR-γ)-dependent switch from lipogenesis to fat oxidation (den Besten et al. 2015). Specifically, propionic and butyric acids have been described to have "anti-obesity" properties by inhibiting lipolysis and insulin-stimulated de novo lipogenesis in rats (Sekirov et al. 2010).

Various studies on rodents suggest that SCFAs, as principal gut bacterial metabolites, can regulate host immune function in ways that may exert beneficial effect in inflammatory disease. SCFAs have been shown to exert multiple effects on immune function (Vinolo et al. 2011). SCFAs act as both pro- and anti-inflammatory; for example, SCFAs influences regulatory T cell homeostasis in the colon (Smith et al. 2013), and dampens cytokine production by intestinal epithelial cells (IEC) (Arpaia 2014; Ishiguro et al. 2014) and human monocytes (Cox et al. 2009), whereas, others have shown that SCFAs promote inflammatory response in mice through activation of G protein-coupled receptors GPR41 and GPR43 (Kim et al. 2013). In addition, butyrate reduces LPS-induced cytokine production by intestinal macrophages, the most abundant immune cell type in the lamina propria (Chang et al. 2014). In liver, acetate, which is also a principal hepatic end product of ethanol metabolism, modulates inflammatory response in acute alcoholic hepatitis (Kendrick et al. 2010). Taken together, these findings suggest that certain bacterial species communicate with the host immune system through their metabolites to persuade both magnitude and quality of the immune response (Arpaia 2014).

6 Secondary Bile Acids Modulate Hepatic Bile Acid-Mediated Metabolic Pathways

Bile acids are synthesized from cholesterol in the liver and released into the intestines via bile secretions. In addition to their role in facilitating intestinal absorption of dietary lipids and fat-soluble vitamins, re-absorbed bile acids via enterohepatic circulation also serve as potent signaling molecules. For instance, glycine- or taurine-conjugated primary bile acids (cholate and chenodeoxycholate) are major ligands for the farnesoid X receptor (FXR), whose activation serves to suppress hepatic de novo bile acid synthesis, lipogenesis and gluconeogenesis (Sinal et al.

2000). However, a fraction of the primary bile acids in the gut can be deconjugated by bacterial bile acid hydrolases (particularly those expressed by bacteria from the genera Bacteroides, Eubacterium and Clostridium), and further metabolized into secondary bile acids (deoxycholate and lithocholate) (Ridlon et al. 2006).

Bacterial-transformed secondary bile acids have increased hydrophobicity, and thus enhance intestinal absorption of dietary lipids and fat-soluble vitamins (Ridlon et al. 2006). Unlike their primary counterparts, secondary bile acids instead activate the hepatic G protein-coupled receptor TGR5, whose signaling promotes bile acid synthesis and energy homeostasis (Thomas et al. 2009; Watanabe et al. 2006). Furthermore, the gut microbiota recently was demonstrated to promote/prevent obesity by controlling the level of FXR antagonist, tauro-beta-muricholic acid (Li et al. 2013a; Sayin et al. 2013) via modulating the composition of primary/secondary bile acid pool.

In one recent study, Yoshimoto et al. reported an association between secondary bile acid deoxycholate and the development of hepatocellular carcinoma (HCC) in mice (Yoshimoto et al. 2013). In their study, the authors used a chemically-induced HCC model, whereby day 4–5 postnatal mice were treated once with carcinogen 7,12-Dimethylbenz(a)anthracene and then placed on a high-fat diet to potentiate HCC. These mice displayed a substantial increase in serum deoxycholate, which can be explained by the increased abundance of Clostridium cluster XI populations (a well-known strain capable of performing 7α-dehydroxylation on primary bile acid to generate deoxycholate) in the gut. Remarkably, the levels of deoxycholate are reflected in the severity of HCC; HCC development is reduced when deoxycholate level decreases (via inhibiting bacterial 7α-dehydroxylation activity), but the opposite is true when deoxycholate levels increase via dietary supplementation (Yoshimoto et al. 2013). Altogether, the capacity to modulate the host bile acid pool allows the gut microbiota to significantly influence hepatic cholesterogenesis, contribute to obesity and possibly promote the development of liver pathologies.

7 Choline Metabolites Influence Liver Functions

Choline is an essential nutrient that is present in most dietary sources as phosphatidylcholine (lecithin). Its importance is highlighted by its requirementin various biological processes including hepatic lipogenesis, as well as in the synthesis of VLDL, neurotransmitter acetylcholine and cell membrane phospholipids (Vance 2008). Dietary choline typically is metabolized in the liver, but gut microbiota are known to convert choline to hepatotoxic dimethylamine and trimetylamine. These toxic methylamines are rapidly metabolized by host hepatic flavin monooxygenase 3 (FMO3) to generate trimethylamine-N-oxide (TMAO). However, bacterial metabolism of choline reduces choline bioavailability, and this can impair phosphatidylcholine synthesis in the liver. In a study by Dumas et al., the obesity-predisposed 129S6 mice, when fed a high-fat diet, exhibit reduced choline bioavailability, impaired VLDL secretion and increased triglycerides accumulation in the liver, all of which are correlated with increased trimethylamine (Dumas et al. 2006). These mice display classic symptoms (*e.g.*, hepatic steatosis) of non-alcoholic fatty liver disease (NALFD), which are often seen in mice fed with choline-deficient diet.

Wang et al. reported an association between the levels of bacterial-derived phosphatidylcholine metabolites with the risks for cardiovascular diseases (Wang et al. 2011). They found that supplementation of TMAO or its precursors choline and betaine upregulates the expression of macrophage scavenger receptors CD36 and SR-A1 in ApoE deficient mice. These receptors potently suppress reverse cholesterol transport and induce the formation of macrophage foam cells, both of which are implicated in driving atherogenesis (Bremer 1983; Wang et al. 2012). Similarly, bacterial-mediated metabolism of dietary L-carnitine from red meat into TMAO has been shown to accelerate atherosclerosis in both mice and humans (Koeth et al. 2013). Incidentally, microbiota ablation protects ApoE deficient mice against elevated plasma TMAO, foam cell formation and atherosclerosis (Wang et al. 2011).

Intriguingly, Warrier et al. recently demonstrate a prominent role of microbiota-regulated hepatic enzyme FMO3 in modulating hepatic cholesterol balance. Deficiency of FMO3 in mice decreases intestinal absorption of cholesterol, hepatic cholesteryl ester levels, biliary secretion of cholesterol and the total bile acid pool; all which are associated with increased expression of sterol regulatory element-binding proteins (SREBPs) target genes (*e.g.*, HMG-CoA reductase, HMG-CoA synthase) and decreased expression of liver X receptor (LXRs) target genes (*e.g.*, SCD1, fatty acid synthase) (Warrier et al. 2015). The same group also found FMO3 to be a potent negative regulator of nonbiliary macrophage reverse cholesterol transport while promoting increased levels of biliary cholesterol and its absorption from the intestines (Warrier et al. 2015). However, further studies are warranted to delineate the independent as well as combined roles of microbiota-associated TMA, FMO3 and TMAO in influencing host cholesterol homeostasis.

8 Microbiotal Metabolites Promote Inflammation-Driven Liver Diseases

The contributions of gut microbiota to liver diseases have been increasingly recognized over the past decades, especially those involving NAFLD. NAFLD typically is characterized histologically by the presence of hepatic steatosis, accompanied with a constellation of metabolic abnormalities including hyperlipidemia, hypertension, insulin resistance and obesity in the absence of significant alcohol consumption. A fraction of patients (~25 %) with simple steatosis may progress into the more severe form of NAFLD, such as non-alcoholic steatohepatitis (NASH) and/or NASH-related cirrhosis (Williams et al. 2011). Although the etiology of NAFLD was initially proposed to be a two-hit process (steatosis followed by inflammation and oxidative stress) (Day and James 1998), recent studies found that NAFLD is more likely to be multifactorial, with microbiota being one of the factors driving NAFLD.

The translocation of bacteria and/or their metabolites to the liver have been strongly associated with small intestinal bacteria overgrowth and increased gut permeability in patients with NAFLD. One human study observed small intestinal bacterial overgrowth in 50 % of NASH patients compared to only 22 % in healthy controls (Wigg et al. 2001). Yet, in another study, 35 patients with biopsy-proven NAFLD were found to exhibit increased gut permeability due to altered tight junctions in the gut (Miele et al. 2009). By using mouse models, Gäbele et al. is able to convincingly demonstrate the gut-liver axis involvement in NASH (Gabele et al. 2011). They found that high-fat diet-fed mice developed NASH, but displayed a more severe NASH when mice are co-treated with dextran sodium sulfate, a known chemical colitogen to cause intestinal inflammation/injury. This observation can be explained by the increased levels of bacterial lipopolysaccharide (LPS) in the hepatic portal vein of these mice, which reaffirms the involvement of microbiota in NAFLD (Gabele et al. 2011).

Among the gut metabolites associated with NAFLD, the bacterial LPS is perhaps the most immunologically potent in inducing chronic inflammation and liver injury. LPS, a component of the outer membrane of gram negative bacteria, routinely gets translocated from the gut and into the liver where it will be detoxified by hepatic lipase (Shao et al. 2007) and/or alkaline phosphatase (Koyama et al. 2002). In pathological conditions, however, the liver fails to mediate efficient clearance of LPS, which then give rise to low-grade chronic inflammation that precedes steatosis in NAFLD. Mechanistically, this metabolite binds to LPS-binding protein, MD2 and CD14, forming a complex that activates host Toll-like receptor 4 (TLR4) expressed on Kupffer cells. Activated TLR4 then upregulates the expression of pro-inflammatory cytokines, such as TNF-α (a well-established proinflammatory cytokine) that induces insulin resistance and inhibits lipoprotein lipase. In addition, activation of LPS-TLR4 signaling on hepatic stellate cells also promotes TGF-β-mediated fibrogenesis, which leads to liver fibrosis and then cirrhosis (Seki et al. 2007).

Cani et al. demonstrate that mice on 4 weeks of continuous subcutaneous infusion of LPS (from *E. coli* 055:B5; 300 μg/kg/day for 4 weeks) are able to recapitulate hallmarks of metabolic syndrome (increased fasting glucose, insulin resistance, obesity and steatosis) observed in high-fat diet-induced NALFD mice (Cani et al. 2007). Their studies also found that high-fat diet significantly reduced Lactobacillus and Bacteroides populations while concomitantly increased the intestinal gram negative bacteria load in mice. This is accompanied by increased levels of cecal and plasma LPS that are available to leak into circulation and drive metabolic endotoxemia. The administration of antibiotics, however, significantly reduces cecal and plasma LPS levels, and improves metabolic syndrome in both genetically obese *ob/ob* mice and high-fat diet-fed wild-type (WT) mice. Mice that are deficient in TLR4 do not exhibit insulin resistance and are protected against microbiota-dependent inflammation and metabolic disorders, further confirming the involvement of LPS in NAFLD (Csak et al. 2011).

Another component of innate immunity, the inflammasomes also are implicated in NAFLD via participation in lipid peroxidation and reactive oxygen species generation in the liver. Inflammasomes are multi-protein complexes composed of proteins with leucine-rich repeats and nucleotide-binding domains that sense microbial ligands (including LPS) and/or damage-associated molecular patterns. Upon activation, inflammasomes induce caspase 1-mediated cleavage of inactive IL-1β and IL-18 to their active pro-inflammatory forms. An increased activation in hepatic NLRP3 inflammasome recently has been associated with liver fibrosis and injury (Imaeda et al. 2009; Watanabe et al. 2009). Vandanmagsar et al. found that ablation of NLRP3 improves hepatic insulin signaling and protects against obesity (Vandanmagsar et al. 2011). Yet in another study, Henao-Mejia et al. demonstrate that mice lacking either NLRP6 or NLRP3 inflammasomes develop microbiotal dysbiosis that resulted in an increased translocation of bacterial metabolites into the hepatic portal vein, enhanced TNF-α expression and exacerbated steastosis (Henao-Mejia et al. 2012). Both studies on TLR4 and inflammasomes strongly implicate the role of host innate immunity in maintaining host-microbiota homeostasis, where the deficiency or excessive stimulation will eventuate metabolic disorders.

Gut microbiota also could contribute to liver injury by producing endogenous alcohol and its metabolites, such as acetate and acetaldehyde, leading to fatty acid synthesis as well as oxidative stress in the liver. Endogenous alcohol also can potentially increase intestinal permeability and contribute to the metabolic endotoxemia implicated in NAFLD. This is reaffirmed in one recent study whereby patients with NASH were reported to have higher blood alcohol concentration associated with increased abundance of alcohol-producing bacteria (*e.g.*, *E. coli*), further suggesting that endogenous alcohol production may be another potential mechanism by which microbiota drive NAFLD (Zhu et al. 2013).

Aside from NAFLD, chronic inflammation in the liver can also potentially lead to hepatocellular carcinoma. In a recent study, Dapito et al. reported that microbiota-induced TLR4 signaling plays a prominent role in HCC promotion, but not required in the initiation of HCC in mice (Dapito et al. 2012). Interestingly, they found that TLR4 activation induces increased hepatocytes proliferation while preventing their apoptosis. HCC tumor number and size are remarkably reduced in germ-free and microbiota-ablated mice, indicating that LPS-induced signaling pathways occupy an important role in inflammation-associated HCC.

9 Microbiota-Mediated Metabolic Diseases in the Absence of Functional Host Immunity

The interplay between gut microbiota and liver metabolism is further complicated by the participation of innate and adaptive arms of host immunity. The host immune system plays important roles in shaping the gut microbiota, mostly by eliminating harmful pathogens, preventing their epithelial breach and restricting pathobiont expansion. Hence, deficiencies in host immunity would dysregulate the delicate balance between gut microbiota and host metabolism, resulting in the onset of metabolic disorders.

The link between host immunity, microbiota and hepatic lipogenesis was demonstrated in mice lacking Toll-like receptor 5 (TLR5), an innate immune receptor that senses the bacterial flagellin. Our group demonstrated that TLR5KO mice display pronounced hallmarks of metabolic syndrome, including increased adiposity, hyperlipidemia, hyperglycemia, hepatic insulin resistance and hepatic steatosis (Vijay-Kumar et al. 2010). TLR5KO mice also exhibit significantly increased bacterial load with altered microbiota composition when compared to their wild-type littermates. Interestingly, the metabolic phenotype of TLR5KO mice absolutely are microbiota dependent, as TLR5KO microbiota transfer alone is sufficient to confer the phenotype to germ-free WT mice. Additionally, microbiota ablation is found to be effective in correcting the metabolic syndrome of TLR5KO mice.

In similar fashion to TLR5KO mice, Caricilli et al. demonstrate that mice deficient in Toll-Like receptor 2 (TLR2) also develop hallmarks of metabolic syndrome (Caricilli et al. 2011). Notably, they reported that TLR2KO mice harbor threefold increase in Firmicutes with a minor increase in Bacteroidetes when compared to WT controls. Alterations in gut microbiota in TLR2KO mice are strongly associated with metabolic endotoxemia, subclinical inflammation, insulin resistance, glucose intolerance and obesity. Further analysis reveals that microbiotal dysbiosis in these mice activate TLR4 signaling, causing ER stress and JNK activation, which underlie development of obesity.

The study by Shulzhenko et al. demonstrate that IgA deficiency indirectly impair metabolic functions of the intestinal epithelium, an effect that is secondary to the onset of microbiotal dysbiosis (Shulzhenko et al. 2011). Mice deficient in B cells (and thus unable to produce IgA) display significantly lowered expression of lipid metabolizing genes with a concomitant upregulation of inflammation-related genes when compared to WT mice (Shulzhenko et al. 2011). Although no change in food intake was observed, B cell-deficient mice exhibit reduced fat and cholesterol absorption that are reflected in decreased perigonadal fat pad mass. The involvement of gut microbiota is affirmed as microbiota transfer alone is sufficient to confer the phenotype to germ-free WT mice. Interestingly, Shulzhenko et al. found that this phenotype is due to impaired function of GATA4, a transcription factor that serves as a ‘switch’ controlling the metabolic versus immune gene expression in the intestinal epithelia. In response to stimulation by altered gut microbiota, intestinal epithelial cells upregulate interferon-inducible genes to compensate for IgA deficiency while simultaneously repress GATA 4-related genes, resulting in reduced accumulation of body fat (Shulzhenko et al. 2011).

Despite this, one should note that immune dysregulation does not lead necessarily to adverse microbiotal alterations that potentiate diseases; instead, downregulation of some components of the host immunity may be counter-intuitively beneficial. Indeed, intestinal epithelial cell-specific deletion of MyD88 is reported to confer partial protection against diet-induced obesity, diabetes and inflammation in mice. MyD88 is the universal adapter protein downstream of TLRs (TLR1-9; except TLR3), and therefore, its absence may serve to down regulate TLR pathways. Everard et al. found that deficiency in MyD88 in murine epithelial cells causes the host to produce more anti-inflammatory endocannabinoids, increase the number of intestinal regulatory T cells, and normalize the production antimicrobial peptides during diet-induced obesity (Everard et al. 2014). Moreover, they also observed improved energy homeostasis with reduced inflammation, fat mass and hepatic steatosis. In short, this study identifies a novel role for MyD88 as a sensor/switch (that is reminiscent of GATA4 discussed earlier) in modulating microbiota-host metabolism, and therefore, presenting a promising therapeutic target for treating obesity and other metabolic disorders.

10 Manipulating the Gut Microbiota, a Potential Treatment for 'New Age' Metabolic Diseases

The increasinglywidespread consumption of Westernized high-fat and high-carbohydrate diets have given rise to the so-called 'New Age' metabolic disorders that include inflammatory bowel diseases and NAFLD (Huang et al. 2013). Although the etiology of 'New Age' disorders are yet to be understood, many studies proposed that the Western diet could mediate its negative effects, at least in part, by significantly altering the gut microbiota. If this is so, then perhaps therapeutic interventions that reshape the gut microbiota may present a promising strategy in correcting metabolic diseases. Indeed, the administration of broad-spectrum antibiotics can prevent diet-induced obesity and improve insulin sensitivity (Carvalho et al. 2012). The long-term use of antibiotics in humans, however, is not recommended because it also depletes other beneficial gut commensals in the gut.

As alternatives, many studies began to explore the potential use of probiotics and prebiotics to reshape the microbiota without extensively harming the gut ecosystem. The administration of probiotic Lactobacillus strains has been shown to decrease fat mass, alleviate insulin resistance and reduce the risk of type II diabetes in humans (Andreasen et al. 2010). Additionally, the probiotic formula VSL#3 (a mixture of Lactobacillus and Bifidobacterium strains) and prebiotic oligofructose have been demonstrated to reduce liver injury and improve liver functions, as characterized by reduced expression of transaminases and pro-inflammatory TNF-α, reduced total cholesterol and improved insulin sensitivity (Ma et al. 2013).

Recently, there has been an increasing interest in the use of fecal microbiota transplantation (FMT) to treat microbiota-mediated diseases. FMT was initially applied to treat *Clostridium difficile* infection (CDI), with recent study showing that FMT can normalize the abnormally high fecal primary bile acids in patients with recurrent CDI (Weingarden et al. 2014). CDI patients receiving FMT undergo rapid change in their microbiotal composition and exhibit increased production of secondary bile acids (which were undetectable prior to FMT) from primary bile acids. The shift in bile acid composition likely may be one of the mechanisms underlying the effectiveness of FMT against CDI. Based on its successful application in treating CDI, FMT is now considered a viable treatment for microbiota-driven metabolic/cardiovascular diseases, inflammatory bowel diseases and autoimmune diseases (Smits et al. 2013). By transplanting fecal microbiota from lean donors to obese recipients, one recent study was able to improve insulin resistance in patients with metabolic syndrome (Vrieze et al. 2012). However, like broad-spectrum antibiotics treatment, the FMT approach is an extreme manipulation of the gut microbiota and therefore, needs to be administered with utmost care. Nevertheless, FMT may prove to be a viable therapeutic strategy to treat microbiota-driven diseases as more studies on the immunological and metabolic consequences of microbiota alterations become available.

11 Summary and Future Perspectives

Recent captivating findings demonstrate that the microbiota are not simply freeloaders, but rather are involved in many functions vital to our health and well-being: microbiota digest food, produce anti-inflammatory chemicals and compounds, and regulate key metabolic processes in the liver. Findings and implications from these studies prompted many new research to elucidate the possible role of microbiota in various kinds of human pathologies. Many studies have shown that gut microbiota not only regulates our metabolism but also guides our immune system, and even our moods and behavior. Accordingly, the microbial community can be exploited in several ways: (1) Isolating microbiota consortium from lean individuals and use it to protect against weight gain in obese individuals with dysregulated microbiota; (2) Gut bacteria that produce anti-inflammatory molecules can be used to provide local relief in the inflamed gut; (3) Microbiota can be manipulated in a way to enhance extraction of energy/nutrient from ingested food, specifically for individuals with compromised digestion.

The liver, a principal metabolic organ in association with the gut, plays a key role in the development of obesity and associated metabolic disorders. Combating the global epidemic of liver-associated pathologies is a major public health issue that requires a multifaceted approach that should encompasses expertise from biochemists, hepatologists, gastroenterologists, immunologists, dietitians and nutritionists. Collaborative research endeavors integrating various platforms of omics technologies (metagenomics, metabolomics, microbiomics and proteomics) may likely help to elucidate the underlying molecular mechanisms and manipulate them in a variety of ways to improve human health.

Glossary

Microbiota Total number of microbes that inhabit our intestines

Microbiome The total number of genes from our internal microbial communities

References

(2015). Your microbes at work: fiber fermenters keep us healthy. Nature 518:S9

Abu-Shanab A, Quigley EM (2010) The role of the gut microbiota in nonalcoholic fatty liver disease. Nat Rev Gastroenterol Hepatol 7:691–701

Andreasen AS, Larsen N, Pedersen-Skovsgaard T, Berg RM, Moller K, Svendsen KD, Jakobsen M, Pedersen BK (2010) Effects of Lactobacillus acidophilus NCFM on insulin sensitivity and the systemic inflammatory response in human subjects. Br J Nutr 104:1831–1838

Anitha, M., Vijay-Kumar, M., Sitaraman, S.V., Gewirtz, A.T., and Srinivasan, S. (2012). Gut microbial products regulate murine gastrointestinal motility via Toll-like Receptor 4 signaling. Gastroenterology 143(4):1006–16.e4

Arpaia N (2014) Keeping peace with the microbiome: acetate dampens inflammatory cytokine production in intestinal epithelial cells. Immunol Cell Biol 92:561–562

Backhed F, Ding H, Wang T, Hooper LV, Koh GY, Nagy A, Semenkovich CF, Gordon JI (2004) The gut microbiota as an environmental factor that regulates fat storage. Proc Natl Acad Sci U S A 101:15718–15723

Backhed F, Ley RE, Sonnenburg JL, Peterson DA, Gordon JI (2005) Host-bacterial mutualism in the human intestine. Science 307:1915–1920

Backhed F, Manchester JK, Semenkovich CF, Gordon JI (2007) Mechanisms underlying the resistance to diet-induced obesity in germ-free mice. Proc Natl Acad Sci U S A 104:979–984

Ballard FJ (1972) Supply and utilization of acetate in mammals. Am J Clin Nutr 25:773–779

Balmer, M.L., Slack, E., de Gottardi, A., Lawson, M.A., Hapfelmeier, S., Miele, L., Grieco, A., Van Vlierberghe, H., Fahrner, R., Patuto, N., et al. (2014). The liver may act as a firewall mediating mutualism between the host and its gut commensal microbiota. Science translational medicine *6*, 237ra266.

Bellahcene M, O'Dowd JF, Wargent ET, Zaibi MS, Hislop DC, Ngala RA, Smith DM, Cawthorne MA, Stocker CJ, Arch JR (2013) Male mice that lack the G-protein-coupled receptor GPR41 have low energy expenditure and increased body fat content. Br J Nutr 109:1755–1764

Berggren AM, Nyman EM, Lundquist I, Bjorck IM (1996) Influence of orally and rectally administered propionate on cholesterol and glucose metabolism in obese rats. Br J Nutr 76:287–294

Bergman EN (1990) Energy contributions of volatile fatty acids from the gastrointestinal tract in various species. Physiol Rev 70:567–590

Bloemen JG, Venema K, van de Poll MC, Olde Damink SW, Buurman WA, Dejong CH (2009) Short chain fatty acids exchange across the gut and liver in humans measured at surgery. Clin Nutr 28:657–661

Bloemen JG, Olde Damink SW, Venema K, Buurman WA, Jalan R, Dejong CH (2010) Short chain fatty acids exchange: Is the cirrhotic, dysfunctional liver still able to clear them? Clin Nutr 29:365–369

Boillot J, Alamowitch C, Berger AM, Luo J, Bruzzo F, Bornet FR, Slama G (1995) Effects of dietary propionate on hepatic glucose production, whole-body glucose utilization, carbohydrate and lipid metabolism in normal rats. Br J Nutr 73:241–251

Bremer J (1983) Carnitine--metabolism and functions. Physiol Rev 63:1420–1480

Cani PD, Amar J, Iglesias MA, Poggi M, Knauf C, Bastelica D, Neyrinck AM, Fava F, Tuohy KM, Chabo C et al (2007) Metabolic endotoxemia initiates obesity and insulin resistance. Diabetes 56:1761–1772

Cani PD, Bibiloni R, Knauf C, Waget A, Neyrinck AM, Delzenne NM, Burcelin R (2008) Changes in gut microbiota control metabolic endotoxemia-induced inflammation in high-fat diet-induced obesity and diabetes in mice. Diabetes 57:1470–1481

Caricilli AM, Picardi PK, de Abreu LL, Ueno M, Prada PO, Ropelle ER, Hirabara SM, Castoldi A, Vieira P, Camara NO et al (2011) Gut microbiota is a key modulator of insulin resistance in TLR 2 knockout mice. PLoS Biol 9, e1001212

Carmody RN, Turnbaugh PJ (2012) Gut microbes make for fattier fish. Cell Host Microbe 12:259–261

Carvalho BM, Guadagnini D, Tsukumo DM, Schenka AA, Latuf-Filho P, Vassallo J, Dias JC, Kubota LT, Carvalheira JB, Saad MJ (2012) Modulation of gut microbiota by antibiotics improves insulin signalling in high-fat fed mice. Diabetologia 55:2823–2834

Chang PV, Hao L, Offermanns S, Medzhitov R (2014) The microbial metabolite butyrate regulates intestinal macrophage function via histone deacetylase inhibition. Proc Natl Acad Sci U S A 111(6):2247–2252

Conterno L, Fava F, Viola R, Tuohy KM (2011) Obesity and the gut microbiota: does up-regulating colonic fermentation protect against obesity and metabolic disease? Genes Nutr 6:241–260

Cox MA, Jackson J, Stanton M, Rojas-Triana A, Bober L, Laverty M, Yang X, Zhu F, Liu J, Wang S et al (2009) Short-chain fatty acids act as antiinflammatory mediators by regulating prostaglandin E(2) and cytokines. World journal of gastroenterology : WJG 15:5549–5557

Csak T, Velayudham A, Hritz I, Petrasek J, Levin I, Lippai D, Catalano D, Mandrekar P, Dolganiuc A, Kurt-Jones E et al (2011) Deficiency in myeloid differentiation factor-2 and toll-like receptor 4 expression attenuates nonalcoholic steatohepatitis and fibrosis in mice. Am J Physiol Gastrointest Liver Physiol 300:G433–G441

Cummings JH, Pomare EW, Branch WJ, Naylor CP, Macfarlane GT (1987) Short chain fatty acids in human large intestine, portal, hepatic and venous blood. Gut 28:1221–1227

Dapito DH, Mencin A, Gwak GY, Pradere JP, Jang MK, Mederacke I, Caviglia JM, Khiabanian H, Adeyemi A, Bataller R et al (2012) Promotion of hepatocellular carcinoma by the intestinal microbiota and TLR4. Cancer Cell 21:504–516

Day CP, James OF (1998) Steatohepatitis: a tale of two "hits"? Gastroenterology 114:842–845

den Besten G, Lange K, Havinga R, van Dijk TH, Gerding A, van Eunen K, Muller M, Groen AK, Hooiveld GJ, Bakker BM et al (2013a) Gut-derived short-chain fatty acids are vividly assimilated into host carbohydrates and lipids. Am J Physiol Gastrointest Liver Physiol 305(12):G900–G910

den Besten G, van Eunen K, Groen AK, Venema K, Reijngoud DJ, Bakker BM (2013b) The role of short-chain fatty acids in the interplay between diet, gut microbiota, and host energy metabolism. J Lipid Res 54:2325–2340

den Besten G, Bleeker A, Gerding A, van Eunen K, Havinga R, van Dijk TH, Oosterveer MH, Jonker JW, Groen AK, Reijngoud DJ et al (2015) Short-Chain Fatty Acids protect against High-Fat Diet-Induced Obesity via a PPARgamma-dependent switch from lipogenesis to fat oxidation. Diabetes 64(7):2398–2408

Dorrestein PC, Mazmanian SK, Knight R (2014) Finding the missing links among metabolites, microbes, and the host. Immunity 40:824–832

Dumas ME, Barton RH, Toye A, Cloarec O, Blancher C, Rothwell A, Fearnside J, Tatoud R, Blanc V, Lindon JC et al (2006) Metabolic profiling reveals a contribution of gut microbiota to fatty liver phenotype in insulin-resistant mice. Proc Natl Acad Sci U S A 103:12511–12516

Eckburg PB, Bik EM, Bernstein CN, Purdom E, Dethlefsen L, Sargent M, Gill SR, Nelson KE, Relman DA (2005) Diversity of the human intestinal microbial flora. Science 308:1635–1638

Everard A, Geurts L, Caesar R, Van Hul M, Matamoros S, Duparc T, Denis RG, Cochez P, Pierard F, Castel J et al (2014) Intestinal epithelial MyD88 is a sensor switching host metabolism towards obesity according to nutritional status. Nat Commun 5:5648

Fernandes J, Su W, Rahat-Rozenbloom S, Wolever TM, Comelli EM (2014) Adiposity, gut microbiota and faecal short chain fatty acids are linked in adult humans. Nutr Diabetes 4, e121

Flint HJ, Bayer EA, Rincon MT, Lamed R, White BA (2008) Polysaccharide utilization by gut bacteria: potential for new insights from genomic analysis. Nat Rev Microbiol 6:121–131

Gabele E, Dostert K, Hofmann C, Wiest R, Scholmerich J, Hellerbrand C, Obermeier F (2011) DSS induced colitis increases portal LPS levels and enhances hepatic inflammation and fibrogenesis in experimental NASH. J Hepatol 55:1391–1399

Grice EA, Segre JA (2012) The human microbiome: our second genome. Annu Rev Genomics Hum Genet 13:151–170

Henao-Mejia J, Elinav E, Jin C, Hao L, Mehal WZ, Strowig T, Thaiss CA, Kau AL, Eisenbarth SC, Jurczak MJ et al (2012) Inflammasome-mediated dysbiosis regulates progression of NAFLD and obesity. Nature 482:179–185

Hodson L, Fielding BA (2013) Stearoyl-CoA desaturase: rogue or innocent bystander? Prog Lipid Res 52:15–42

Hosseini E, Grootaert C, Verstraete W, Van de Wiele T (2011) Propionate as a health-promoting microbial metabolite in the human gut. Nutr Rev 69:245–258

Huang EY, Devkota S, Moscoso D, Chang EB, Leone VA (2013) The role of diet in triggering human inflammatory disorders in the modern age. Microbes and infection / Institut Pasteur 15:765–774

Imaeda AB, Watanabe A, Sohail MA, Mahmood S, Mohamadnejad M, Sutterwala FS, Flavell RA, Mehal WZ (2009) Acetaminophen-induced hepatotoxicity in mice is dependent on Tlr9 and the Nalp3 inflammasome. J Clin Invest 119:305–314

Ishiguro K, Ando T, Maeda O, Watanabe O, Goto H (2014) Suppressive action of acetate on interleukin-8 production via tubulin-alpha acetylation. Immunol Cell Biol 92:624–630

Kendrick SF, O'Boyle G, Mann J, Zeybel M, Palmer J, Jones DE, Day CP (2010) Acetate, the key modulator of inflammatory responses in acute alcoholic hepatitis. Hepatology 51:1988–1997

Kim MH, Kang SG, Park JH, Yanagisawa M, Kim CH (2013) Short-chain fatty acids activate GPR41 and GPR43 on intestinal epithelial cells to promote inflammatory responses in mice. Gastroenterology 145(396–406):e391–310

Kimura I, Ozawa K, Inoue D, Imamura T, Kimura K, Maeda T, Terasawa K, Kashihara D, Hirano K, Tani T et al (2013) The gut microbiota suppresses insulin-mediated fat accumulation via the short-chain fatty acid receptor GPR43. Nat Commun 4:1829

Koeth RA, Wang Z, Levison BS, Buffa JA, Org E, Sheehy BT, Britt EB, Fu X, Wu Y, Li L et al (2013) Intestinal microbiota metabolism of L-carnitine, a nutrient in red meat, promotes atherosclerosis. Nat Med 19:576–585

Koyama I, Matsunaga T, Harada T, Hokari S, Komoda T (2002) Alkaline phosphatases reduce toxicity of lipopolysaccharides in vivo and in vitro through dephosphorylation. Clin Biochem 35:455–461

Laffel L (1999) Ketone bodies: a review of physiology, pathophysiology and application of monitoring to diabetes. Diabetes Metab Res Rev 15:412–426

Ley RE, Turnbaugh PJ, Klein S, Gordon JI (2006) Microbial ecology: human gut microbes associated with obesity. Nature 444:1022–1023

Li F, Jiang C, Krausz KW, Li Y, Albert I, Hao H, Fabre KM, Mitchell JB, Patterson AD, Gonzalez FJ (2013a) Microbiome remodelling leads to inhibition of intestinal farnesoid X receptor signalling and decreased obesity. Nat Commun 4:2384

Li X, Chen H, Guan Y, Li X, Lei L, Liu J, Yin L, Liu G, Wang Z (2013b) Acetic acid activates the AMP-activated protein kinase signaling pathway to regulate lipid metabolism in bovine hepatocytes. PLoS One 8, e67880

Lin HV, Frassetto A, Kowalik EJ Jr, Nawrocki AR, Lu MM, Kosinski JR, Hubert JA, Szeto D, Yao X, Forrest G et al (2012) Butyrate and propionate protect against diet-induced obesity and regulate gut hormones via free fatty acid receptor 3-independent mechanisms. PLoS One 7, e35240

Lyte M (2013) Microbial endocrinology in the microbiome-gut-brain axis: how bacterial production and utilization of neurochemicals influence behavior. PLoS Pathog 9, e1003726

Ma YY, Li L, Yu CH, Shen Z, Chen LH, Li YM (2013) Effects of probiotics on nonalcoholic fatty liver disease: a meta-analysis. World journal of gastroenterology : WJG 19:6911–6918

Macfarlane S, Macfarlane GT (2003) Regulation of short-chain fatty acid production. Proc Nutr Soc 62:67–72

Madsbad S (2014) The role of glucagon-like peptide-1 impairment in obesity and potential therapeutic implications. Diabetes Obes Metab 16:9–21

Maslowski KM, Vieira AT, Ng A, Kranich J, Sierro F, Yu D, Schilter HC, Rolph MS, Mackay F, Artis D et al (2009) Regulation of inflammatory responses by gut microbiota and chemoattractant receptor GPR43. Nature 461:1282–1286

McNeil NI (1984) The contribution of the large intestine to energy supplies in man. Am J Clin Nutr 39:338–342

Miele L, Valenza V, La Torre G, Montalto M, Cammarota G, Ricci R, Masciana R, Forgione A, Gabrieli ML, Perotti G et al (2009) Increased intestinal permeability and tight junction alterations in nonalcoholic fatty liver disease. Hepatology 49:1877–1887

Musso G, Gambino R, Cassader M (2010) Obesity, diabetes, and gut microbiota: the hygiene hypothesis expanded? Diabetes Care 33:2277–2284

Rabot S, Membrez M, Bruneau A, Gerard P, Harach T, Moser M, Raymond F, Mansourian R, Chou CJ (2010) Germ-free C57BL/6 J mice are resistant to high-fat-diet-induced insulin resistance and have altered cholesterol metabolism. FASEB journal : official publication of the Federation of American Societies for Experimental Biology 24:4948–4959

Ridlon JM, Kang DJ, Hylemon PB (2006) Bile salt biotransformations by human intestinal bacteria. J Lipid Res 47:241–259

Riva MA, Riva E, Spicci M, Strazzabosco M, Giovannini M, Cesana G (2011) "The city of Hepar": rituals, gastronomy, and politics at the origins of the modern names for the liver. J Hepatol 55:1132–1136

Royall D, Wolever TM, Jeejeebhoy KN (1990) Clinical significance of colonic fermentation. Am J Gastroenterol 85:1307–1312

Said HM (2011) Intestinal absorption of water-soluble vitamins in health and disease. Biochem J 437:357–372

Samuel BS, Shaito A, Motoike T, Rey FE, Backhed F, Manchester JK, Hammer RE, Williams SC, Crowley J, Yanagisawa M et al (2008) Effects of the gut microbiota on host adiposity are modulated by the short-chain fatty-acid binding G protein-coupled receptor, Gpr41. Proc Natl Acad Sci U S A 105:16767–16772

Sayin SI, Wahlstrom A, Felin J, Jantti S, Marschall HU, Bamberg K, Angelin B, Hyotylainen T, Oresic M, Backhed F (2013) Gut microbiota regulates bile acid metabolism by reducing the levels of tauro-beta-muricholic acid, a naturally occurring FXR antagonist. Cell Metab 17:225–235

Seki E, De Minicis S, Osterreicher CH, Kluwe J, Osawa Y, Brenner DA, Schwabe RF (2007) TLR4 enhances TGF-beta signaling and hepatic fibrosis. Nat Med 13:1324–1332

Sekirov I, Russell SL, Antunes LC, Finlay BB (2010) Gut microbiota in health and disease. Physiol Rev 90:859–904

Semova I, Carten JD, Stombaugh J, Mackey LC, Knight R, Farber SA, Rawls JF (2012) Microbiota regulate intestinal absorption and metabolism of fatty acids in the zebrafish. Cell Host Microbe 12:277–288

Shao B, Lu M, Katz SC, Varley AW, Hardwick J, Rogers TE, Ojogun N, Rockey DC, Dematteo RP, Munford RS (2007) A host lipase detoxifies bacterial lipopolysaccharides in the liver and spleen. J Biol Chem 282:13726–13735

Shoaie S, Karlsson F, Mardinoglu A, Nookaew I, Bordel S, Nielsen J (2013) Understanding the interactions between bacteria in the human gut through metabolic modeling. Sci Rep 3:2532

Shulzhenko N, Morgun A, Hsiao W, Battle M, Yao M, Gavrilova O, Orandle M, Mayer L, Macpherson AJ, McCoy KD et al (2011) Crosstalk between B lymphocytes, microbiota and the intestinal epithelium governs immunity versus metabolism in the gut. Nat Med 17:1585–1593

Sinal CJ, Tohkin M, Miyata M, Ward JM, Lambert G, Gonzalez FJ (2000) Targeted disruption of the nuclear receptor FXR/BAR impairs bile acid and lipid homeostasis. Cell 102:731–744

Skutches CL, Holroyde CP, Myers RN, Paul P, Reichard GA (1979) Plasma acetate turnover and oxidation. J Clin Invest 64:708–713

Smith PM, Howitt MR, Panikov N, Michaud M, Gallini CA, Bohlooly YM, Glickman JN, Garrett WS (2013) The microbial metabolites, short-chain fatty acids, regulate colonic Treg cell homeostasis. Science 341:569–573

Smits LP, Bouter KE, de Vos WM, Borody TJ, Nieuwdorp M (2013) Therapeutic potential of fecal microbiota transplantation. Gastroenterology 145:946–953

Tang C, Ahmed K, Gille A, Lu S, Grone HJ, Tunaru S, Offermanns S (2015) Loss of FFA2 and FFA3 increases insulin secretion and improves glucose tolerance in type 2 diabetes. Nat Med 21:173–177

Thomas C, Gioiello A, Noriega L, Strehle A, Oury J, Rizzo G, Macchiarulo A, Yamamoto H, Mataki C, Pruzanski M et al (2009) TGR5-mediated bile acid sensing controls glucose homeostasis. Cell Metab 10:167–177

Topping DL, Clifton PM (2001) Short-chain fatty acids and human colonic function: roles of resistant starch and nonstarch polysaccharides. Physiol Rev 81:1031–1064

Turnbaugh PJ, Ley RE, Mahowald MA, Magrini V, Mardis ER, Gordon JI (2006) An obesity-associated gut microbiome with increased capacity for energy harvest. Nature 444:1027–1031

Turnbaugh PJ, Backhed F, Fulton L, Gordon JI (2008) Diet-induced obesity is linked to marked but reversible alterations in the mouse distal gut microbiome. Cell Host Microbe 3:213–223

Vance DE (2008) Role of phosphatidylcholine biosynthesis in the regulation of lipoprotein homeostasis. Curr Opin Lipidol 19:229–234

Vandanmagsar B, Youm YH, Ravussin A, Galgani JE, Stadler K, Mynatt RL, Ravussin E, Stephens JM, Dixit VD (2011) The NLRP3 inflammasome instigates obesity-induced inflammation and insulin resistance. Nat Med 17:179–188

Velagapudi VR, Hezaveh R, Reigstad CS, Gopalacharyulu P, Yetukuri L, Islam S, Felin J, Perkins R, Boren J, Oresic M et al (2010) The gut microbiota modulates host energy and lipid metabolism in mice. J Lipid Res 51:1101–1112

Vijay-Kumar M, Aitken JD, Carvalho FA, Cullender TC, Mwangi S, Srinivasan S, Sitaraman SV, Knight R, Ley RE, Gewirtz AT (2010) Metabolic syndrome and altered gut microbiota in mice lacking Toll-like receptor 5. Science 328:228–231

Vinolo MA, Rodrigues HG, Nachbar RT, Curi R (2011) Regulation of inflammation by short chain fatty acids. Nutrients 3:858–876

Vrieze A, Van Nood E, Holleman F, Salojarvi J, Kootte RS, Bartelsman JF, Dallinga-Thie GM, Ackermans MT, Serlie MJ, Oozeer R et al (2012) Transfer of intestinal microbiota from lean donors increases insulin sensitivity in individuals with metabolic syndrome. Gastroenterology 143(913–916), e917

Wall R, Cryan JF, Ross RP, Fitzgerald GF, Dinan TG, Stanton C (2014) Bacterial neuroactive compounds produced by psychobiotics. Adv Exp Med Biol 817:221–239

Wang Z, Klipfell E, Bennett BJ, Koeth R, Levison BS, Dugar B, Feldstein AE, Britt EB, Fu X, Chung YM et al (2011) Gut flora metabolism of phosphatidylcholine promotes cardiovascular disease. Nature 472:57–63

Wang D, Xia M, Yan X, Li D, Wang L, Xu Y, Jin T, Ling W (2012) Gut microbiota metabolism of anthocyanin promotes reverse cholesterol transport in mice via repressing miRNA-10b. Circ Res 111:967–981

Warrier, M., Shih, D.M., Burrows, A.C., Ferguson, D., Gromovsky, A.D., Brown, A.L., Marshall, S., McDaniel, A., Schugar, R.C., Wang, Z., et al. (2015). The TMAO-generating enzyme flavin monooxygenase 3 is a central regulator of cholesterol balance. Cell Rep. pii: S2211-1247(14)01065-1

Watanabe M, Houten SM, Mataki C, Christoffolete MA, Kim BW, Sato H, Messaddeq N, Harney JW, Ezaki O, Kodama T et al (2006) Bile acids induce energy expenditure by promoting intracellular thyroid hormone activation. Nature 439:484–489

Watanabe A, Sohail MA, Gomes DA, Hashmi A, Nagata J, Sutterwala FS, Mahmood S, Jhandier MN, Shi Y, Flavell RA et al (2009) Inflammasome-mediated regulation of hepatic stellate cells. Am J Physiol Gastrointest Liver Physiol 296:G1248–G1257

Weingarden AR, Chen C, Bobr A, Yao D, Lu Y, Nelson VM, Sadowsky MJ, Khoruts A (2014) Microbiota transplantation restores normal fecal bile acid composition in recurrent Clostridium difficile infection. Am J Physiol Gastrointest Liver Physiol 306:G310–G319

Wigg AJ, Roberts-Thomson IC, Dymock RB, McCarthy PJ, Grose RH, Cummins AG (2001) The role of small intestinal bacterial overgrowth, intestinal permeability, endotoxaemia, and tumour necrosis factor alpha in the pathogenesis of non-alcoholic steatohepatitis. Gut 48:206–211

Williams CD, Stengel J, Asike MI, Torres DM, Shaw J, Contreras M, Landt CL, Harrison SA (2011) Prevalence of nonalcoholic fatty liver disease and nonalcoholic steatohepatitis among a largely middle-aged population utilizing ultrasound and liver biopsy: a prospective study. Gastroenterology 140:124–131

Wolever TM, Brighenti F, Royall D, Jenkins AL, Jenkins DJ (1989) Effect of rectal infusion of short chain fatty acids in human subjects. Am J Gastroenterol 84:1027–1033

Wong JM, de Souza R, Kendall CW, Emam A, Jenkins DJ (2006) Colonic health: fermentation and short chain fatty acids. J Clin Gastroenterol 40:235–243

Yamashita H, Kaneyuki T, Tagawa K (2001) Production of acetate in the liver and its utilization in peripheral tissues. Biochim Biophys Acta 1532:79–87

Yamashita H, Fujisawa K, Ito E, Idei S, Kawaguchi N, Kimoto M, Hiemori M, Tsuji H (2007) Improvement of obesity and glucose tolerance by acetate in Type 2 diabetic Otsuka Long-Evans Tokushima Fatty (OLETF) rats. Biosci Biotechnol Biochem 71:1236–1243

Yoshimoto S, Loo TM, Atarashi K, Kanda H, Sato S, Oyadomari S, Iwakura Y, Oshima K, Morita H, Hattori M et al (2013) Obesity-induced gut microbial metabolite promotes liver cancer through senescence secretome. Nature 499:97–101

Zambell KL, Fitch MD, Fleming SE (2003) Acetate and butyrate are the major substrates for de novo lipogenesis in rat colonic epithelial cells. J Nutr 133:3509–3515

Zhu L, Baker SS, Gill C, Liu W, Alkhouri R, Baker RD, Gill SR (2013) Characterization of gut microbiomes in nonalcoholic steatohepatitis (NASH) patients: a connection between endogenous alcohol and NASH. Hepatology 57:601–609

Chapter 10
Hepatic Lipogenesis: Nutritional Control and Pathophysiological Relevance

Shih-Lung Woo, Ting Guo, and Chaodong Wu

Abstract Hepatic de novo lipogenesis (DNL) is important in generating various fatty acids to meet the needs of cellular functions such as cell membrane structural composition and molecular signaling. Like many cellular processes, hepatic DNL is tightly regulated by different dietary nutrient composition and hormonal fluctuation. For example, dietary glucose and fructose could act as a substrate for DNL; insulin together with the stimulation of glucose could induce signaling to enhance hepatic DNL, while glucagon has an inhibitory effect on hepatic DNL. However, the dysregulation of hepatic DNL could lead to metabolic disorders such as non-alcoholic fatty liver disease (NAFLD), and cardiovascular disease (CVD). In this article, nutrient composition and hormonal regulation of hepatic DNL, as well as the implication of hepatic DNL in contributing to metabolic diseases will be discussed in detail.

Keywords Nutrition • Nutrients • Liver • Lipogenesis • Fatty liver disease • Dyslipidemia

1 Introduction

Lipogenesis is the synthesis of fatty acids from a two-carbon precursor, acetyl-CoA. Before the late 1950s, it was thought that fatty acid synthesis was a mere reversible reaction of β-oxidation (Lynen 1953). Until the late 1950s, Wakil et al. announced their discovery that fatty acid biosynthesis in pigeon liver extracts is greatly increased under the presence of carbon dioxide, and that the condensation of two acetyl-CoA molecules is greatly increased in the presence of the newly discovered enzyme, acetyl-CoA carboxylase (ACC) (Wakil et al. 1957). Approximately twenty years later, another milestone of fatty acid biosynthesis was achieved when

S.-L. Woo • T. Guo • C. Wu (✉)
Department of Nutrition and Food Science, Texas A&M University,
College Station, TX, USA
e-mail: cdwu@tamu.edu

J.M. Ntambi (ed.), *Hepatic De Novo Lipogenesis and Regulation of Metabolism*,
DOI 10.1007/978-3-319-25065-6_10

Lynen et al. announced the characterization and purification of a multi-enzyme complex, known as fatty acid synthetase (FAS) (Lynen et al. 1961). Alongside, that was the first study to describe the highly efficient mechanism of palmitoyl-CoA synthesis from acetyl-CoA and malonyl-CoA.

Fatty acid de novo synthesis is present in almost all tissues but major sites are considered to be liver, adipose tissue, and lactating mammary gland (Pearce 1983). However, the quantitative differences of fatty acid synthesis vary across tissues within and between species. For example, adipose tissue is considered as the main lipogenic tissue than liver in mice (Feller 1954), rats (Hausberger et al. 1954), sheep (Ingle et al. 1972), goat (Liepa et al. 1978), and cattle; while human liver was shown to be the major site of lipogenesis (Shrago et al. 1971). Fatty acid synthesis is crucial in generating various fatty acids as essential structural components of membranes, signaling molecules, and metabolic mediators. However, dysregulation of the lipogenesis pathway could contribute to various metabolic diseases including obesity, non-alcoholic fatty liver disease, and cardiovascular disease (Ameer et al. 2014). Here, we will discuss the nutritional and hormonal regulation of hepatic de novo lipogenesis (DNL). Also, the pathophysiological significance of dysregulation in lipogenesis pathway will be reviewed.

2 Regulation of Hepatic DNL

The summary of the biochemical pathways of hepatic DNL is shown below (Fig. 10.1), with acetyl-CoA being the principal substrate for ACC in the first step of DNL. The product of ACC is malonyl-CoA, an important suppressor for carnitine palmitoyl transferase-1 (CPT-1), which is responsible for transporting fatty acid across mitochondrial membrane for β-oxidation. Therefore, fatty acid synthesis antagonizes fatty acid oxidation. Malonyl-CoA then undergoes a series of condensation process with acetyl-CoA, catalyzed by the multi-complex FAS through a three-step ketoreduction-dehydration-enoylreduction process (Wakil 1989), to form the C16:0 final product, palmitic acid. The reaction is summarized as:

$$\begin{aligned}&\text{Acetyl CoA} + 7\text{Malonyl CoA} + 14\text{NADPH} + 14\text{H}^{+}\\&\rightarrow \text{Palmitic acid} + 7\text{HCO}_{3^-} + 8\text{CoA} + 14\text{NADP}^{+} + 6\text{H}_2\text{O}\end{aligned}$$

Then, palmitic acid can undergo desaturation by stearoyl-CoA desaturase 1 (SCD-1) and other desaturases or elongation by elongase to form various fatty acids such as palmitoleic acid (16:1), and oleic acid (C18:1).

The two isoforms of ACC—ACC1 and ACC2 have different localization, and thus have slightly different role in lipid metabolism. ACC1, which is predominantly expressed in cytosol of lipogenic tissues such as liver, adipose, and mammary gland, plays a role in synthesizing malonyl-CoA for fatty acid synthesis (Abu-Elheiga et al. 2005). In contrast, ACC2, which is localized on the outer mitochondrial membrane, is expressed mainly in tissues with low lipogenic capacity such as the

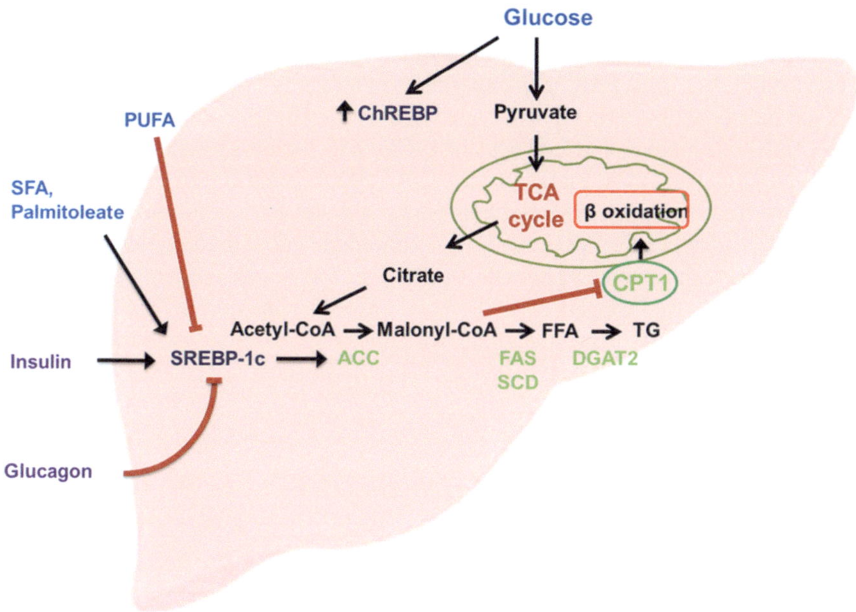

Fig. 10.1 Hepatic de novo lipogenesis (DNL) is tightly regulated by different dietary nutrient composition and hormonal fluctation. In response to feeding, dietary glucose and fructose act as a substrate for DNL. Upon fasting, glucagon has an inhibitory effect on hepatic DNL. Also, fatty acids critically regulate hepatic DNL. However, the regulatory effects of fatty acids vary depending on types, i.e. saturated, monosaturated, and polyunsaturated fatty acids, and the composition of fatty acids. *CHREBP* carbohydrate response element-binding protein, *SREBP-1c* sterol regulatory element-binding protein-1c, *ACC* acetyl-CoA carboxylase, *CPT1* carnitine palmitoyl transferase 1, *DGAT2* diacylglycerol transferase 2, *FAS* fatty acid synthetase, *SCD* stearoyl-CoA desaturase, *FFA* free fatty acid, *PUFA* polyunsaturated fatty acid, *SFA* saturated fatty acid and *TG* triglyceride

heart and skeletal muscle. The fact that β-oxidation occurs mainly in the mitochondria, function of ACC2 is generally considered to inhibit β-oxidation. This is achieved by the production of malonyl-CoA produced by ACC2 (Oh et al. 2005).

ACC is regulated by short-term and long-term mechanisms while FAS is regulated mainly by long-term regulation. Short-term regulation involves both allosteric regulation and multisite phosphorylation. ACC1 and ACC2 activity is regulated by allosteric activation of citrate while deregulated by long-chain acyl-CoA through allosteric inhibition (Barber et al. 2005; Brownsey et al. 2006). In addition, ACC is also regulated by phosphorylation, where phosphorylation of ACC reduces its activity, by kinases in response to hormonal and nutritional stimuli (Munday 2002). For example, glucagon could induce phosphorylation of ACC via protein kinase A (PKA) and AMP-activated protein kinase (AMPK). It is known that ACC has multiple phosphorylation sites, however, it seems that the phosphorylation at Ser77 and Ser79 has inhibitory effects on ACC (Munday 2002). In contrast, FAS is not regulated by this short-term regulation, and is considered to alter rate of DNL through long-term regulation by a change in gene transcription of this enzyme. In fact, both ACC and FAS may regulate long-term DNL via changes in gene expression.

Other than ACC and FAS, it is important to know a few other enzymes that are involved indirectly, but crucial, in lipogenesis. These enzymes include ATP-citrate lyase (ACL) (Kornacker and Lowenstein 1965), which cleaves the cytosolic citrate into acetyl-CoA and OAA; malic enzyme (ME) (Walker and Bailey 1969), which is responsible for converting cytosolic malate to pyruvate; glucose-6-phosphate dehydrogenase (Rudack et al. 1971a) and 6-phosphogluconate dehydrogenase (Rudack et al. 1971b), both of which generate NADPH in the phosphate pentose pathway that is necessary for lipogenesis. All these enzymes, including ACC and FAS, respond to nutritional/hormonal signals, which will increase/decrease the activities and/or concentration of the enzymes.

2.1 Nutritional Regulation of Hepatic DNL

Hepatic DNL is generally most activated by high-carbohydrate feeding but inhibited by high-fat feeding and fasting. The rate of DNL varies markedly in response to different dietary compositions. For example, the relative quantities of carbohydrate in the diet and the form of carbohydrate have significant effects on hepatic DNL.

It should be clearly distinguished that DNL may be significant during dietary surplus of carbohydrate but not as significant during high-carbohydrate (low-fat) eucaloric diets (Hellerstein et al. 1996).

Overfeeding of carbohydrate enhances the conversion of carbohydrate into fatty acids. Perhaps one very good example of this phenomenon is the clinical observation of fatty infiltration in liver induced by high dextrose content of total parenteral nutrition (Baker and Rosenberg 1987). Numerous studies have also shown the lipogenic effect of carbohydrate overfeeding (Masoro et al. 1949; Lammert et al. 2000). Particularly, during massive overfeeding of a carbohydrate-rich diet in the Guru Walla model—a Cameroon ritual where teenage boys deliberately overfeed a 7000 kcal/day, high carbohydrate diet (70 %) but just a total of 4 kg fat consumption in over 10 weeks, which they gain 12 kg of fat (Pasquet et al. 1992). However, during less extreme carbohydrate feeding, DNL is less significant. As shown by Acheson et al., DNL becomes significant after maximum whole-body glycogen storage of approximately 700–1000 g is reached. In other words, the excess carbohydrate during short-term ingestion of a large amount of carbohydrate should go towards saturation of glycogen storage before DNL becomes significant (Acheson et al. 1984, 1988).

While the above given examples did not explain the lipogenic effect related to the composition of carbohydrate, it is generally recognized that glucose is the main lipogenic substrate in monogastric animals. The lipogenic effect of glucose is further discussed by mechanistic studies of the lipogenic transcription factor such as carbohydrate response element-binding protein (ChREBP) and sterol regulatory element-binding protein-1c (SREBP-1c), both of which will be discussed further in the sections below. Fructose is also now emerging as a strong lipogenic stimulator, and will be discussed further.

2.2 *Glucose*

The role for glucose in stimulating DNL has long been recognized. This is because glucose, through its metabolism, provides substrates that are needed for DNL. Furthermore, as a signaling molecule, glucose has a stimulatory effect on the expression of several key lipogenic enzymes, in particular ACC (Ma et al. 2005; Towle 2005). However, this role should be interpreted as a role for glucose metabolism, but not glucose, per se, in stimulating DNL. Physiologically, glucose metabolism is tightly coupled to insulin. Thus, it is very difficult to separate the effect of glucose from that of insulin on DNL. After the identification and functional characterization of ChREBP in controlling the expression of lipogenic enzymes (Ishii et al. 2004), it becomes largely accepted that ChREBP and SREPB1c mediate the effect of glucose and insulin, respectively, on stimulating lipogenic enzyme expression.

The stimulatory effect of glucose, strictly speaking glucose metabolism, on DNL is manifested by the finding that glucokinase overexpression in rodents led to hepatic steatosis (O'Doherty et al. 1999; Wu et al. 2005) A simple explanation is that enhancing glycolysis provides substrates for DNL. Because generation of fructose-1,6-bisphosphate from fructose-6-phosphate is a rate-determining step of glycolysis, overexpression of a kinase-dominant 6-phosphofructo-2-kinase was used to enhance glycolysis (Wu et al. 2005). Of interest, DNL was not increased; although glycolysis was enhanced. Further mechanistic investigation demonstrated the importance of pentose phosphate shunt in activating ChREBP, and, likely, in generating NADPH, another required substrate of DNL, to enable DNL (Wu et al. 2005). In that study, insulin levels were decreased in response to glucokinase overexpression. This, however, does not imply a stimulatory effect of glucose (metabolism) on DNL that is independent of insulin. Indeed, enhancing glycolysis through overexpression of glucokinase generates effects on glycolysis that are almost identical to the combined effect of glucose and insulin.

2.3 *Fructose*

Fructose has received much attention recently due to its association with extensive dietary consumption in the modern society, its close relationship with metabolic diseases such as non-alcoholic fatty liver disease (NAFLD), insulin resistance, and dyslipidemia, due mostly to the lipogenic capacity of fructose.

Fructose is a monosaccharide (other than glucose and galactose) that occurs in the form of sucrose, a disaccharide composed of fructose and glucose, or in the form fructose alone. After absorption, fructose is rapidly phosphorylated by fructokinase into fructose-1-phosphate in hepatocytes. Fructose-1-phosphate is then converted to glyceraldehyde and dihydroxyacetone, which could feed into the regular glycolytic pathway. This is of importance because the fructose metabolic pathway is able to bypass the main regulatory step of glycolysis, which is 6PFK1. In addition, fructose uptake is not regulated by insulin nor does fructose stimulate insulin secretion as much as glucose does, and fructokinase has almost 10 times higher enzyme activity than glucokinase (Zakim 1973). Moreover, Havel et al. showed that an equivalent dose of glucose and fructose (1 g/kg dose) produced dissimilarities in their levels reflected in blood. Blood fructose level increases minimally to approximately 0.5 mM, which is relatively low when compared to 10 mM postprandial blood glucose level (Havel 2005). This suggests that fructose is highly metabolized upon the first pass to the liver.

There is no doubt that fructose stimulates the mRNA expression of lipogenic enzymes such as ACC and FAS (Miyazaki et al. 2004). However, this does not mean that all the fructose is metabolized in favor of lipogenesis. In fact, in a single-blinded, randomized, crossover study, it was found that 35 % of the fructose was oxidized, approximately 38 % appeared as synthesized glycerol in the form of VLDL-triglyceride, 0.4 % appeared as de novo fatty acids in the form of VLDL-triglyceride, and some likely remained storage as liver glycogen after a fructose load of 0.75 g/kg (Chong et al. 2007).

The contemporary issue nowadays is that the over-consumption of sucrose and high-fructose corn syrup in our daily food products likely is amplifying the significance of the lipogenic effect of fructose (Marriott et al. 2010). Moreover, fructose could induce hepatic and extrahepatic insulin resistance as shown by Dirlewanger et al. (2000). Also, high fructose consumption has been highly correlated with visceral obesity (Pollock et al. 2012; Stanhope et al. 2009). This is important as insulin resistance and obesity are highly related to metabolic diseases such as NAFLD and dyslipidemia (Basciano et al. 2005; Vos and Lavine 2013; Bray 2013), which will be discussed further.

2.4 Fatty Acid Composition

The earliest report of the inhibitory effect of dietary lipid on hepatic lipogenesis is in 1950 when Masoro et al. showed that there were less labelled [^{14}C] fatty acids generated from [^{14}C] glucose in rat livers after 3 days of a diet containing 60 % butter, compared to rats that were fed a 60 % glucose diet (Herzberg 1983). Due to the heterogeneity of dietary lipid composition, such as saturated, mono-, and polyunsaturated fatty acids, numerous subsequent studies then aim to clarify if all dietary fats have equal effects on diminishing DNL. Saturated fatty acids promotes fatty acid synthesis (Sampath et al. 2007), whereas polyunsaturated fatty acids (PUFA), both n-3 and n-6 PUFAs, are able to suppress the hepatic lipogenic enzymes (Musch et al. 1974; Toussant et al. 1981; Flick et al. 1977; Clarke and Jump 1996). Interestingly, palmitoleate, a monounsaturated fatty acid, seems to controversially stimulate DNL as observed in the inducing effect of palmitoleate in increasing FAS mRNA expression in mice liver, as well as an increase in hepatic lipid accumulation (Guo et al. 2012).

2.5 *Palmitoleate*

As mentioned above, palmitoleate controversially increases DNL. In two lines of mice studies (Cao et al. 2008; Yang et al. 2011), palmitoleate was thought to decrease hepatic lipid accumulation due to inhibition of FAS expression in liver. In contrast, Petit et al. reported that circulating palmitoleate level might positively correlate with the degree of hepatic steatosis in human subjects (Petit et al. 2012). A recent mechanism study by Guo et al. reported similar effects of palmitoleate on inducing hepatic DNL and hepatic steatosis in mice (Guo et al. 2012). This is supported by an increase in expression of SREBP-1c and FAS upon palmitoleate treatment in hepatocytes cell line. Interestingly, palmitoleate supplementation in mice also improved insulin sensitivity and glucose tolerance, thus dissociating the association between insulin resistance and hepatic steatosis. In particular, these effects are found to be activated via liver Akt signaling by palmitoleate (Guo et al. 2012), which is important as SREBP-1c is known to be activated via the Akt-mTORC1 pathway and gluconeogenesis is known to be downregulated via Akt-FOXO1 pathway. Also, the ability of palmitoleate to reduce liver and macrophage inflammatory response in this study might also help explain the improvement in insulin resistance (Guo et al. 2012).

The findings of this study provide a new insight of how different lipid composition could have profound effects on human physiology; as palmitoleic acid is a monounsaturated fatty acid that could be a dietary component and more importantly, could also be endogenously produced by adipocytes. The latter effect has important implication for further investigation on elucidating the cross-talk between adipose tissue and liver in the pathophysiology of metabolic diseases such as insulin resistance and NAFLD.

2.6 *Hormonal Regulation of Hepatic DNL*

Hormonal regulation of DNL could be short-term, which takes approximately 15–30 min to become operative; and long-term, which takes approximately 6 h to become effective (Beynen et al. 1979).

The involvement of insulin in DNL, specifically increasing FAS activity in rat liver in response to a high-carbohydrate, fat-free diet was first described by

Lakshmanan et al. (1972). Subsequent studies confirmed the effect of insulin in DNL by showing similar results that insulin stimulates important DNL enzymes, ACC and FAS in hepatocyte cell cultures and isolated hepatocytes (Geelen et al. 1978; Geelen and Gibson 1975; Muller et al. 1976; Geelen and Vaartjes 1977). Concurrently, most of these studies reported opposite effect of glucagon on DNL, and that the hormonal effects of both insulin and glucagon on DNL were noticeable within 30 min of administration.

2.7 *Insulin*

It is well established that insulin has a stimulatory effect on DNL, which explains perfectly the lipogenic effect of glucose. It is not until the discovery of the transcription factor sterol regulatory element-binding proteins (SREBPs) that sheds more light to the stimulatory mechanism of insulin on DNL. SREBPs are transcription factors that were described to regulate genes encoding enzymes that are responsible for cholesterol and fatty acid biosynthesis (Brown and Goldstein 1997).

SREBP-1c is one of the isoforms that is expressed in most tissues of humans and mice with specially high levels in liver, skeletal muscle, white adipose tissue, adrenal gland, and brain (Shimomura et al. 1997). The expression of SREBP-1c is known to be regulated by liver X receptor (LXR), a nuclear hormone receptor, upon activation by insulin (Chen et al. 2004). The SREBP-1c protein precursor, which has a C-terminal domain bound to SREBP-cleavage-activating protein (SCAP), is newly synthesized and located on the endoplasmic reticulum membrane. Upon activation by insulin, the SREBP-1c precursor is transported to the Golgi apparatus to undergo proteolytic splicing by specific proteases at Site 1 and Site 2 so as to release the N-terminal of the mature, truncated protein (Horton et al. 2002). The mature form of SREBP-1c can then translocate to the nucleus, binding to sterol-regulating-element-like or E-Box of the promoter region to act as a transcription factor of lipogenic genes such as ACC and FAS (Amemiya-Kudo et al. 2002). Therefore, the physiological function of SREBP-1c is recognized as the main regulator of lipogenesis, especially in liver.

In a normal liver, insulin-binding to insulin receptor results in auto-phosphorylation of the β subunit intracellularly by the receptor's own tyrosine kinase domain. This then recruits several substrates including insulin receptor substrates-1 and -2 (IRS-1 and IRS-2), which are important in transducing the cascade for glucose metabolism. Activation of IRSs are important for the activation of phosphatidylinositide-3 kinase (PI-3 K), which then generates phosphatidylinositol (3,4,5)-triphosphate (PIP3) that activates 3-phosphoinositide dependent kinase-1 and -2 (PDK1, and PDK2 also known as mTORC2). PDK1 and PDK2 then phosphorylate protein kinase B (PKB)/Akt at Thr^{308} and Ser^{473} residues respectively, thus activating Akt. On one hand, Akt phosphorylates (thus inactivates) Forkhead box protein O1 (FOXO1), which is a transcription factor that is necessary for transcription of gluconeogenic enzymes such as glucose-6-phosphatase (G6Pase) and phosphoenolpyruvate carboxylase (PEPCK) (Matsumoto et al. 2006; Kubota et al. 2008). This results in the inhibitory effect of insulin on hepatic gluconeogenesis. On the other hand, Akt is able to activate mammalian target of rapamycin complex 1 (mTORC1), which activates p70 S6 kinase (S6K) (Li et al. 2010; Ma and Blenis 2009). This then, leads to an increase in the truncated, activated SREBP-1c in the nucleus, thus, able to activate the transcription of lipogenic genes.

2.8 Glucagon

The antagonizing effect of insulin and glucagon on glucose metabolism has long been known. This is not any different in hepatic DNL. As already discussed, glycolysis is crucial for generating carbon flux into fatty acid biosynthesis. Insulin is known to stimulate lactate and pyruvate production through glycolytic flux while glucagon effect is the opposite due to the inhibitory effect of glucagon on pyruvate kinase and phosphofructokinase-2. This means that conditions which inhibit lactate and pyruvate production should inhibit hepatic DNL during the absence of another source of acetyl-CoA (Beynen et al. 1980, 1981; Harris et al. 1979). In fact, the diminishing effect of glucagon on pyruvate kinase and phosphofructokinase-2 is affected by the increase levels of cyclic adenosine monophosphate (cAMP), which is due to increased adenylate cyclase activation, thereby increasing the activity of cAMP-dependent protein kinase (PKA) (Muller et al. 1988; Ljungstrom and Ekman 1977). Activated PKA then phosphorylates both pyruvate kinase and phosphofructokinase-2, thus inactivation of these two enzymes. In fact, that glucagon effects on the downstream cascades through PKA signaling is now well-recognized to be the canonical pathway of glucagon signaling pathway. In addition, glucagon could directly decrease DNL via the action of PKA phosphorylation (thus inactivation) on ACC in rat liver (Holland et al. 1984; Tipper and Witters 1982). While this cAMP-dependent pathway of glucagon has long been discovered, a cAMP-independent pathway of the glucagon action has also been reported. This is supported by the evidence of glucagon effect on AMPK (Sim and Hardie 1988).

Several studies have also shown that hepatic glucagon action could cause an energy-depleted state as characterized by elevated AMP: ATP ratio that is sufficient to activate AMPK (Berglund et al. 2009; Kimball et al. 2004; Longuet et al. 2008). AMPK is an energy-sensing switch that is involved in glucose, lipid and protein metabolisms. It was originally discovered by its ability to inhibit fatty acid synthesis (Carlson and Kim 1973) and cholesterol synthesis (Beg et al. 1973) through phosphorylation of both ACC and HMG-CoA reductase respectively. Extensive studies have then showed that AMPK plays more roles in human metabolism and diseased states than just lipid metabolism (Grahame Hardie 2014). But, they will not be covered here due to space limitations.

It is worth noting that AMPK has an inhibitory effect on SREBP-1c. In a recent study (Li et al. 2011), Li et al. showed that AMPK was able to directly phosphorylate SREBP-1c at Ser 372 residue so as to reduce the proteolytic cleavage of

SREBP-1c, thus reducing the nuclear accumulation of the mature SREBP-1c, resulting in a decrease in transcription of lipogenic gene such as ACC, FAS, and SCD-1in in vitro essay using HepG2 cell lines. In vivo data also showed that AMPK-induced phosphorylation of SREBP-1c renders a decrease in hepatic lipid accumulation (Li et al. 2011).

3 Regulation of Hepatic DNL by Exercise

It is important to bear in mind that physical exercise involves skeletal muscle contraction, which in turn may generate beneficial metabolic changes that places physical exercise on the top hierarchical position of disease prevention. One that is of great emphasis nowadays is the beneficial effect of physical exercise on improving metabolic diseases (Lancaster and Febbraio 2014). Indeed, physical exercise is important for health promotion and prevention of metabolic diseases such as non-alcoholic fatty liver disease and dyslipidemia that are highly associated with insulin resistance (Lee et al. 2004). Most of the molecular mechanistic studies of how exercise improves insulin resistance focus on skeletal muscle. These molecular signaling changes in skeletal muscle arise from alteration in energy status (i.e., elevated AMP to ATP ratio), increase in intracellular calcium concentration, increase in ROS, and increase in protein kinase C (PKC) due to mechanical contraction (Stanford and Goodyear 2014). There is evidence that reversal of muscle insulin resistance with exercise could indirectly reduce postprandial hepatic DNL due to divergence of ingested carbohydrate into glycogen synthesis in the muscle instead of de novo lipogenesis in the liver during a state of insulin resistance (Rabol et al. 2011).

Indeed, there is a good amount of studies pointing that exercise could down-regulate enzymes involved in hepatic DNL (Gorski et al. 1990; Garrido et al. 1996). Though, there is evidence that exercise may inhibit hepatic lipid synthesis and stimulate β-oxidation via activation of AMPK (Carlson and Winder 1999; Park et al. 2002; Ruderman et al. 2003). As recalled, activation of AMPK responds to a state of energy deprivation (high AMP:ATP ratio). There are studies that suggest exercise could increase AMPK activation via sensitization of leptin signaling induced by exercising. In his study (Ruderman et al. 2003), Ruderman et al. reported that liver AMPK was activated during exercise and AMPK was still activated 30-min post-exercise. However, AMP and ATP levels cannot likely be altered at 30-min post-exercise, and this could not be the factor anymore that leads to AMPK activation. This suggests that there might other factors that could lead to AMPK activation. Since leptin has been shown to inhibit fatty acid oxidation through AMPK activation (Minokoshi et al. 2002), Yi et al. (2013) hypothesized that acute and chronic exercise was able to activate the liver leptin-AMPK signaling cascade. By using an insulin- and leptin-resistant rat model, Yi et al. showed that acute and chronic exercise was able to improve leptin signaling by increasing liver expression of leptin receptor, and an increased in AMPK phosphorylation. Consistently, phosphorylation of ACC was increased. It is also worth noting that liver lipid accumulation was decreased by acute and chronic exercise in the diabetic rats (Yi et al. 2013). In fact, in patients with NAFLD, exercise programs consisted of 2-3 sessions per week of 30–60 min was able to reduce liver lipid accumulation (Shojaee-Moradie et al. 2007; Johnson et al. 2009; van der Heijden et al. 2010). The reversal of insulin resistance and hepatic lipid accumulation is important as we further discuss the sections below.

4 Pathophysiological Role for Hepatic DNL

The homeostasis of hepatic lipid metabolism involves a balance between fatty acid/triglyceride synthesis/input and fatty acid/triglyceride oxidation/output. And, each of these two components involves several pathways. For example, sources of hepatic fatty acid/triglyceride include hepatic DNL, dietary fatty acids derived from chylomicron remnants, fatty acids esterification, and plasma non-esterified fatty acids (NEFA) from adipose tissue lipolysis; whereas hepatic fatty acid/triglyceride should be balanced by β-oxidation, and triglyceride export in the form of very-low density lipoprotein (VLDL). Dysregulation on any of these pathways could pose significant physiological consequences that could lead to metabolic disorders. In particular, fatty liver disease (both alcoholic and non-alcoholic fatty liver disease), as well as cardiovascular disease (CVD)-related dyslipidemia are closely related to dysregulation of DNL. Here, the physiological significance of DNL dysregulation will be focused.

4.1 Implication of DNL Dysregulation in Non-alcoholic Fatty Liver Disease

Non-alcoholic fatty liver disease (NAFLD) is a clinical manifestation which encompasses the whole spectrum of liver diseases including hepatic steatosis, non-alcoholic steatohepatitis (NASH), and cirrhosis without significant alcohol consumption (Chalasani et al. 2012). While simple lipid accumulation in the liver is generally considered as histologically benign, it could progress to NASH during overt liver necroinflammation, and could eventually progress to cirrhosis, liver failure and liver cancer (Chalasani et al. 2012; Tilg and Moschen 2010). The estimated prevalence of NAFLD ranges from 6 to 35 % with a median of 20 % worldwide in the general population (Chalasani et al. 2012; Vernon et al. 2011). It is reported that NASH is becoming a more common cause for liver transplantation in the United States, and is on the path of becoming the most common (Wattacheril and Chalasani 2012).

The pathogenesis of NAFLD is still not well understood. However, the general pathogenesis of NAFLD could stem from a "two hits" model that was first proposed by Day and James (Day and James 1998). The first "hit" requires the production of hepatic steatosis. Factors that contribute to hepatic steatosis include increased hepatic DNL, decreased hepatic β-oxidation, increased free fatty acid supply from adipose, and decreased very-low density lipoprotein (VLDL) triglyceride output (Tilg and Moschen 2010; Browning and Horton 2004; Fabbrini et al. 2008). The second "hit" requires a source of oxidative stress capable of initiating significant lipid peroxidation, leading to histological damage (Day and James 1998); though nowadays, there is more and more evidence showing that the second "hit" could be promoted by a chronic proinflammatory environment induced by obesity-related adipose tissue dysfunction and obesity-induced insulin resistance. However, the sequence of these "two hits" has been challenged in the sense that inflammation could precede hepatic steatosis and the metabolic events present in NAFLD are suggested to occur in a parallel rather than a consecutive manner. In addition, emerging evidence suggests that there are multiple factors contributing to NAFLD concurrently. These factors that lead to liver inflammation include gut-derived mediators, adipose-derived mediators, and endoplasmic reticulum stress. Therefore, a "multiple parallel hits" concept might be a more precise reflection of the current knowledge of NAFLD (Tilg and Moschen 2010).

Though, it is important to point out that NAFLD is nearly always associated with metabolic disorders such as insulin resistance, obesity, dyslipidemia, and others (Farrell and Larter 2006). Studies in rodents and humans have revealed that the excessive accumulation of triglyceride in the liver is mainly caused by an overflow of fatty acids from hypertrophied and insulin resistant adipose tissue, and from increased DNL in the liver; but is less contributed by fatty acid disposal through β-oxidation or VLDL export (Lewis et al. 2002). In fact, isotopic quantitative analysis in NAFLD patients showed that labelled triglyceride in the liver was distributed around 59 % derived mainly from adipose lipolysis, 26 % derived from DNL, and 15 % derived from the diet (Donnelly et al. 2005). Importantly, the lipogenic rate, which is normally inhibited during fasted state, was already elevated.

4.2 Hepatic Insulin Resistance and Elevated DNL

Appropriate insulin signaling in the liver suppresses gluconeogenesis and promotes lipogenesis. However, during a state of hepatic insulin resistance, insulin signaling fails to suppress gluconeogenesis but insulin-induced lipogenesis remains functional. This "selective insulin resistance" paradox was proposed by Brown and Goldstein (Brown and Goldstein 2008), and that impairment of insulin signaling diverges after Akt, suggesting that insulin-mediated suppression of FOXO1 is impaired while insulin-mediated SREBP-1c remains sensitive (Li et al. 2010). Moreover, hyperinsulinemia that usually accompanies insulin resistance, as a compensation mechanism of the pancreas due to hyperglycemia, could further amplifies the insulin-mediated lipogenesis. What causes the impaired suppression of gluconeogenesis but intact lipogenesis remains poorly understood. Accordingly, many theories have arisen to explain the pathology of this insulin signaling paradox (Haeusler and Accili 2008; Capeau 2008; Sajan et al. 2015). Of the very recent, there is evidence suggesting that over-nutrition-induced overproduction of triglyceride metabolites could result in hyper-activation of atypical protein kinase C (aPKC), and that this might be the key in selectively impairing FOXO1 phosphorylation while stimulation of SREBP-1c remain elevated (Sajan et al. 2015). However, to be on the conservative side, more studies are still required to fully elucidate the molecular mechanisms of this selective insulin resistance.

It is important to recognize that there hasn't been a cause-and-effect relationship established between NAFLD and hepatic insulin resistance. This could be evidenced by the work of Samuel et al. (Samuel et al. 2007). The authors showed that knocking down the expression of protein kinase Cε (PKCε), which is known to impair insulin signaling, was able to reverse insulin resistance despite hepatic steatosis. They further showed that liver accumulation of diacylglycerol, which is a lipid metabolite, could activate PKCε, and that PKCε could directly bind and inhibit the activity of insulin receptor tyrosine kinase, thus impairing insulin signaling (Samuel et al. 2007). In addition, hepatic steatosis is not always associated with insulin resistance, as evidenced by the work of Monetti et al., showing that the hepatic steatosis induced by overexpression of diacylglycerol acyltransferase 2 (DGAT2), the committed step in triglyceride synthesis, did not result in insulin resistance (Monetti et al. 2007). So, this begs the question: what accounts for the discrepancies of the existence between hepatic steatosis and insulin resistance? It has been suggested that when steatosis is associated with an accumulation of specific lipid component, fatty acid, or lipid derivative, these component(s) could improve or impair insulin sensitivity (Monetti et al. 2007). Indeed, a recent animal study by Huo et al. found an association between hepatic steatosis and an increase in palmitoleate level of the lipid content in the steatotic liver (Huo et al. 2012). To further verify the significance of this finding, Guo et al. demonstrated that palmitoleate was able to induce hepatic steatosis but improve insulin sensitivity at both the hepatic and systemic level. More interestingly is that palmitoleate was able to suppress liver inflammation. This indicates that the improvement of insulin sensitivity by palmitoleate is, at least partly, due to the suppression of liver inflammation (Guo et al. 2012).

Clearly, hepatic steatosis does not necessarily correlate with insulin resistance. Nonetheless, when both hepatic steatosis and insulin resistance do exist concurrently, there seems to be no clear cut pathophysiology of which one contributes to one another. This is because either hepatic steatosis or insulin resistance has the potential to lead to one another. It remains important to factor in other "players" into the equation. This includes inflammation that is also often associated with insulin resistance and hepatic steatosis. However, the physiological significance of inflammation is out of the scope of discussion here.

4.3 Alcoholic Fatty Liver Disease Has Similar Features of NAFLD

The characteristics of alcoholic fatty liver disease (AFLD) is similar to NAFLD in the sense that AFLD also include hepatic steatosis, steatohepatitis, fibrosis, however, AFLD is caused by over-consumption of alcohol (ethanol) instead of over-nutrition. Of course, the factors that determine the degree of liver injury depends on amount, duration, type of alcohol, drinking patterns, gender, genetic variance, and others. According to the World Health Organization's Global Alcohol database, which is used to estimate the patterns of alcohol consumption and compare the alcohol-related morbidity and mortality, the burden of alcohol-related liver injury is most prominent in developed countries; however, the burden is increasing in developing countries (O'Shea et al. 2010). Most of the patients with heavy alcohol intake have some degree of hepatic steatosis, which is usually non-symptomatic and is reversible with abstinence (Lefkowitch 2005). Like NAFLD, the genesis of lipid accumulation in the liver is also caused by an imbalance of fatty acid or triglyceride synthesis and disposal (Sozio et al. 2010).

Highly metabolized in the liver, ethanol is converted by alcohol dehydrogenase to acetaldehyde, which is then converted to acetyl-CoA by acetaldehyde dehydrogenase. More importantly, these two steps involve the conversion of NAD^+ to NADH. One of the earliest mechanisms for alcoholic steatosis is that the accumulation of NADH acts as a product inhibition of mitochondrial β-oxidation dehydrogenases, resulting in hepatic steatosis (Sozio and Crabb 2008). However, as new studies evolve, there are additional implications of how ethanol induces hepatic steatosis, including an increase in hepatic lipid synthesis induced by alcohol. Though, the molecular mechanisms have been unclear.

Nonetheless, numerous studies suggested that one of the lipogenic effects of ethanol is at the level of AMPK. Through a series of studies, You et al. reported that ethanol is able to inhibit AMPK activity in rat hepatoma cells, followed by a increase in transcription of SREBP-1c, which is a downstream of AMPK (You et al. 2004). Further in vitro experiments using rat hepatoma cell lines, You et al. showed that acetaldehyde, which is a metabolite of ethanol, could increase the level of active, mature form of SREBP-1c, an effect that was further confirmed by in vivo study using mice chronically fed with ethanol to induce fatty liver (You et al. 2002). Consistently, chronic ethanol feeding in mice also decreased the expression of SREBP-1c target genes such as ACC, FAS, SCD, ACL, and malic enzyme (Yin et al. 2007), whereas SREBP-1c knockout mice partially reduced ethanol-induced fatty liver (Kaplowitz and Ji 2006). Also, the lipogenic effects of ethanol seem to involve ChREBP (Liangpunsakul et al. 2013).

Other mechanisms that are involved in ethanol-induced hepatic steatosis include upregulating factors that stimulate SREBP-1c expression. These factors include endoplasmic reticulum stress, endocannabinoids, lipolysaccharide (LPS) signaling due to increased gut permeability. Also, alcohol has been reported to downregulate factors that reduce SREBP-1c expression include sirtuin1, adiponectin, and signal transducer and activator of transcription 3 (STAT3). Moreover, alcohol has been reported to inhibit fatty acid oxidation via inactivation of the peroxisome proliferator-activated receptor-α (PPAR- α), a nuclear receptor that activates the transcription of genes involved in fatty acid oxidation (Gao and Bataller 2011).

Undoubtedly, many new mechanisms have been unraveled to explain molecular players in mediating DNL in fatty liver (both AFLD and NAFLD). However, there remain missing links of the molecular cascade to be further clarified; for example, in the case of insulin resistance-associated NAFLD, the molecular mechanism that leads to intact lipogenesis during insulin resistance is still not fully understood.

4.4 Implication of DNL in Cardiovascular Disease

Cardiovascular disease (CVD) has long been the leading cause of death worldwide. The World Health Organization (WHO) predicted that in 2015, CVD will be accounting for approximately 30 % of all deaths worldwide (Institute of Medicine 2010). CVD is a range of diseases that include coronary artery disease, heart failure, stroke, and others that affects the heart and the vascular system. Among the cluster of risk factors for CVD, metabolic syndrome, which includes dyslipidemia and hypertension, increases the risk of CVD two- to threefold (Isomaa et al. 2001). Despite the commonality of metabolic syndrome among the U.S. population, it has been difficult to characterize the common molecular mechanisms between those syndromes and CVD pathophysiology (Kahn et al. 2005; Zimmet et al. 2005). Dyslipidemia in the clinical setting, which could be induced by insulin resistance (Biddinger et al. 2008), is characterized by increased triglyceride level, increased low-density lipoprotein (LDL) cholesterol, and decreased high-density lipoprotein (HDL) cholesterol (Grundy 1998). All of these components of dyslipidemia are now known to be contributed by hepatic overproduction of triglyceride-rich very-low-density lipoprotein 1 (VLDL1) (Adiels et al. 2008).

4.5 Metabolism of VLDL

The metabolism (synthesis and degradation) of VLDL require the integrative function of several proteins that are mostly mediated by insulin. The assembly of VLDL in the hepatocytes starts from the lipidation of apoprotein B100 (apoB100) by microsomal triglyceride transfer protein (MTP) in the rough endoplasmic reticulum, forming the triglyceride-poor VLDL particles (VLDL2), which then undergo further lipidation to form the triglyceride-rich VLDL1 (Olofsson and Boren 2005; Stillemark-Billton et al. 2005). VLDL is then secreted from the liver to the circulation, forming intermediate-density lipoprotein (IDL) as triglycerides are hydrolyzed by lipoprotein lipase in the peripheral tissues. As IDL further travels through the circulation, it can then undergo further hydrolysis by lipoprotein lipase, forming low-density lipoproteins (LDL), which can then be removed by LDL-receptor mainly present in the liver (Goldstein and Brown 2009).

4.6 Role for DNL in VLDL Overproduction

In the 1990s, Gibbons et al. reported that fatty acids synthesized via hepatic DNL (during a period of high DNL rate) appear to be preferentially secreted as VLDL rather than stored as cytosolic triglyceride pool (Gibbons and Burnham 1991; Gibbons 1990). These were shown by rats after normal-fed and sucrose-fed (where DNL should be higher than starvation state), the capacity for VLDL secretion via DNL is higher than via extracellular NEFAs (Schonfeld and Pfleger 1971; Salam et al. 1988). Hepatic DNL could increase VLDL-triglyceride secretion by two ways. Firstly, the newly synthesized fatty acids from DNL could be directly made available for esterification of triglyceride. Secondly, the lipogenic product, malonyl-CoA, could indirectly inhibit fatty acid oxidation, therefore, circulating fatty acids are more available for triglyceride reesterification. Consequently, by these two mechanisms, elevated hepatic DNL could contribute to an increased rate of VLDL-triglyceride production, thus dyslipidemia.

5 What Could Cause Elevated DNL That Contributes to Increased VLDL-Related Dyslipidemia?

Firstly, we would like to emphasize the importance of hyperinsulinemia, one of the characteristics of insulin resistance, in contributing to increased hepatic DNL and VLDL-triglyceride secretion. Indeed, numerous studies have shown contribution of DNL to VLDL secretion from the liver of obese and insulin resistant subjects (Hellerstein 1999; Schwarz et al. 2003; Shimomura et al. 2000). In this particular study, Schwarz et al. (2003) showed that hyperinsulinemic obese subjects had significantly higher DNL rate and VLDL-triglyceride than obese subjects without hyperinsulinemia. In addition, the hyperinsulinemic obese subjects who ate a non-lipogenic diet have a significantly higher DNL than the normoinsulinemic obese subjects. These observations again confirm the importance of hyperinsulinemia in contributing to increased hepatic DNL and VLDL output. The observation that hyperinsulinemia-induced DNL in the obese subjects could also match the finding in (Shimomura et al. 2000) when Shimomura et al. reported hyperinsulinemic and insulin resistant animals showed increased lipogenesis. Therefore, insulin resistance may well be the key component in causing elevated DNL-related VLDL overproduction, thus dyslipidemia. As for the molecular mechanism for the upregulation of hepatic DNL by insulin resistance, the defective insulin signaling at the hepatic level and hyperinsulinemia could contribute to over-active lipogenic and glycolytic pathway, such as ACC, FAS, and GK, via increasing expression and activity of SREBP-1c; and this was thoroughly explained in the NAFLD section.

Of course, there are other mechanisms of which insulin resistance induces dyslipidemia. These include reduced hepatic uptake of VLDL, IDL, and LDL; reduced LDL receptors; and increased apoB secretion, thus increased VLDL assembly (Meshkani and Adeli 2009). However, these mechanisms will not be discussed here.

6 Targeting DNL for Management of Metabolic Diseases

Although DNL is important in generating fatty acids for biological needs, the dysregulation of hepatic DNL could induce various metabolic diseases as were discussed above, which are fatty liver disease and CVD. The dysregulation of DNL in these diseases mainly arise from a variety of sources, including hepatic insulin resistance, increased substrate availability such as glucose and fructose, and perhaps induction of hepatic DNL by palmitoleate as well. Interestingly, palmitoleate content in VLDL has just recently been shown to positively correlate with hepatic DNL and hepatic lipid accumulation in clinical patients (Lee et al. 2015). This indicates that palmitoleate has the potential to be utilized as a biomarker in identifying increased hepatic DNL and perhaps predicting fatty liver. This is important because patients with pure fatty liver do not usually demonstrate noticeable symptoms, and that early detection is crucial in reducing progression of the disease and further liver damage.

While it remains to be explored if hepatic DNL could be targeted for the treatment of fatty liver disease and its associated diseases, it is clear that inhibiting hepatic DNL accounts for, at least in part, the beneficial effects of a number of therapeutic agents for treatment of metabolic disease. For example, metformin, the first-class anti-diabetic medicine, has recently been considered as an effective treatment for fatty liver disease. As supportive evidence, metformin treatment caused a decrease in the levels of hepatic steatosis and liver inflammation, as well as an increase in insulin sensitivity in mice fed a high-fat diet (Woo et al. 2014). Notably, metformin was shown to decrease the activation of ACC and suppress the mRNA expression of FAS in the liver. This indicates that inhibition of ACC and FAS contributes to reduction of hepatic steatosis, thus could prevent fat deposition-associated further liver inflammatory damage. Considering that increased hepatic DNL also greatly increases VLDL secretion, leading to increased risk of CVD, further research directions should be emphasized on finding ways to suppress hepatic DNL as a mean of preventative intervention to interfere with the progression of metabolic diseases and associated complications.

References

Abu-Elheiga L, Matzuk MM, Kordari P, Oh W, Shaikenov T, Gu Z, Wakil SJ (2005) Mutant mice lacking acetyl-CoA carboxylase 1 are embryonically lethal. Proc Natl Acad Sci U S A 102:12011–12016

Acheson KJ, Schutz Y, Bessard T, Ravussin E, Jequier E, Flatt JP (1984) Nutritional influences on lipogenesis and thermogenesis after a carbohydrate meal. Am J Physiol 246:E62–E70

Acheson KJ, Schutz Y, Bessard T, Anantharaman K, Flatt JP, Jequier E (1988) Glycogen storage capacity and de novo lipogenesis during massive carbohydrate overfeeding in man. Am J Clin Nutr 48:240–247

Adiels M, Olofsson SO, Taskinen MR, Boren J (2008) Overproduction of very low-density lipoproteins is the hallmark of the dyslipidemia in the metabolic syndrome. Arterioscler Thromb Vasc Biol 28:1225–1236

Ameer F, Scandiuzzi L, Hasnain S, Kalbacher H, Zaidi N (2014) De novo lipogenesis in health and disease. Metabolism 63:895–902

Amemiya-Kudo M, Shimano H, Hasty AH, Yahagi N, Yoshikawa T, Matsuzaka T, Okazaki H, Tamura Y, Iizuka Y, Ohashi K, Osuga J, Harada K, Gotoda T, Sato R, Kimura S, Ishibashi S, Yamada N (2002) Transcriptional activities of nuclear SREBP-1a, -1c, and -2 to different target promoters of lipogenic and cholesterogenic genes. J Lipid Res 43:1220–1235

Baker AL, Rosenberg IH (1987) Hepatic complications of total parenteral nutrition. Am J Med 82:489–497

Barber MC, Price NT, Travers MT (2005) Structure and regulation of acetyl-CoA carboxylase genes of metazoa. Biochim Biophys Acta 1733:1–28

Basciano H, Federico L, Adeli K (2005) Fructose, insulin resistance, and metabolic dyslipidemia. Nutr Metab (Lond) 2:5

Beg ZH, Allmann DW, Gibson DM (1973) Modulation of 3-hydroxy-3-methylglutaryl coenzyme A reductase activity with cAMP and wth protein fractions of rat liver cytosol. Biochem Biophys Res Commun 54:1362–1369

Berglund ED, Lee-Young RS, Lustig DG, Lynes SE, Donahue EP, Camacho RC, Meredith ME, Magnuson MA, Charron MJ, Wasserman DH (2009) Hepatic energy state is regulated by glucagon receptor signaling in mice. J Clin Invest 119:2412–2422

Beynen AC, Vaartjes WJ, Geelen MJ (1979) Opposite effects of insulin and glucagon in acute hormonal control of hepatic lipogenesis. Diabetes 28:828–835

Beynen AC, Vaartjes WJ, Geelen MJ (1980) Acute effects of insulin on fatty acid metabolism in isolated rat hepatocytes. Horm Metab Res 12:425–430

Beynen AC, Haagsman HP, Van Golde LM, Geelen MJ (1981) The effects of insulin and glucagon on the release of triacylglycerols by isolated rat hepatocytes are mere reflections of the hormonal effects on the rate of triacylglycerol synthesis. Biochim Biophys Acta 665:1–7

Biddinger SB, Hernandez-Ono A, Rask-Madsen C, Haas JT, Aleman JO, Suzuki R, Scapa EF, Agarwal C, Carey MC, Stephanopoulos G, Cohen DE, King GL, Ginsberg HN, Kahn CR (2008) Hepatic insulin resistance is sufficient to produce dyslipidemia and susceptibility to atherosclerosis. Cell Metab 7:125–134

Bray GA (2013) Energy and fructose from beverages sweetened with sugar or high-fructose corn syrup pose a health risk for some people. Adv Nutr 4:220–225

Brown MS, Goldstein JL (1997) The SREBP pathway: regulation of cholesterol metabolism by proteolysis of a membrane-bound transcription factor. Cell 89:331–340

Brown MS, Goldstein JL (2008) Selective versus total insulin resistance: a pathogenic paradox. Cell Metab 7:95–96

Browning JD, Horton JD (2004) Molecular mediators of hepatic steatosis and liver injury. J Clin Invest 114:147–152

Brownsey RW, Boone AN, Elliott JE, Kulpa JE, Lee WM (2006) Regulation of acetyl-CoA carboxylase. Biochem Soc Trans 34:223–227

Cao H, Gerhold K, Mayers JR, Wiest MM, Watkins SM, Hotamisligil GS (2008) Identification of a lipokine, a lipid hormone linking adipose tissue to systemic metabolism. Cell 134:933–944

Capeau J (2008) Insulin resistance and steatosis in humans. Diabetes Metab 34:649–657

Carlson CA, Kim KH (1973) Regulation of hepatic acetyl coenzyme A carboxylase by phosphorylation and dephosphorylation. J Biol Chem 248:378–380

Carlson CL, Winder WW (1999) Liver AMP-activated protein kinase and acetyl-CoA carboxylase during and after exercise. J Appl Physiol 86:669–674

Chalasani N, Younossi Z, Lavine JE, Diehl AM, Brunt EM, Cusi K, Charlton M, Sanyal AJ (2012) The diagnosis and management of non-alcoholic fatty liver disease: practice Guideline by the American Association for the Study of Liver Diseases, American College of Gastroenterology, and the American Gastroenterological Association. Hepatology 55:2005–2023

Chen G, Liang G, Ou J, Goldstein JL, Brown MS (2004) Central role for liver X receptor in insulin-mediated activation of Srebp-1c transcription and stimulation of fatty acid synthesis in liver. Proc Natl Acad Sci U S A 101:11245–11250

Chong MF, Fielding BA, Frayn KN (2007) Mechanisms for the acute effect of fructose on postprandial lipemia. Am J Clin Nutr 85:1511–1520

Clarke SD, Jump DB (1996) Polyunsaturated fatty acid regulation of hepatic gene transcription. J Nutr 126:1105S–1109S

Day CP, James OF (1998) Steatohepatitis: a tale of two "hits"? Gastroenterology 114:842–845

Dirlewanger M, Schneiter P, Jequier E, Tappy L (2000) Effects of fructose on hepatic glucose metabolism in humans. Am J Physiol Endocrinol Metab 279:E907–E911

Donnelly KL, Smith CI, Schwarzenberg SJ, Jessurun J, Boldt MD, Parks EJ (2005) Sources of fatty acids stored in liver and secreted via lipoproteins in patients with nonalcoholic fatty liver disease. J Clin Invest 115:1343–1351

Fabbrini E, Mohammed BS, Magkos F, Korenblat KM, Patterson BW, Klein S (2008) Alterations in adipose tissue and hepatic lipid kinetics in obese men and women with nonalcoholic fatty liver disease. Gastroenterology 134:424–431

Farrell GC, Larter CZ (2006) Nonalcoholic fatty liver disease: from steatosis to cirrhosis. Hepatology 43:S99–S112

Feller DD (1954) Metabolism of adipose tissue. I. Incorporation of acetate carbon into lipides by slices of adipose tissue. J Biol Chem 206:171–180

Flick PK, Chen J, Vagelos PR (1977) Effect of dietary linoleate on synthesis and degradation of fatty acid synthetase from rat liver. J Biol Chem 252:4242–4249

Gao B, Bataller R (2011) Alcoholic liver disease: pathogenesis and new therapeutic targets. Gastroenterology 141:1572–1585

Garrido G, Guzman M, Odriozola JM (1996) Effect of different types of high carbohydrate diets on glycogen metabolism in liver and skeletal muscle of endurance-trained rats. Eur J Appl Physiol Occup Physiol 74:91–99

Geelen MJ, Gibson DM (1975) Lipogenesis in maintenance cultures of rat hepatocytes. FEBS Lett 58:334–339

Geelen MJ, Vaartjes WJ (1977) Levels of cyclic 3′-5′-adenosine monophosphate (cAMP) in maintenance cultures of rat hepatocytes in response to insulin and glucagon. Lipids 12:577–580

Geelen MJ, Beynen AC, Christiansen RZ, Lepreau-Jose MJ, Gibson DM (1978) Short-term effects of insulin and glucagon on lipid synthesis in isolated rat hepatocytes. Covariance of acetyl-CoA carboxylase activity and the rate of 3H2O incorporation into fatty acids. FEBS Lett 95:326–330

Gibbons GF (1990) Assembly and secretion of hepatic very-low-density lipoprotein. Biochem J 268:1–13

Gibbons GF, Burnham FJ (1991) Effect of nutritional state on the utilization of fatty acids for hepatitic triacylglycerol synthesis and secretion as very-low-density lipoprotein. Biochem J 275(Pt 1):87–92

Goldstein JL, Brown MS (2009) The LDL receptor. Arterioscler Thromb Vasc Biol 29:431–438

Gorski J, Oscai LB, Palmer WK (1990) Hepatic lipid metabolism in exercise and training. Med Sci Sports Exerc 22:213–221

Grahame Hardie D (2014) AMP-activated protein kinase: a key regulator of energy balance with many roles in human disease. J Intern Med 276:543–559

Grundy SM (1998) Hypertriglyceridemia, atherogenic dyslipidemia, and the metabolic syndrome. Am J Cardiol 81:18B–25B

Guo X, Li H, Xu H, Halim V, Zhang W, Wang H, Ong KT, Woo SL, Walzem RL, Mashek DG, Dong H, Lu F, Wei L, Huo Y, Wu C (2012) Palmitoleate induces hepatic steatosis but suppresses liver inflammatory response in mice. PLoS One 7, e39286

Haeusler RA, Accili D (2008) The double life of Irs. Cell Metab 8:7–9

Harris RA, Mapes JP, Ochs RS, Crabb DW, Stropes L (1979) Hormonal control of hepatic lipogenesis. Adv Exp Med Biol 111:17–42

Hausberger FX, Milstein SW, Rutman RJ (1954) The influence of insulin on glucose utilization in adipose and hepatic tissues in vitro. J Biol Chem 208:431–438

Havel PJ (2005) Dietary fructose: implications for dysregulation of energy homeostasis and lipid/carbohydrate metabolism. Nutr Rev 63:133–157
Hellerstein MK (1999) De novo lipogenesis in humans: metabolic and regulatory aspects. Eur J Clin Nutr 53(Suppl 1):S53–S65
Hellerstein MK, Schwarz JM, Neese RA (1996) Regulation of hepatic de novo lipogenesis in humans. Annu Rev Nutr 16:523–557
Herzberg GR (1983) The influence of dietary fatty acid composition on lipogenesis. Adv Nutr Res 5:221–253
Holland R, Witters LA, Hardie DG (1984) Glucagon inhibits fatty acid synthesis in isolated hepatocytes via phosphorylation of acetyl-CoA carboxylase by cyclic-AMP-dependent protein kinase. Eur J Biochem 140:325–333
Horton JD, Goldstein JL, Brown MS (2002) SREBPs: activators of the complete program of cholesterol and fatty acid synthesis in the liver. J Clin Invest 109:1125–1131
Huo Y, Guo X, Li H, Xu H, Halim V, Zhang W, Wang H, Fan YY, Ong KT, Woo SL, Chapkin RS, Mashek DG, Chen Y, Dong H, Lu F, Wei L, Wu C (2012) Targeted overexpression of inducible 6-phosphofructo-2-kinase in adipose tissue increases fat deposition but protects against diet-induced insulin resistance and inflammatory responses. J Biol Chem 287:21492–21500
Ingle DL, Bauman DE, Garrigus US (1972) Lipogenesis in the ruminant: in vivo site of fatty acid synthesis in sheep. J Nutr 102:617–623
Institute of Medicine (US) Committee on Preventing the Global Epidemic of Cardiovascular Disease: Meeting the Challenges in Developing Countries, Fuster V, Kelly BB (eds) (2010) Promoting cardiovascular health in the developing world: a critical challenge to achieve global health. National Academies Press, Washington DC
Ishii S, IIzuka K, Miller BC, Uyeda K (2004) Carbohydrate response element binding protein directly promotes lipogenic enzyme gene transcription. Proc Natl Acad Sci U S A 101:15597–15602
Isomaa B, Almgren P, Tuomi T, Forsen B, Lahti K, Nissen M, Taskinen MR, Groop L (2001) Cardiovascular morbidity and mortality associated with the metabolic syndrome. Diabetes Care 24:683–689
Johnson NA, Sachinwalla T, Walton DW, Smith K, Armstrong A, Thompson MW, George J (2009) Aerobic exercise training reduces hepatic and visceral lipids in obese individuals without weight loss. Hepatology 50:1105–1112
Kahn, R., Buse, J., Ferrannini, E., Stern, M., American Diabetes, A., and European Association for the Study of, D (2005) The metabolic syndrome: time for a critical appraisal: joint statement from the American Diabetes Association and the European Association for the Study of Diabetes. Diabetes Care 28:2289–2304
Kaplowitz N, Ji C (2006) Unfolding new mechanisms of alcoholic liver disease in the endoplasmic reticulum. J Gastroenterol Hepatol 21(Suppl 3):S7–S9
Kimball SR, Siegfried BA, Jefferson LS (2004) Glucagon represses signaling through the mammalian target of rapamycin in rat liver by activating AMP-activated protein kinase. J Biol Chem 279:54103–54109
Kornacker MS, Lowenstein JM (1965) Citrate and the conversion of carbohydrate into fat. The activities of citrate-cleavage enzyme and acetate thiokinase in livers of starved and re-fed rats. Biochem J 94:209–215
Kubota N, Kubota T, Itoh S, Kumagai H, Kozono H, Takamoto I, Mineyama T, Ogata H, Tokuyama K, Ohsugi M, Sasako T, Moroi M, Sugi K, Kakuta S, Iwakura Y, Noda T, Ohnishi S, Nagai R, Tobe K, Terauchi Y, Ueki K, Kadowaki T (2008) Dynamic functional relay between insulin receptor substrate 1 and 2 in hepatic insulin signaling during fasting and feeding. Cell Metab 8:49–64
Lakshmanan MR, Nepokroeff CM, Porter JW (1972) Control of the synthesis of fatty-acid synthetase in rat liver by insulin, glucagon, and adenosine 3′:5′cyclic monophosphate. Proc Natl Acad Sci U S A 69:3516–3519
Lammert O, Grunnet N, Faber P, Bjornsbo KS, Dich J, Larsen LO, Neese RA, Hellerstein MK, Quistorff B (2000) Effects of isoenergetic overfeeding of either carbohydrate or fat in young men. Br J Nutr 84:233–245

Lancaster GI, Febbraio MA (2014) The immunomodulating role of exercise in metabolic disease. Trends Immunol 35:262–269
Lee IM, Sesso HD, Oguma Y, Paffenbarger RS Jr (2004) The "weekend warrior" and risk of mortality. Am J Epidemiol 160:636–641
Lee JJ, Lambert JE, Hovhannisyan Y, Ramos-Roman MA, Trombold JR, Wagner DA, Parks EJ (2015) Palmitoleic acid is elevated in fatty liver disease and reflects hepatic lipogenesis. Am J Clin Nutr 101:34–43
Lefkowitch JH (2005) Morphology of alcoholic liver disease. Clin Liver Dis 9:37–53
Lewis GF, Carpentier A, Adeli K, Giacca A (2002) Disordered fat storage and mobilization in the pathogenesis of insulin resistance and type 2 diabetes. Endocr Rev 23:201–229
Li S, Brown MS, Goldstein JL (2010) Bifurcation of insulin signaling pathway in rat liver: mTORC1 required for stimulation of lipogenesis, but not inhibition of gluconeogenesis. Proc Natl Acad Sci U S A 107:3441–3446
Li Y, Xu S, Mihaylova MM, Zheng B, Hou X, Jiang B, Park O, Luo Z, Lefai E, Shyy JY, Gao B, Wierzbicki M, Verbeuren TJ, Shaw RJ, Cohen RA, Zang M (2011) AMPK phosphorylates and inhibits SREBP activity to attenuate hepatic steatosis and atherosclerosis in diet-induced insulin-resistant mice. Cell Metab 13:376–388
Liangpunsakul S, Ross RA, Crabb DW (2013) Activation of carbohydrate response element-binding protein by ethanol. J Investig Med 61:270–277
Liepa GU, Beitz DC, Linder JR (1978) Fatty acid synthesis in ruminating and nonruminating goats. J Nutr 108:1733–1739
Ljungstrom O, Ekman P (1977) Glucagon-induced phosphorylation of pyruvate kinase (type L) in rat liver slices. Biochem Biophys Res Commun 78:1147–1155
Longuet C, Sinclair EM, Maida A, Baggio LL, Maziarz M, Charron MJ, Drucker DJ (2008) The glucagon receptor is required for the adaptive metabolic response to fasting. Cell Metab 8:359–371
Lynen F (1953) Functional group of coenzyme A and its metabolic relations, especially in the fatty acid cycle. Fed Proc 12:683–691
Lynen F, Knappe J, Lorch E, Juetting G, Ringelmann E, Lachance JP (1961) On the biochemical function of biotin. II. Purification and mode of action of beta-methyl-crotonyl-carboxylase. Biochem Z 335:123–167
Ma XM, Blenis J (2009) Molecular mechanisms of mTOR-mediated translational control. Nat Rev Mol Cell Biol 10:307–318
Ma L, Tsatsos NG, Towle HC (2005) Direct role of ChREBP.Mlx in regulating hepatic glucose-responsive genes. J Biol Chem 280:12019–12027
Marriott BP, Olsho L, Hadden L, Connor P (2010) Intake of added sugars and selected nutrients in the United States, National Health and Nutrition Examination Survey (NHANES) 2003-2006. Crit Rev Food Sci Nutr 50:228–258
Masoro EJ, Chaikoff IL, Dauben WG (1949) Lipogenesis from glucose in the normal and liverless animal as studied with C14-labeled glucose. J Biol Chem 179:1117–1125
Matsumoto M, Han S, Kitamura T, Accili D (2006) Dual role of transcription factor FoxO1 in controlling hepatic insulin sensitivity and lipid metabolism. J Clin Invest 116:2464–2472
Meshkani R, Adeli K (2009) Hepatic insulin resistance, metabolic syndrome and cardiovascular disease. Clin Biochem 42:1331–1346
Minokoshi Y, Kim YB, Peroni OD, Fryer LG, Muller C, Carling D, Kahn BB (2002) Leptin stimulates fatty-acid oxidation by activating AMP-activated protein kinase. Nature 415:339–343
Miyazaki M, Dobrzyn A, Man WC, Chu K, Sampath H, Kim HJ, Ntambi JM (2004) Stearoyl-CoA desaturase 1 gene expression is necessary for fructose-mediated induction of lipogenic gene expression by sterol regulatory element-binding protein-1c-dependent and -independent mechanisms. J Biol Chem 279:25164–25171
Monetti M, Levin MC, Watt MJ, Sajan MP, Marmor S, Hubbard BK, Stevens RD, Bain JR, Newgard CB, Farese RV Sr, Hevener AL, Farese RV Jr (2007) Dissociation of hepatic steatosis and insulin resistance in mice overexpressing DGAT in the liver. Cell Metab 6:69–78

Muller P, Singh A, Orci L, Jeanrenaud B (1976) Secretory processes, carbohydrate and lipid metabolism in isolated mouse hepatocytes. Aspects of regulation by glucagon and insulin. Biochim Biophys Acta 428:480–494

Muller A, Unthan-Fechner K, Probst I (1988) Activation of phosphofructokinase 2 by insulin in cultured hepatocytes without accompanying changes of effector levels or cAMP-stimulated protein kinase activity ratios. Eur J Biochem 176:415–420

Munday MR (2002) Regulation of mammalian acetyl-CoA carboxylase. Biochem Soc Trans 30:1059–1064

Musch K, Ojakian MA, Williams MA (1974) Comparison of alpha-linolenate and oleate in lowering activity of lipogenic enzymes in rat liver: evidence for a greater effect of dietary linoleinate independent of food and carbohydrate intake. Biochim Biophys Acta 337:343–348

O'Doherty RM, Lehman DL, Telemaque-Potts S, Newgard CB (1999) Metabolic impact of glucokinase overexpression in liver: lowering of blood glucose in fed rats is accompanied by hyperlipidemia. Diabetes 48:2022–2027

O'Shea RS, Dasarathy S, McCullough AJ, Practice Guideline Committee of the American Association for the Study of Liver Diseases, Practice Parameters Committee of the American College of Gastroenterology. (2010) Alcoholic liver disease. Hepatology 51:307–328

Oh W, Abu-Elheiga L, Kordari P, Gu Z, Shaikenov T, Chirala SS, Wakil SJ (2005) Glucose and fat metabolism in adipose tissue of acetyl-CoA carboxylase 2 knockout mice. Proc Natl Acad Sci U S A 102:1384–1389

Olofsson SO, Boren J (2005) Apolipoprotein B: a clinically important apolipoprotein which assembles atherogenic lipoproteins and promotes the development of atherosclerosis. J Intern Med 258:395–410

Park H, Kaushik VK, Constant S, Prentki M, Przybytkowski E, Ruderman NB, Saha AK (2002) Coordinate regulation of malonyl-CoA decarboxylase, sn-glycerol-3-phosphate acyltransferase, and acetyl-CoA carboxylase by AMP-activated protein kinase in rat tissues in response to exercise. J Biol Chem 277:32571–32577

Pasquet P, Brigant L, Froment A, Koppert GA, Bard D, de Garine I, Apfelbaum M (1992) Massive overfeeding and energy balance in men: the Guru Walla model. Am J Clin Nutr 56:483–490

Pearce J (1983) Fatty acid synthesis in liver and adipose tissue. Proc Nutr Soc 42:263–271

Petit JM, Guiu B, Duvillard L, Jooste V, Brindisi MC, Athias A, Bouillet B, Habchi M, Cottet V, Gambert P, Hillon P, Cercueil JP, Verges B (2012) Increased erythrocytes n-3 and n-6 polyunsaturated fatty acids is significantly associated with a lower prevalence of steatosis in patients with type 2 diabetes. Clin Nutr 31(4):520–525

Pollock NK, Bundy V, Kanto W, Davis CL, Bernard PJ, Zhu H, Gutin B, Dong Y (2012) Greater fructose consumption is associated with cardiometabolic risk markers and visceral adiposity in adolescents. J Nutr 142:251–257

Rabol R, Petersen KF, Dufour S, Flannery C, Shulman GI (2011) Reversal of muscle insulin resistance with exercise reduces postprandial hepatic de novo lipogenesis in insulin resistant individuals. Proc Natl Acad Sci U S A 108:13705–13709

Rudack D, Chisholm EM, Holten D (1971a) Rat liver glucose 6-phosphate dehydrogenase. Regulation by carbohydrate diet and insulin. J Biol Chem 246:1249–1254

Rudack D, Gozukara EM, Chisholm EM, Holten D (1971b) The effect of dietary carbohydrate and fat on the synthesis of rat liver 6-phosphogluconate dehydrogenase. Biochim Biophys Acta 252:305–313

Ruderman NB, Park H, Kaushik VK, Dean D, Constant S, Prentki M, Saha AK (2003) AMPK as a metabolic switch in rat muscle, liver and adipose tissue after exercise. Acta Physiol Scand 178:435–442

Sajan MP, Ivey RA, Lee MC, Farese RV (2015) Hepatic insulin resistance in ob/ob mice involves increases in ceramide, aPKC activity, and selective impairment of Akt-dependent FoxO1 phosphorylation. J Lipid Res 56:70–80

Salam WH, Wilcox HG, Heimberg M (1988) Effects of oleic acid on the biosynthesis of lipoprotein apoproteins and distribution into the very-low-density lipoprotein by the isolated perfused rat liver. Biochem J 251:809–816

Sampath H, Miyazaki M, Dobrzyn A, Ntambi JM (2007) Stearoyl-CoA desaturase-1 mediates the pro-lipogenic effects of dietary saturated fat. J Biol Chem 282:2483–2493
Samuel VT, Liu ZX, Wang A, Beddow SA, Geisler JG, Kahn M, Zhang XM, Monia BP, Bhanot S, Shulman GI (2007) Inhibition of protein kinase Cepsilon prevents hepatic insulin resistance in nonalcoholic fatty liver disease. J Clin Invest 117:739–745
Schonfeld G, Pfleger B (1971) Utilization of exogenous free fatty acids for the production of very low density lipoprotein triglyceride by livers of carbohydrate-fed rats. J Lipid Res 12:614–621
Schwarz JM, Linfoot P, Dare D, Aghajanian K (2003) Hepatic de novo lipogenesis in normoinsulinemic and hyperinsulinemic subjects consuming high-fat, low-carbohydrate and low-fat, high-carbohydrate isoenergetic diets. Am J Clin Nutr 77:43–50
Shimomura I, Shimano H, Horton JD, Goldstein JL, Brown MS (1997) Differential expression of exons 1a and 1c in mRNAs for sterol regulatory element binding protein-1 in human and mouse organs and cultured cells. J Clin Invest 99:838–845
Shimomura I, Matsuda M, Hammer RE, Bashmakov Y, Brown MS, Goldstein JL (2000) Decreased IRS-2 and increased SREBP-1c lead to mixed insulin resistance and sensitivity in livers of lipodystrophic and ob/ob mice. Mol Cell 6:77–86
Shojaee-Moradie F, Baynes KC, Pentecost C, Bell JD, Thomas EL, Jackson NC, Stolinski M, Whyte M, Lovell D, Bowes SB, Gibney J, Jones RH, Umpleby AM (2007) Exercise training reduces fatty acid availability and improves the insulin sensitivity of glucose metabolism. Diabetologia 50:404–413
Shrago E, Glennon JA, Gordon ES (1971) Comparative aspects of lipogenesis in mammalian tissues. Metabolism 20:54–62
Sim AT, Hardie DG (1988) The low activity of acetyl-CoA carboxylase in basal and glucagon-stimulated hepatocytes is due to phosphorylation by the AMP-activated protein kinase and not cyclic AMP-dependent protein kinase. FEBS Lett 233:294–298
Sozio M, Crabb DW (2008) Alcohol and lipid metabolism. Am J Physiol Endocrinol Metab 295:E10–E16
Sozio MS, Liangpunsakul S, Crabb D (2010) The role of lipid metabolism in the pathogenesis of alcoholic and nonalcoholic hepatic steatosis. Semin Liver Dis 30:378–390
Stanford KI, Goodyear LJ (2014) Exercise and type 2 diabetes: molecular mechanisms regulating glucose uptake in skeletal muscle. Adv Physiol Educ 38:308–314
Stanhope KL, Schwarz JM, Keim NL, Griffen SC, Bremer AA, Graham JL, Hatcher B, Cox CL, Dyachenko A, Zhang W, McGahan JP, Seibert A, Krauss RM, Chiu S, Schaefer EJ, Ai M, Otokozawa S, Nakajima K, Nakano T, Beysen C, Hellerstein MK, Berglund L, Havel PJ (2009) Consuming fructose-sweetened, not glucose-sweetened, beverages increases visceral adiposity and lipids and decreases insulin sensitivity in overweight/obese humans. J Clin Invest 119:1322–1334
Stillemark-Billton P, Beck C, Boren J, Olofsson SO (2005) Relation of the size and intracellular sorting of apoB to the formation of VLDL 1 and VLDL 2. J Lipid Res 46:104–114
Tilg H, Moschen AR (2010) Evolution of inflammation in nonalcoholic fatty liver disease: the multiple parallel hits hypothesis. Hepatology 52:1836–1846
Tipper JP, Witters LA (1982) In vitro phosphorylation and inactivation of rat liver acetyl-CoA carboxylase purified by avidin affinity chromatography. Biochim Biophys Acta 715:162–169
Toussant MJ, Wilson MD, Clarke SD (1981) Coordinate suppression of liver acetyl-CoA carboxylase and fatty acid synthetase by polyunsaturated fat. J Nutr 111:146–153
Towle HC (2005) Glucose as a regulator of eukaryotic gene transcription. Trends Endocrinol Metab 16:489–494
van der Heijden GJ, Wang ZJ, Chu ZD, Sauer PJ, Haymond MW, Rodriguez LM, Sunehag AL (2010) A 12-week aerobic exercise program reduces hepatic fat accumulation and insulin resistance in obese, Hispanic adolescents. Obesity (Silver Spring) 18:384–390
Vernon G, Baranova A, Younossi ZM (2011) Systematic review: the epidemiology and natural history of non-alcoholic fatty liver disease and non-alcoholic steatohepatitis in adults. Aliment Pharmacol Ther 34:274–285

Vos MB, Lavine JE (2013) Dietary fructose in nonalcoholic fatty liver disease. Hepatology 57:2525–2531

Wakil SJ (1989) Fatty acid synthase, a proficient multifunctional enzyme. Biochemistry 28:4523–4530

Wakil SJ, Porter JW, Gibson DM (1957) Studies on the mechanism of fatty acid synthesis. I. Preparation and purification of an enzymes system for reconstruction of fatty acid synthesis. Biochim Biophys Acta 24:453–461

Walker PR, Bailey E (1969) Role of "malic enzyme" in lipogenesis. Biochim Biophys Acta 187:591–593

Wattacheril J, Chalasani N (2012) Nonalcoholic fatty liver disease (NAFLD): is it really a serious condition? Hepatology 56:1580–1584

Woo SL, Xu H, Li H, Zhao Y, Hu X, Zhao J, Guo X, Guo T, Botchlett R, Qi T, Pei Y, Zheng J, Xu Y, An X, Chen L, Chen L, Li Q, Xiao X, Huo Y, Wu C (2014) Metformin ameliorates hepatic steatosis and inflammation without altering adipose phenotype in diet-induced obesity. PLoS One 9, e91111

Wu C, Kang JE, Peng L, Li H, Khan SA, Hillard CJ, Okar DA, Lange AJ (2005) Enhancing hepatic glycolysis reduces obesity: differential effects on lipogenesis depend on site of glycolytic modulation. Cell Metab 2:131–140

Yang Z-H, Miyahara H, Hatanaka A (2011) Chronic administration of palmitoleic acid reduces insulin resistance and hepatic lipid accumulation in KK-Ay Mice with genetic type 2 diabetes. Lipids Health Dis 10:120

Yi X, Cao S, Chang B, Zhao D, Gao H, Wan Y, Shi J, Wei W, Guan Y (2013) Effects of acute exercise and chronic exercise on the liver leptin-AMPK-ACC signaling pathway in rats with type 2 diabetes. J Diabetes Res 2013:946432

Yin HQ, Kim M, Kim JH, Kong G, Kang KS, Kim HL, Yoon BI, Lee MO, Lee BH (2007) Differential gene expression and lipid metabolism in fatty liver induced by acute ethanol treatment in mice. Toxicol Appl Pharmacol 223:225–233

You M, Fischer M, Deeg MA, Crabb DW (2002) Ethanol induces fatty acid synthesis pathways by activation of sterol regulatory element-binding protein (SREBP). J Biol Chem 277:29342–29347

You M, Matsumoto M, Pacold CM, Cho WK, Crabb DW (2004) The role of AMP-activated protein kinase in the action of ethanol in the liver. Gastroenterology 127:1798–1808

Zakim D (1973) Influence of fructose of hepatic synthesis of lipids. Prog Biochem Pharmacol 8:161–188

Zimmet P, Magliano D, Matsuzawa Y, Alberti G, Shaw J (2005) The metabolic syndrome: a global public health problem and a new definition. J Atheroscler Thromb 12:295–300

Chapter 11
Metabolic Rhythm of Hepatic Lipogenesis: Regulation and Roles in Metabolism

David Jacobi, Hyunjii J. Cho, Ryan K. Alexander, and Chih-Hao Lee

Abstract Hepatic lipid metabolism is modulated by multiple pathways, including hormones (e.g., insulin and glucagon), energy/nutrient-sensing signaling, and circadian rhythm. The latter constitutes a pre-programmed transcriptional mechanism in anticipation of upcoming feeding/fasting metabolic cycles. Although the central clock is controlled by light, the peripheral clock, such as that in the liver, is very sensitive to the nutrient status. As such, studies in mice and humans have demonstrated that disrupted circadian rhythm is linked to metabolic diseases. This chapter will describe roles of the molecular clock and downstream nuclear receptors in the control of liver lipid metabolism. Potential mechanisms through which hepatic lipogenesis may affect peripheral metabolic homeostasis via lipid metabolites will also be discussed.

Keywords Circadian rhythm • Lipid metabolism • Inter-organ communication • Hepatic de novo lipogenesis • Lipid mediator

1 Introduction

The liver maintains blood glucose and lipids at physiological levels. As such, it is central to the metabolic flexibility of the body to utilize appropriate energy substrates at a given time of the day. The ability to switch between glucose and fatty acid utilization during physiological feeding/fasting is achieved by several hormonal signaling pathways, notably insulin and glucagon. De novo lipogenesis, occurring primarily in the liver, plays a key role in maintaining metabolic flexibility, as it converts unused glucose at the fed state to fatty acids for storage in adipose tissue that are released at the fasted state as a main energy source. Not surprisingly, dysregulation in liver lipogenic pathway often leads to hepatic and systemic

D. Jacobi • H.J. Cho • R.K. Alexander • C.-H. Lee (✉)
Division of Biological Sciences, Department of Genetics and Complex Diseases, Harvard T.H. Chan School of Public Health, 665 Huntington Ave, Bldg1, Rm 409, Boston, MA 02115, USA
e-mail: clee@hsph.harvard.edu

J.M. Ntambi (ed.), *Hepatic De Novo Lipogenesis and Regulation of Metabolism*, DOI 10.1007/978-3-319-25065-6_11 metabolic dysfunction, as demonstrated in mouse genetic models. In this chapter, we will discuss regulatory mechanisms governing de novo lipogenesis in the liver with an emphasis on transcriptional regulation. The first layer of control (in addition to insulin) is mediated by pre-programmed, rhythmic mRNA expression of lipogenic genes regulated by the circadian clock transcription factors. A second layer is mediated by several nuclear hormone receptors that are downstream of the molecular clock and activated by dietary lipids, thereby integrating nutrient status with energy metabolism. We will also summarize metabolic phenotypes in mice with loss- or gain-of-function studies of lipogenic genes. These mouse genetic models raise the possibility of liver-derived lipid metabolites that serve as mediators for crosstalk between the liver and other peripheral tissues in a coordinated effort to maintain metabolic homeostasis. Two examples of such lipid metabolites will be discussed.

2 Metabolic Rhythm of Hepatic Lipogenesis

2.1 Physiology of Feeding/Fasting Responses

As the first organ flooded with dietary nutrients carried by the portal vein, the liver is key in controlling energy substrate homeostasis. A central mechanism in the process is hepatic de novo lipogenesis, whereby energy from dietary sources can be efficiently repackaged. Following a meal, elevation in glycemia stimulates insulin secretion by pancreatic beta cells. Energy homeostasis is maintained in part by insulin's direct action on the liver, including inhibition of hepatic gluconeogenesis and glycogenolysis and stimulation of glucose storage in the form of glycogen. In addition, insulin induces de novo lipogenesis, allowing for the transfer of the energy contained in excess dietary carbohydrates toward exportable and energy dense lipids.

In short, the canonical signaling pathway of insulin commences with its binding to the insulin receptor (IR), a cell surface receptor tyrosine kinase. Subsequent autophosphorylation and activation of IR follows and initiates a cascade of phosphorylation events through the insulin receptor substrate (IRS) family of proteins 1–6, which then act through the phosphatidylinositol-3-kinase (PI3K)/Protein kinase B (PKB or AKT) signaling pathway to regulate metabolism (Boucher et al. 2014). The action of insulin triggers many metabolic pathways downstream of AKT. For instance, AKT inhibits Tuberous Sclerosis Complex (TSC) 1/2, which leads to activation of Sterol regulatory element-binding transcription factor (SREBP) 1 and upregulation of lipogenic gene expression through the mammalian target of rapamycin complex 1 (mTORC1) (Yecies et al. 2011).

Storage of glucose as glycogen is limited and the liver converts excess glucose to fatty acids through de novo lipogenesis. In the liver, glucose exists as glucose-6-phosphate that is processed to pyruvate through glycolysis. Pyruvate decarboxylation generates acetyl coenzyme A (acetyl-CoA), a substrate for acetyl-CoA carboxylase α and β (ACACA/B or Acc1/2) to synthesize malonyl-CoA, which enters the committed step for lipogenesis by fatty acid synthase (FASN). Additionally, malonyl-CoA accumulation inhibits hepatic carnitine palmitoyltransferase I (CPT1), a rate-limiting enzyme regulating fatty acids transport to the mitochondria for β oxidation. In addition, dietary fatty acids taken-up by the liver are re-esterified to form triglycerides in the liver and exported as very low-density lipoproteins (VLDLs) for peripheral use and storage. At the postprandial state, insulin levels drop, allowing the release of fatty acids from adipose tissue to be used as a primary energy substrate. The action of glucagon also increases hepatic gluconeogenesis to maintain blood glucose concentrations.

In addition to hormonal and substrate-driven metabolic regulation, transcriptomic studies have illustrated the presence of circadian control in metabolism. In the liver, ~15 % of genes transcripts oscillate rhythmically (Panda et al. 2002; Vollmers et al. 2009), many of which play a role in glucose and lipid metabolism. This anticipatory metabolic pre-programming is thought to promote metabolic efficiency and is coordinately controlled by several "molecular clocks". This will be discussed in detail below.

2.2 *Circadian Regulation of Liver Metabolism*

The observation of endogenous biological rhythms dates back to the 1700s, when French scientist Jean-Jacques d'Ortous De Mairan discovered that the leaf movements of a plant retained a periodicity of 24-h in complete darkness. This suggested that cycling biological phenomena could be not only a simple response to external stimuli, but also a consequence of autonomous internal clocks (Somers 1999). Much later in the 1950s, studies in Drosophila by Pittendrigh and in humans by Aschoff, suggested the existence of innate circadian rhythms (Aschoff 1960; Pittendrigh 1954). Throughout the years, the reliance of systemic metabolic homeostasis on the precision and coordination of the molecular processes within tissues and cells has become increasingly evident.

The master regulator of the mammalian circadian clock is expressed within the pacemaker neurons of the supra chiasmic nucleus (SCN) in the hypothalamus (Inouye and Kawamura 1979; Stephan and Zucker 1972). The rhythm is entrained by daylight through neuronal connection to photoreceptors in the retina. This establishes a close functional coupling of the light and dark cycle. The SCN in turn, installs a "standard time" for a number of basic physiological functions such as alertness, blood pressure, or body temperature. Peripheral organs, although synchronized under the influence of the brain central clock, possess their own circadian machinery to control organ-specific cellular processes. With regard to the liver, the brain clock directly influences feeding and fasting, which then entrains the hepatic clock to control lipid and glucose metabolism (Yamazaki et al. 2000; Yoo et al. 2004).

2.3 The Molecular Clock

At the core of the circadian clock machinery is a well-synchronized transcription-translation feedback loop generated by transcription factors aryl hydrocarbon receptor nuclear translocator-like (ARNTL or BMAL1) and circadian locomotor output cycles kaput (CLOCK), together with co-regulators Period (PER1, PER2, and PER3) and Cryptochrome (CRY1 and CRY2) (Asher and Schibler 2011; Bass and Takahashi 2010; Zhang and Kay 2010). The BMAL:CLOCK heterodimer binds to the E-box response element in the promoter of its target genes, which include the clock repressors, *PER* and *CRY*, and genes involved in the regulation glucose and lipid metabolism (Fig. 11.1). At the peak of BMAL:CLOCK activity, PER and CRY begin to accumulate, forming an inhibitory complex that binds and suppresses BMAL:CLOCK activity. The cycle is then completed by releasing the inhibition of BMAL:CLOCK with phosphorylation-dependent proteolytic degradation of PER through casein kinases (CK1ε and CK1δ) and CRY through AMP-activated protein kinase (AMPK) (Gallego and Virshup 2007).

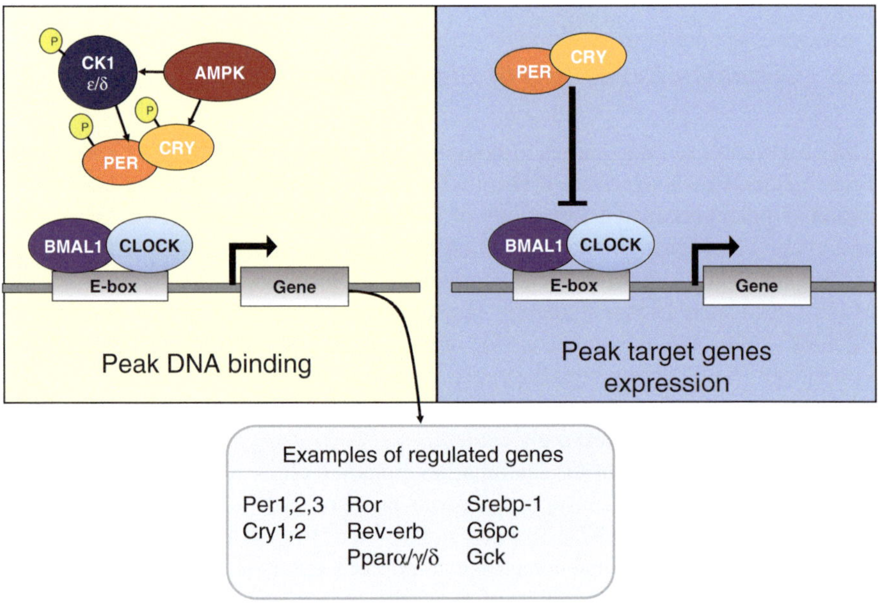

Fig. 11.1 The transcriptional network of the molecular clock and its downstream effectors. The circadian clock is a coordinated transcription-translation feedback loop generated by transcriptional activators, BMAL1 and CLOCK, and repressors PER and CRY. BMAL1 and CLOCK heterodimerize and bind to the E-box response element in the promoter of their target genes, including PER and CRY. As PER and CRY accumulate, they form a complex that inhibits the BMAL1:CLOCK dimer. The cycle renews with the proteolytic degradation of PER and CRY through phosphorylation by CK1ε/δ and AMPK. In mice, the peak DNA binding of BMAL1:CLOCK is in the light cycle, while the peak expression of target genes is in the dark cycle

As mentioned previously, studies have demonstrated the existence of cell/tissue specific autonomous clock machinery. Interestingly, the liver, even after 16 days of a shifted light/dark cycle, does not completely adapt to the change (Yamazaki et al. 2000). Correspondingly, using a PER2:Luciferase protein as a reporter of circadian activity, it was revealed that peripheral tissues in isolation were able to self-sustain circadian oscillations for more than 20 days (Yoo et al. 2004). Furthermore, lesions in the SCN did not abolish peripheral circadian oscillations, but instead resulted in phase desynchrony among tissues (Yoo et al. 2004). This suggests that the brain clock is necessary for phase control of clocks in peripheral organs but not required for tissue-specific circadian oscillations. Instead, there appears to be tissue-specific cues that are more effective in synchronizing peripheral clocks. In the liver, the inversion of the feeding cycle rapidly alters the hepatic expression of metabolic genes, suggesting that the liver metabolic program is more rapidly entrained to rhythmic feeding and fasting, even in the presence of a functioning central pacemaker (Damiola et al. 2000; Stokkan et al. 2001), providing a link between the circadian clock machinery and metabolic regulation.

2.4 Clock Regulators and Metabolic Homeostasis

The important role of the circadian machinery in metabolism became evident through genetic deletion or mutations of individual core clock components. Specifically, mice with loss-of-function of *Bmal1* or *Clock* developed metabolic disorders. *Clock* knockout mice are hyperphagic and obese and display a range of abnormal metabolic parameters such as hyperlipidemia, β cell dysfunction, hepatic steatosis, and hyperglycemia (Pan et al. 2010; Rudic et al. 2004; Turek et al. 2005). Additional studies have indicated a potential role of *Clock* in lipid uptake, absorption, synthesis, and degradation. Microsomal triglyceride transfer protein (MTP) is an essential component to the hepatic secretion of apolipoprotein B-containing lipoproteins, such as chylomicrons and VLDLs. Clock regulates *Mtp* and plasma triglycerides circadian changes via Shp, which binds to Hnf4α/Lrh-1 at the *Mtp* promoter. *Clock* loss of function and *Shp* knockdown mice lose diurnal expression of *Mtp* and have high plasma triglycerides at all times (Pan et al. 2010). *Clock* knockout mice also display altered rhythmic expression of enzymes involved in lipid biosynthesis and fatty acid degradation, including *Acsl4* and *Fabp1* (Kudo et al. 2007).

Similarly, *Bmal1* knockout mice are arrhythmic in complete darkness, develop hyperlipidemia and hepatic steatosis, and are unable to maintain oscillations in plasma glucose and triglycerides (Rudic et al. 2004; Shimba et al. 2011). Bmal1 regulates de novo lipogenesis in the liver via insulin-mTORC1-Akt signaling (Zhang et al. 2014), and Bmal1 overexpression in mouse liver significantly elevated the mRNA level of *Srebp-1c* and of lipogenic enzymes such as *Scd1*, *Fasn*, and *Acaca* (Zhang et al. 2014). Furthermore, *Bmal1* KO primary hepatocytes displayed impaired expression of lipogenic enzymes, *Acaca*, *Fasn*, *Scd1*, glycerol-3-phosphate acyltransferase (*Gpat*), and lipogenic transcription factors, *Chrebp*, *Srebp1c*, and *Pgc1β* (Zhang et al. 2014), which highlights the critical function of Bmal1 in hepatic lipogenesis. In addition, abolishment of the negative regulators, *Per* and *Cry*, has also resulted in altered lipid metabolism (Grimaldi et al. 2010; Rudic et al. 2004; Turek et al. 2005) with reduced plasma triglyceride level in mice lacking *Per1*/*Per2* (Grimaldi et al. 2010). These studies collectively have suggested a critical role of the clock machinery as a novel regulator that couples circadian rhythms to metabolism.

2.5 Circadian Regulation of Liver Metabolism via Nuclear Receptors

While the molecular clock directly regulates gene expression, it also "outsources" metabolic regulation to several downstream transcriptional regulators. Notable among these are nuclear receptors, which are transcription factors that contain both DNA and ligand binding domains (Chawla et al. 2001). They can be activated by a range of metabolic ligands including steroid hormones, fatty acids, oxysterols, bile acids, and heme. Several studies have suggested that tissue-specific expression and rhythmicity of nuclear receptors link peripheral circadian clocks to tissue-specific metabolic outputs. Pertinently, 20 of the 41 nuclear receptors expressed in the mouse liver display a rhythmic expression pattern. Indeed several of these receptors have been implicated in hepatic lipid metabolism (Yang et al. 2006). Moreover, endogenous ligands for some of these nuclear receptorsare known to oscillate in a circadian fashion, thus adding another layer of integration between the molecular clock and nuclear receptors in regulating metabolic homeostasis.

RORs and REV-ERBs. In addition to the core molecular clock described earlier, the nuclear receptors RAR-related orphan receptors (RORs) and REV-ERBs act as auxiliary clock components that directly regulate clock gene expression. ROR and REV-ERB are clock output proteins that bind ROR response elements (ROREs) in target enhancer regions and act as constitutive transcriptional activators and repressors, respectively (Dumas et al. 1994; Forman et al. 1994; Retnakaran et al. 1994). Pertinently, Rors and Rev-erbs directly regulate *Bmal1* gene expression, thus acting as accessory clock components that facilitate rhythmic *Bmal1* expression. The expression and rhythmicity of *Ror* (α, β, and γ) and *Rev-erb* (α and β) isoforms is tissue-specific. Specifically, *Rorγ* and *Rev-erbα/β* display rhythmic expression in the liver (Yang et al. 2006). In circadian terminology, a standard of time is defined based on the period of a cue given by the environment (called a zeitgeber). Under standard light-dark cycles, the time of lights on defines zeitgeber time zero (ZT0) and the time of lights off defines zeitgeber time twelve (ZT12). For nocturnal animals, ZT0 and ZT12 represent the beginning of a physiological fasting and feeding phases, respectively. In the mouse liver, mRNA levels of *Rev-erbs* α and β peak during the light/fasting cycle at ZT4 and ZT8, respectively, while *Rorγ* expression peaks during the dark cycle at ZT16. Furthermore, crosstalk exists between Rev-Erb

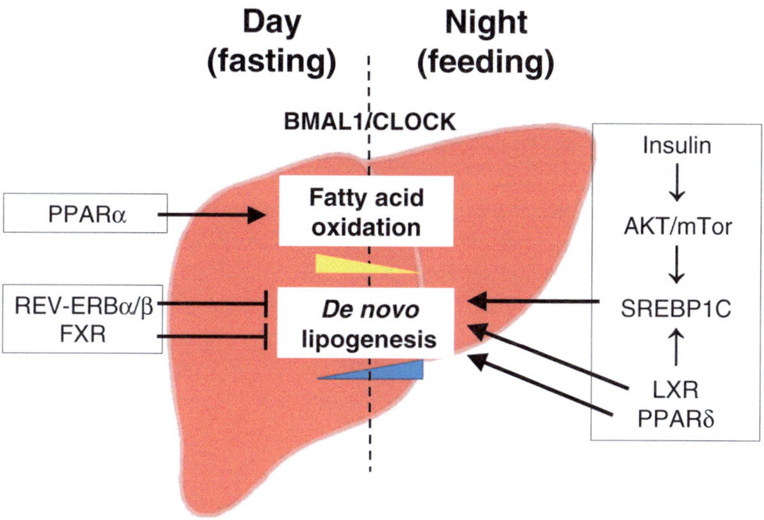

Fig. 11.2 Circadian partitioning of hepatic lipid synthesis and catabolism. The circadian clock orchestrates the transcriptional partitioning of hepatic lipid usage through several downstream regulators. The gene expression patterns of PPARα and PPARδ, REV-ERBα and REV-ERBβ, and SREBP1C all display circadian rhythmicity. Further, the activating ligands for these nuclear receptors as well as for FXR and LXR also display rhythmic accumulation throughout the fasting/feeding cycle. This transcriptional coordination of de novo lipogenesis by the molecular clock and its downstream transcriptional regulators coincides with post-prandial regulation of lipogenesis by insulin. Diagram illustrates liver metabolism in mice

and Ror proteins and occurs in a context-dependent manner. For example, Rev-Erbα has an RORE in its promoter region, and thus it represses its own expression and is induced by Rors (Delerive et al. 2002; Raspe et al. 2002).

Several studies have established that Rev-erb α and β play crucial and direct roles in regulating hepatic lipogenesis (Fig. 11.2). Cistromic analyses have shown that Rev-erbα co-localizes with the histone deacetylase Hdac3 and its associated co-repressor NCoR at several loci encoding genes involved in lipogenesis (Feng et al. 2011). Consistently, hepatic deletion of either *Hdac3* or *Rev-Erbα* in mice results in dramatic hepatosteatosis, thus consolidating their roles in the repression of lipogenesis. The cistromic analyses further revealed that Rev-Erbα and Hdac3 occupancy displayed a circadian pattern inversely related to histone acetylation and RNA polymerase II recruitment. Therefore, during the active/feeding period Hdac3 occupancy decreases allowing for increased expression of lipogenic genes. In fact, Rev-Erbα appears to be necessary for Hdac3 recruitment to metabolic genes. These data suggest that the circadian epigenomic remodeling controlled by Hdac3 and Rev-erbα is essential for homeostasis of the lipogenic process in the liver. Furthermore, similar cistromic analysis has shown that loci that are mutually bound by both Rev-erbs α and β show a significant enrichment in genes involved in lipid and lipoprotein metabolism (Bugge et al. 2012), suggesting functional redundancy of the β isoform in the diurnal regulation of lipogenesis.

While the negative feedback loop generated by the core clock (namely BMAL1, CLOCK, CRY, and PER) is sufficient to form a circadian oscillator, the establishment of a secondary, or auxiliary, feedback loop via the ROR and REV-ERB proteins is thought to make the circadian oscillator robust and tunable (Stricker et al. 2008; Tigges et al. 2009). Thus, it has been proposed that endogenous ligands may tune molecular clock systems via the ROR and REV-ERB nuclear receptors in response to metabolic cues. As of now, heme is the only known endogenous ligand of REV-ERBα and β, and heme levels are known to oscillate in a circadian fashion (Kaasik and Lee 2004; Raghuram et al. 2007; Yin et al. 2007). Heme is a large heterocyclic prosthetic group that is involved in a range of biological functions including diatomic gas transfer, electron transfer, and chemical catalysis. How this and other ligands of the REV-ERB and ROR nuclear receptors may act to coordinate molecular clock function with dynamic metabolic needs in peripheral tissues has yet to be characterized in depth.

PPARs. Peroxisome Proliferator-activated Receptors (PPARs) play multi-faceted roles in the regulation of lipid metabolism in a variety of tissues and cell types. PPARs form heterodimers with an obligate interaction partner, Retinoid X Receptor (RXR). Endogenous ligands of the three PPAR isoforms (α, β/δ, and γ) include mono- and polyunsaturated fatty acids and eicosanoids. Several synthetic ligands have also been generated for these receptors. All three isoforms show rhythmic expression in the mouse liver, with expression of *Pparγ* and *Pparα* peaking during the light cycle at ZT8 and ZT12, respectively, and *Pparβ/δ* peaking during the dark cycle at ZT20 (Yang et al. 2006). *Pparγ* is most abundantly expressed in adipose tissue and plays an indispensable role in adipocyte differentiation (Grimaldi et al. 2010); however, its role in the liver is less well understood. PPARα is a master regulator of fatty acid catabolism (Kersten et al. 1999; Leone et al. 1999), while PPARδ promotes de novo lipogenesis (Liu et al. 2011).

PPARα mediates metabolic responses to fasting including fatty acid oxidation and ketogenesis. It regulates the expression mitochondrial enzymes involved in fatty acid import and oxidation, including medium-chain acyl-CoA dehydrogenase and carnitine palmitoyltransferase I, as well as extramitochondrial enzymes such as acyl-CoA oxidase, cytochrome P450 4A3, and Abcd2 and Abcd3 which mediate peroxisomal fatty acid oxidation (Fourcade et al. 2001; Leone et al. 1999). PPARα also regulates the expression of fatty acid binding proteins which chaperone intracellular fatty acids and transport PPARα ligands to the nucleus (Wolfrum et al. 2001), thus enacting a feed-forward process.

PPARδ is a positive regulator of hepatic lipogenesis. In contrast to its counterpart Rev-erbα, which suppresses lipogenesis during the light/fasting cycle in mice, Pparδ promotes lipogenesis during the dark/feeding cycle (Fig. 11.2). Pparδ directly regulates the expression of the rate-limiting enzyme of de novo lipogenesis, acetyl-coA carboxylase 1 (*Acc1*) (Liu et al. 2013). Thus, Rev-erbα and Pparδ coordinate the circadian rhythmicity of lipid synthesis in the liver. Furthermore, recent research has implicated Pparδ-dependent hepatic lipogenesis in interorgan communication via lipid signaling molecules, which will be discussed in detail later.

LXR and FXR. The Liver X Receptors (LXRα and LXRβ) promote de novo lipogensis at least in part through activating the expression of *SREBP1C*, a direct transcriptional activator of lipogenic enzymes (Repa et al. 2000). Like the PPARs, RXR is the obligate LXR binding partner. On the other hand, the Farnesoid X Receptor (FXR) represses de novo lipogenesis and promotes triglyceride clearance. Fxr indirectly represses the expression of several genes involved in lipid metabolism including *Srebp1c* through induction of Shp (Watanabe et al. 2004).

While *Lxr* and *Fxr*are constitutively expressed in mouse liver (Yang et al. 2006), studies have suggested that the molecular clock regulates the activity of these nuclear receptors through indirect mechanisms. First, the clock regulates Lxr and Fxr protein stability via the NAD-dependent protein deacetylase sirtuin-1 (Sirt1) (Kemper et al. 2009; Li et al. 2007; Nakahata et al. 2009). The Bmal1-Clock heterodimer drives the rhythmic accumulation of the redox substrate NAD+ partly by controlling the transcription of the enzyme nicotinamide phosphoribosyltransferase (Nampt), which is rate-limiting in the NAD+ salvage pathway (Ramsey et al. 2009). This poises the cyclic activation of Sirt1, which in turn deacetylates and destabilizes Lxr and Fxr during fasting.

LXR activity can also be regulated by the rhythmic accumulation of is putative ligands, cholesterol and oxysterols, throughout fasting/feeding cycles (Janowski et al. 1996). Furthermore, LXRs promote the conversion of cholesterol to bile acids through up-regulation of the CYP7A1 enzyme (Peet et al. 1998). Bile acids are known FXR ligands (Wang et al. 1999). In turn, FXR generates a negative feedback loop by repressing CYP7A1 expression through SHP (Goodwin et al. 2000). As such, the protein stability and ligand activation of LXRs and FXR are coordinately regulated by fasting and feeding.

3 Hepatic Lipogenesis and Peripheral Metabolism

In the first part of this chapter, we discussed critical regulatory nodes controlling metabolic rhythm of hepatic metabolism. The remainder of the chapter will focus on the impact of dysregulation in the hepatic lipogenic pathway on systemic metabolic homeostasis.

3.1 The Lipogenic Pathway

Lipogenesis involves several enzymatic steps, from the mitochondrial synthesis of citrate and its transport to the cytosol for generation of acetyl-CoA, to malonyl CoA synthesis followed by production of palmitate that serves as a building block for synthesis of complex lipids (Fig. 11.3). The steps involved in the generation of the shared intermediate acetyl-CoA directly link lipid and carbohydrate metabolism and dictate the capacity to synthesize fatty acids from carbohydrates. De novo lipogenesis is essential for embryonic development. Whole-body knockouts of ATP citrate lyase (Acly, catalyzes cytosolic formation of acetyl-CoA from citrate), ACACA, or FASN (the latter two are rate-limiting enzymes of de novo lipogenesis),

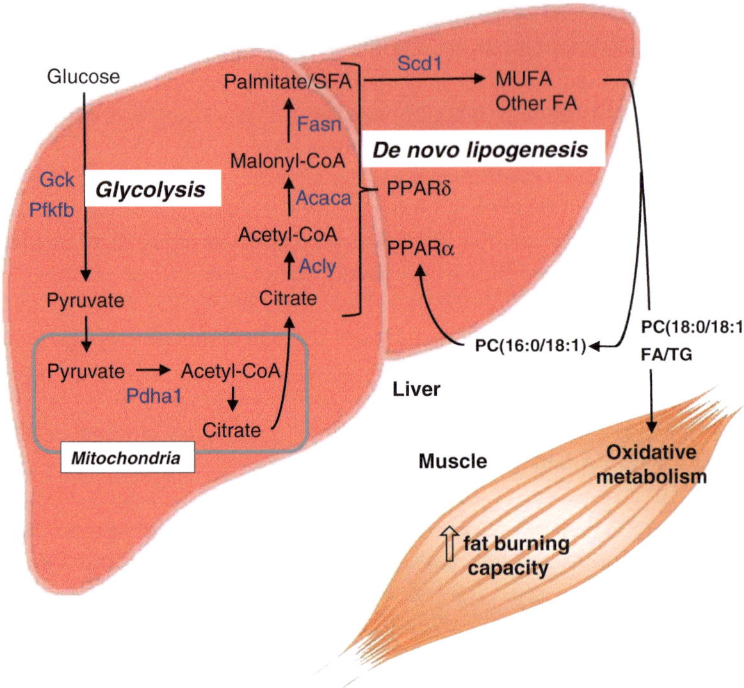

Fig. 11.3 Hepatic de novo lipogenesis and the generation of lipid mediators. Glycolysis is a major pathway providing substrates for de novo lipogenesis in the fed state. Mitochondrial citrate is exported to cytosol and converted to acetyl-CoA. Malonyl CoA synthesis is followed by production of palmitate that serves as a building block for synthesis of complex lipids. Products of hepatic de novo lipogenesis, such as PC(16:0/18:1) and PC(18:0/18:1), have been implicated in mediating postprandial fat utilization in the liver and skeletal muscle, respectively. *Gck* glucokinase, *Pfkfb* 6-phosphofructo-2-kinase/fructose-2,6-bisphosphatase, *Pdha1* puruvate dehydrogenase, *Acly* ATP citrate lyase, *Acaca* acetyl-Coenzyme A carboxylase alpha, *Fasn* fatty acid synthase, *Scd1* stearoyl-Coenzyme A desaturase 1, *SFA* saturated fatty acids, *MUFA* mono unsaturated fatty acids

are embryonic lethal (Abu-Elheiga et al. 2005; Beigneux et al. 2004; Chirala et al. 2003). In the case of liver-specific loss- or gain-of-function, mice are viable but appear to have both hepatic and systemic metabolic phenotypes. A summary of the mouse genetic models, focusing on liver specific modulation, is provided in Table 11.1. As discussed earlier, most enzymes involved in liver de novo lipogenesis display a 24-h oscillation in mRNA transcript (Hughes et al. 2009; Miller et al. 2007; Panda et al. 2002). The studies reported herein do not necessarily focus on time-dependent effects. They remain consistent with the notion that hepatic lipogenesis has strong regulatory effects on metabolism in peripheral organs and on whole body energy balance.

Table 11.1 Metabolic phenotypes of mouse models with liver specific modulation of de novo lipogenesis

Enzymes	Role in lipogenesis	Genetic model	Effect of liver specific modulation: Hepatic	Effect of liver specific modulation: Extra-hepatic	Reference
Glucokinase (Gck)	Participate in glycolysis therefore allowing for mitochondrial citrate production	Adenovirus mediated overexpression of Gck or Pfkfb	Increased hepatic glycolysis	Increase in fatty acid oxidation in skeletal muscle	Wu et al. (2005)
6-phosphofructo-2-kinase/fructose-2,6-bisphosphatase (Pfkfb)			Only Gck caused liver steatosis and high plasma triglycerides and FFA		
Puruvate dehydrogenase (Pdha1)	Pyruvate decarboxylation into acetyl-CoA	Liver-specific Pdha1KO	Improved liver insulin sensitivity	Reduction in fat (33 %) and lean mass (9 %)	Choi et al. (2010)
				Increased lipogenic capacity in epididymal adipose tissue	
				Improved peripheral insulin sensitivity	
ATP citrate lyase (Acly)	Synthesis of acetyl-CoA from citrate	Adenovirus-mediated RNA silencing	Reduced liver acetyl-CoA and malonyl-CoA	Improved GTT	Wang et al. (2009)
			Inhibited hepatic lipogenesis	Improved muscle insulin sensitivity	
			Decreased steatosis in db/db mice		
Acetyl-Coenzyme A carboxylase alpha (Acaca)	Malonyl-CoA synthesis from acetyl-CoA	Liver-specific Acaca KO	Decreased hepatic lipids	Decreased serum NEFA under chow or fat free diet	Mao et al. (2006)
			Up regulation of Acacb		

(continued)

Table 11.1 (continued)

Enzymes	Role in lipogenesis	Genetic model	Effect of liver specific modulation		Reference
			Hepatic	Extra-hepatic	
Fatty acid synthase (Fasn)	Palmitate synthesis	Liver-specific Fasn KO	Fatty liver under a fat-free diet	Hypoglycemia under a fat-free diet	Chakravarthy et al. (2005)
Stearoyl-Coenzyme A desaturase 1 (Scd1)	Synthesis of unsaturated fatty acids from saturated fatty acids	Liver-specific Scd1 KO	Protection from high-carbohydrate (but not high fat) diet induced hepatic steatosis	Protection from high-carbohydrate (but not high fat) diet induced adiposity	Flowers et al. (2012), Miyazaki et al. (2007)
			Hepatic MUFA are decreased.	MUFA composition in serum and WAT is altered	

3.2 Metabolic Phenotypes Associated with Gain- or Loss-of Function Mouse Models

Gck: Carbohydrate metabolism is an important source of acetyl-CoA through glycolysis and pyruvate oxidation. Glucokinase (Gck) and 6-phosphofructo-2-kinase/fructose-2,6-bisphosphatase (Pfkfb) are rate limiting enzymes for glycolysis which are also central in driving glucose flux for hepatic lipogenesis. Adenovirus mediated hepatic overexpression of Gck or Pfkfb increases glycolysis (Wu et al. 2005). However, only Gck overexpression causes lipid accumulation in liver and circulation. Interestingly, both hepatic Gck and Pfkfb dramatically increase fatty acid oxidation in skeletal muscle.

Pdha1. The pyruvate dehydrogenase complex is at the crossroads of glucose and fatty acid metabolism, as it catalyzes mitochondrial pyruvate decarboxylation into acetyl-CoA. Liver specific knockout of the *Pdha1* subunit blocks the incorporation of glucose carbon into fatty acids by hepatocytes (Choi et al. 2010). These mice displayed reduced fat (33 %) and lean mass (9 %), increased lipogenic capacity in epididymal adipose tissue, and improved peripheral insulin sensitivities during a hyperinsulinemic euglycemic clamp.

Acly. ATP citrate lyase (Acly) regulates the synthesis of acetyl-CoA from citrate. Liver-specific *Acly* knockdown via adenovirus-mediated RNA interference reduces hepatic contents of acetyl-CoA and malonyl-CoA, inhibits hepatic de novo lipogenesis, and allaviates hepatic steatosis in db/db mice (Wang et al. 2009). Interestingly, these mice display improved glucose tolerance (GTT) and increased muscle insulin sensitivity.

Acaca/b: Acetyl-CoA carboxylase catalyzes the synthesis of malonyl-CoA, which serves as the donor of 2-carbon units to fatty acid synthesis. The two isoforms of acetyl-CoA carboxylase, Acaca and Acacb, are encoded by separate genes and show distinct tissue distributions. Acaca is highly expressed in the liver, whereas Acacb is predominantly expressed in the skeletal muscle and heart, and to a lesser extent in the liver. In addition, Acacb is localized in the mitochondria while Acaca is cytosolic. Liver-specific Acaca knockout mice display a mild phenotype of decreased hepatic lipid content and serum non-esterified fatty acids (NEFA) under chow or fat free diet (Mao et al. 2006). It is not clear whether the up-regulation of Acacb observed in this model alleviates the phenotype by compensating for Acaca loss. Whole body Acacb knockout mice (Abu-Elheiga et al. 2001) have lowered liver fat content, an effect of increased β-oxidation as a result of the decrease in malonyl-CoA (a negative regulator of the mitochondrial carnitine palmitoyl-CoA shuttle system).

Fasn. Fatty acid synthase (Fasn) catalyzes the first committed step in fatty-acid biosynthesis. Liver *Fasn* knockout mice develop fatty liver and hypoglycemia under a fat-free lipogenic diet or prolonged fasting (Chakravarthy et al. 2005), a phenotype that resemble *Ppara* deficiency in the liver. Indeed, the hypoglycemia/steatohepatitis phenotype could be corrected by pharmacological activation of PPARα. It is suggested that that Fasn is required for the production of endogenous PPARα ligands under fat-free conditions, which will be discussed below.

Scd1. Liver-specific knockout of stearoyl-CoA dehydrogynase (Scd1), an enzyme that catalyzes a rate-limiting step in the synthesis of monounsaturated fatty acids (MUFA) from saturated fatty acids, protects mice from high carbohydrate diet-induced metabolic disorders (Cohen et al. 2002; Miyazaki et al. 2007) as a consequence of suppressed hepatic lipid accumulation and gluconeogenesis and reduced adipose tissue weight. In *Scd1* knockout mice fed a high-sucrose, very low-fat diet, oleate feeding rescued the defective hepatic triglyceride secretion and hepatic lipogenesis. These results are consistent with the notion that hepatic SCD1 is required for carbohydrate-induced adiposity. In a follow-up study, the authors showed that lack of *Scd1* in the liver lowered the MUFA levels of white adipose tissue (Flowers et al. 2012). Furthermore, liver and plasma triglycerides showed similar alterations in fatty acid composition, indicating that fatty acid content of plasma triglycerides is predictive of hepatic Scd1 activity. The data therefore support the existence of crosstalk between liver and adipose tissue, which also raises the possibility that blood-borne lipid metabolites can serve as signaling molecules for inter organ communication to achieve coordinated energy substrate utilization. In the following section, we discuss recent studies that identify bioactive lipids synthesized de novo or derived from dietary fats implicated in such communication.

3.3 De Novo *Lipogenesis and Tissue Crosstalk*

Despite ample genetic evidence demonstrating a clear role for hepatic de novo lipogenesis in metabolic homeostasis, the underlying mechanisms remain unclear. A simple explanation is that hepatic lipogenesis modulates energy substrate availability, which could exert metabolic consequences at the whole-body level. Recent studies also suggest that lipiogenic products may function as metabolic signaling molecules. A notable finding came from hepatic *Fasn* knockout mice. As discussed above, the hypoglycemia and fatty liver phenotype under a fat-free diet can be rescued with a synthetic PPARα agonist, suggesting that Fasn participates in the production of endogenous PPARα ligands (Chakravarthy et al. 2005). Mass spectrometry profiling of liver extracts from wild type or Fasn knockout mice was performed to screen for lipids bound by PPARα (Chakravarthy et al. 2009). Phosphatidylcholine (PC) (16:0/18:1) was identified as a putative PPARα ligand (Fig. 11.3). In fact, portal vein infusion of PC(16:0/18:1) increased fatty acid oxidation in a PPARα dependent manner. Of note, in mice Fasn and lipogenesis are most active in the dark (feeding) cycle, whereas Pparα is known to control fat catabolism in the light cycle. It is likely that postprandial production of PC(16:0/18:1) is in anticipation of the upcoming fasting state when Pparα-controlled β-oxidation and ketogenesis are critical for energy metabolism.

PPARδ is best known for its activity in the control of muscle oxidative metabolism in type I fibers (Evans et al. 2004). Previous work has shown that *Pparδ* knockout mice are glucose intolerant, whereas PPARδ activation by an agonist in db/db mice improves insulin sensitivity in insulin-responsive tissues (Lee et al. 2006). Molecular and functional analyses suggested that PPARδ activation reduces hepatic glucose production by increasing glycolysis and the pentose phosphate shunt to promote fatty acid synthesis in the liver. In fact, DNA array analysis showed that genes involved in fatty acid synthesis were up-regulated in the liver by PPARδ agonist treatment. As expected, PPARδ ligand treatment increased the expression of genes involved in fatty acid catabolism in muscle. These results indicated a coordinated regulation of glucose/lipid metabolism in the liver-muscle axis. Follow-up studies revealed that Pparδ regulates the diurnal expression of several lipogenic genes in the dark cycle, which correlates well with rhythmic alterations in serum lipid profiles and the activity of fatty acid uptake in muscle (Liu et al. 2013). Genetic evidence further suggests that hepatic Pparδ activity is sufficient to drive the observed change in diurnal muscle fatty acid utilization. Unbiased lipidomic profiling of serum or liver samples from mice with hepatic *Pparδ* activation/deletion or *Acaca* knockdown showed that PC(18:0/18:1), or 1-stearoyl-2-oleoyl-sn-glycero-3-phosphocholine (SOPC) was correlated with altered muscle fatty acid uptake phenotypes (Fig. 11.3). In vitro or in vivo delivery of this lipid enhanced muscle cell fatty acid uptake. Interestingly, this effect also requires intact PPARα signaling in the muscle, indicating that PC(16:0/18:1) and PC(18:0/18:1) share similar biological effects on PPARα activation. Therefore, functional identification of PC(16:0/18:1) and PC(18:0/18:1) suggests that these lipogenic products may provide signals for postprandial fat utilization in the liver-muscle axis.

4 Concluding Remarks

With the increasing prevalence of obesity and its associated metabolic pathologies becomes pandemic, many researchers have started to focus on identifying lipid metabolites that are involved in metabolic regulation or dysregulation. De novo lipogenesis, a process essential for embryonic development, is undoubtedly involved in both processes. As an important biosynthetic pathway in metabolic flexibility, hepatic de novo *lipogenesis* is regulated by several signaling pathways and a network of transcription factors, including the molecular clock and lipid-sensing nuclear receptors. These regulatory mechanisms integrate temporal, nutritional, and hormonal controls. Recent studies support the notion that the lipogenic pathway may generate bioactive lipids, which serve as intracellular or long-range metabolic mediators that coordinate fat utilization corresponding to feeding or fasting states. In addition to PC(16:0/18:1) and PC(18:0/18:1) discussed above, palmitoleate (C16:1n7) (Kuda et al. 2009) and branched fatty acid esters of hydroxy fatty acids (FAHFAs) (Yore et al. 2014) have been described to be associated with adipose tissue lipogenesis and to exert systemic insulin sensitizing effects. Therefore, understanding both the upstream signals and downstream effectors of the lipogenic pathway may provide new therapeutic opportunities for drug development to treat metabolic diseases.

Acknowledgements Work in the laboratory of the authors is supported by National Institutes of Health grant R01DK075046 and American Diabetes Association grant 1-14-BS-122 (C-.H.L). H.J.C and R.K.A are supported by Herchel Smith Graduate Fellowship.

References

Abu-Elheiga L, Matzuk MM, Abo-Hashema KA, Wakil SJ (2001) Continuous fatty acid oxidation and reduced fat storage in mice lacking acetyl-CoA carboxylase 2. Science 291:2613–2616

Abu-Elheiga L, Matzuk MM, Kordari P, Oh W, Shaikenov T, Gu Z, Wakil SJ (2005) Mutant mice lacking acetyl-CoA carboxylase 1 are embryonically lethal. Proc Natl Acad Sci U S A 102:12011–12016

Aschoff J (1960) Exogenous and endogenous components in circadian rhythms. Cold Spring Harb Symp Quant Biol 25:11–28

Asher G, Schibler U (2011) Crosstalk between components of circadian and metabolic cycles in mammals. Cell Metab 13:125–137

Bass J, Takahashi JS (2010) Circadian integration of metabolism and energetics. Science 330:1349–1354

Beigneux AP, Kosinski C, Gavino B, Horton JD, Skarnes WC, Young SG (2004) ATP-citrate lyase deficiency in the mouse. J Biol Chem 279:9557–9564

Boucher J, Kleinridders A, Kahn CR (2014) Insulin receptor signaling in normal and insulin-resistant states. Cold Spring Harb Perspect Biol 6

Bugge A, Feng D, Everett LJ, Briggs ER, Mullican SE, Wang F, Jager J, Lazar MA (2012) Rev-erbalpha and Rev-erbbeta coordinately protect the circadian clock and normal metabolic function. Genes Dev 26:657–667

Chakravarthy MV, Pan Z, Zhu Y, Tordjman K, Schneider JG, Coleman T, Turk J, Semenkovich CF (2005) "New" hepatic fat activates PPARalpha to maintain glucose, lipid, and cholesterol homeostasis. Cell Metab 1:309–322

Chakravarthy MV, Lodhi IJ, Yin L, Malapaka RR, Xu HE, Turk J, Semenkovich CF (2009) Identification of a physiologically relevant endogenous ligand for PPARalpha in liver. Cell 138:476–488

Chawla A, Repa JJ, Evans RM, Mangelsdorf DJ (2001) Nuclear receptors and lipid physiology: opening the X-files. Science 294:1866–1870

Chirala SS, Chang H, Matzuk M, Abu-Elheiga L, Mao J, Mahon K, Finegold M, Wakil SJ (2003) Fatty acid synthesis is essential in embryonic development: fatty acid synthase null mutants and most of the heterozygotes die in utero. Proc Natl Acad Sci U S A 100:6358–6363

Choi CS, Ghoshal P, Srinivasan M, Kim S, Cline G, Patel MS (2010) Liver-specific pyruvate dehydrogenase complex deficiency upregulates lipogenesis in adipose tissue and improves peripheral insulin sensitivity. Lipids 45:987–995

Cohen P, Miyazaki M, Socci ND, Hagge-Greenberg A, Liedtke W, Soukas AA, Sharma R, Hudgins LC, Ntambi JM, Friedman JM (2002) Role for stearoyl-CoA desaturase-1 in leptin-mediated weight loss. Science 297:240–243

Damiola F, Le Minh N, Preitner N, Kornmann B, Fleury-Olela F, Schibler U (2000) Restricted feeding uncouples circadian oscillators in peripheral tissues from the central pacemaker in the suprachiasmatic nucleus. Genes Dev 14:2950–2961

Delerive P, Chin WW, Suen CS (2002) Identification of Reverb(alpha) as a novel ROR(alpha) target gene. J Biol Chem 277:35013–35018

Dumas B, Harding HP, Choi HS, Lehmann KA, Chung M, Lazar MA, Moore DD (1994) A new orphan member of the nuclear hormone receptor superfamily closely related to Rev-Erb. Mol Endocrinol 8:996–1005

Evans RM, Barish GD, Wang YX (2004) PPARs and the complex journey to obesity. Nat Med 10:355–361

Feng D, Liu T, Sun Z, Bugge A, Mullican SE, Alenghat T, Liu XS, Lazar MA (2011) A circadian rhythm orchestrated by histone deacetylase 3 controls hepatic lipid metabolism. Science 331:1315–1319

Flowers MT, Ade L, Strable MS, Ntambi JM (2012) Combined deletion of SCD1 from adipose tissue and liver does not protect mice from obesity. J Lipid Res 53:1646–1653

Forman BM, Chen J, Blumberg B, Kliewer SA, Henshaw R, Ong ES, Evans RM (1994) Cross-talk among ROR alpha 1 and the Rev-erb family of orphan nuclear receptors. Mol Endocrinol 8:1253–1261

Fourcade S, Savary S, Albet S, Gauthe D, Gondcaille C, Pineau T, Bellenger J, Bentejac M, Holzinger A, Berger J et al (2001) Fibrate induction of the adrenoleukodystrophy-related gene (ABCD2): promoter analysis and role of the peroxisome proliferator-activated receptor PPARalpha. Eur J Biochem 268:3490–3500

Gallego M, Virshup DM (2007) Post-translational modifications regulate the ticking of the circadian clock. Nat Rev Mol Cell Biol 8:139–148

Goodwin B, Jones SA, Price RR, Watson MA, McKee DD, Moore LB, Galardi C, Wilson JG, Lewis MC, Roth ME et al (2000) A regulatory cascade of the nuclear receptors FXR, SHP-1, and LRH-1 represses bile acid biosynthesis. Mol Cell 6:517–526

Grimaldi B, Bellet MM, Katada S, Astarita G, Hirayama J, Amin RH, Granneman JG, Piomelli D, Leff T, Sassone-Corsi P (2010) PER2 controls lipid metabolism by direct regulation of PPARgamma. Cell Metab 12:509–520

Hughes ME, DiTacchio L, Hayes KR, Vollmers C, Pulivarthy S, Baggs JE, Panda S, Hogenesch JB (2009) Harmonics of circadian gene transcription in mammals. PLoS Genet 5, e1000442

Inouye ST, Kawamura H (1979) Persistence of circadian rhythmicity in a mammalian hypothalamic "island" containing the suprachiasmatic nucleus. Proc Natl Acad Sci U S A 76:5962–5966

Janowski BA, Willy PJ, Devi TR, Falck JR, Mangelsdorf DJ (1996) An oxysterol signalling pathway mediated by the nuclear receptor LXR alpha. Nature 383:728–731

Kaasik K, Lee CC (2004) Reciprocal regulation of haem biosynthesis and the circadian clock in mammals. Nature 430:467–471

Kemper JK, Xiao Z, Ponugoti B, Miao J, Fang S, Kanamaluru D, Tsang S, Wu SY, Chiang CM, Veenstra TD (2009) FXR acetylation is normally dynamically regulated by p300 and SIRT1 but constitutively elevated in metabolic disease states. Cell Metab 10:392–404

Kersten S, Seydoux J, Peters JM, Gonzalez FJ, Desvergne B, Wahli W (1999) Peroxisome proliferator-activated receptor alpha mediates the adaptive response to fasting. J Clin Invest 103:1489–1498

Kuda O, Stankova B, Tvrzicka E, Hensler M, Jelenik T, Rossmeisl M, Flachs P, Kopecky J (2009) Prominent role of liver in elevated plasma palmitoleate levels in response to rosiglitazone in mice fed high-fat diet. J Physiol Pharmacol 60:135–140

Kudo T, Tamagawa T, Kawashima M, Mito N, Shibata S (2007) Attenuating effect of clock mutation on triglyceride contents in the ICR mouse liver under a high-fat diet. J Biol Rhythms 22:312–323

Lee CH, Olson P, Hevener A, Mehl I, Chong LW, Olefsky JM, Gonzalez FJ, Ham J, Kang H, Peters JM et al (2006) PPARdelta regulates glucose metabolism and insulin sensitivity. Proc Natl Acad Sci U S A 103:3444–3449

Leone TC, Weinheimer CJ, Kelly DP (1999) A critical role for the peroxisome proliferator-activated receptor alpha (PPARalpha) in the cellular fasting response: the PPARalpha-null mouse as a model of fatty acid oxidation disorders. Proc Natl Acad Sci U S A 96:7473–7478

Li X, Zhang S, Blander G, Tse JG, Krieger M, Guarente L (2007) SIRT1 deacetylates and positively regulates the nuclear receptor LXR. Mol Cell 28:91–106

Liu S, Hatano B, Zhao M, Yen CC, Kang K, Reilly SM, Gangl MR, Gorgun C, Balschi JA, Ntambi JM et al (2011) Role of peroxisome proliferator-activated receptor {delta}/{beta} in hepatic metabolic regulation. J Biol Chem 286:1237–1247

Liu S, Brown JD, Stanya KJ, Homan E, Leidl M, Inouye K, Bhargava P, Gangl MR, Dai L, Hatano B et al (2013) A diurnal serum lipid integrates hepatic lipogenesis and peripheral fatty acid use. Nature 502:550–554

Mao J, DeMayo FJ, Li H, Abu-Elheiga L, Gu Z, Shaikenov TE, Kordari P, Chirala SS, Heird WC, Wakil SJ (2006) Liver-specific deletion of acetyl-CoA carboxylase 1 reduces hepatic triglyceride accumulation without affecting glucose homeostasis. Proc Natl Acad Sci U S A 103:8552–8557

Miller BH, McDearmon EL, Panda S, Hayes KR, Zhang J, Andrews JL, Antoch MP, Walker JR, Esser KA, Hogenesch JB et al (2007) Circadian and CLOCK-controlled regulation of the mouse transcriptome and cell proliferation. Proc Natl Acad Sci U S A 104:3342–3347

Miyazaki M, Flowers MT, Sampath H, Chu K, Otzelberger C, Liu X, Ntambi JM (2007) Hepatic stearoyl-CoA desaturase-1 deficiency protects mice from carbohydrate-induced adiposity and hepatic steatosis. Cell Metab 6:484–496

Nakahata Y, Sahar S, Astarita G, Kaluzova M, Sassone-Corsi P (2009) Circadian control of the NAD+ salvage pathway by CLOCK-SIRT1. Science 324:654–657

Pan X, Zhang Y, Wang L, Hussain MM (2010) Diurnal regulation of MTP and plasma triglyceride by CLOCK is mediated by SHP. Cell Metab 12:174–186

Panda S, Antoch MP, Miller BH, Su AI, Schook AB, Straume M, Schultz PG, Kay SA, Takahashi JS, Hogenesch JB (2002) Coordinated transcription of key pathways in the mouse by the circadian clock. Cell 109:307–320

Peet DJ, Turley SD, Ma W, Janowski BA, Lobaccaro JM, Hammer RE, Mangelsdorf DJ (1998) Cholesterol and bile acid metabolism are impaired in mice lacking the nuclear oxysterol receptor LXR alpha. Cell 93:693–704

Pittendrigh CS (1954) On temperature independence in the clock system controlling emergence time in Drosophila. Proc Natl Acad Sci U S A 40:1018–1029

Raghuram S, Stayrook KR, Huang P, Rogers PM, Nosie AK, McClure DB, Burris LL, Khorasanizadeh S, Burris TP, Rastinejad F (2007) Identification of heme as the ligand for the orphan nuclear receptors REV-ERBalpha and REV-ERBbeta. Nat Struct Mol Biol 14:1207–1213

Ramsey KM, Yoshino J, Brace CS, Abrassart D, Kobayashi Y, Marcheva B, Hong HK, Chong JL, Buhr ED, Lee C et al (2009) Circadian clock feedback cycle through NAMPT-mediated NAD+ biosynthesis. Science 324:651–654

Raspe E, Mautino G, Duval C, Fontaine C, Duez H, Barbier O, Monte D, Fruchart J, Fruchart JC, Staels B (2002) Transcriptional regulation of human Rev-erbalpha gene expression by the orphan nuclear receptor retinoic acid-related orphan receptor alpha. J Biol Chem 277:49275–49281

Repa JJ, Liang G, Ou J, Bashmakov Y, Lobaccaro JM, Shimomura I, Shan B, Brown MS, Goldstein JL, Mangelsdorf DJ (2000) Regulation of mouse sterol regulatory element-binding protein-1c gene (SREBP-1c) by oxysterol receptors, LXRalpha and LXRbeta. Genes Dev 14:2819–2830

Retnakaran R, Flock G, Giguere V (1994) Identification of RVR, a novel orphan nuclear receptor that acts as a negative transcriptional regulator. Mol Endocrinol 8:1234–1244

Rudic RD, McNamara P, Curtis AM, Boston RC, Panda S, Hogenesch JB, Fitzgerald GA (2004) BMAL1 and CLOCK, two essential components of the circadian clock, are involved in glucose homeostasis. PLoS Biol 2, e377

Shimba S, Ogawa T, Hitosugi S, Ichihashi Y, Nakadaira Y, Kobayashi M, Tezuka M, Kosuge Y, Ishige K, Ito Y et al (2011) Deficient of a clock gene, brain and muscle Arnt-like protein-1 (BMAL1), induces dyslipidemia and ectopic fat formation. PLoS One 6, e25231

Somers DE (1999) The physiology and molecular bases of the plant circadian clock. Plant Physiol 121:9–20

Stephan FK, Zucker I (1972) Circadian rhythms in drinking behavior and locomotor activity of rats are eliminated by hypothalamic lesions. Proc Natl Acad Sci U S A 69:1583–1586

Stokkan KA, Yamazaki S, Tei H, Sakaki Y, Menaker M (2001) Entrainment of the circadian clock in the liver by feeding. Science 291:490–493

Stricker J, Cookson S, Bennett MR, Mather WH, Tsimring LS, Hasty J (2008) A fast, robust and tunable synthetic gene oscillator. Nature 456:516–519

Tigges M, Marquez-Lago TT, Stelling J, Fussenegger M (2009) A tunable synthetic mammalian oscillator. Nature 457:309–312

Turek FW, Joshu C, Kohsaka A, Lin E, Ivanova G, McDearmon E, Laposky A, Losee-Olson S, Easton A, Jensen DR et al (2005) Obesity and metabolic syndrome in circadian Clock mutant mice. Science 308:1043–1045

Vollmers C, Gill S, DiTacchio L, Pulivarthy SR, Le HD, Panda S (2009) Time of feeding and the intrinsic circadian clock drive rhythms in hepatic gene expression. Proc Natl Acad Sci U S A 106:21453–21458

Wang H, Chen J, Hollister K, Sowers LC, Forman BM (1999) Endogenous bile acids are ligands for the nuclear receptor FXR/BAR. Mol Cell 3:543–553

Wang Q, Jiang L, Wang J, Li S, Yu Y, You J, Zeng R, Gao X, Rui L, Li W et al (2009) Abrogation of hepatic ATP-citrate lyase protects against fatty liver and ameliorates hyperglycemia in leptin receptor-deficient mice. Hepatology 49:1166–1175

Watanabe M, Houten SM, Wang L, Moschetta A, Mangelsdorf DJ, Heyman RA, Moore DD, Auwerx J (2004) Bile acids lower triglyceride levels via a pathway involving FXR, SHP, and SREBP-1c. J Clin Invest 113:1408–1418

Wolfrum C, Borrmann CM, Borchers T, Spener F (2001) Fatty acids and hypolipidemic drugs regulate peroxisome proliferator-activated receptors alpha - and gamma-mediated gene expression via liver fatty acid binding protein: a signaling path to the nucleus. Proc Natl Acad Sci U S A 98:2323–2328

Wu C, Kang JE, Peng LJ, Li H, Khan SA, Hillard CJ, Okar DA, Lange AJ (2005) Enhancing hepatic glycolysis reduces obesity: differential effects on lipogenesis depend on site of glycolytic modulation. Cell Metab 2:131–140

Yamazaki S, Numano R, Abe M, Hida A, Takahashi R, Ueda M, Block GD, Sakaki Y, Menaker M, Tei H (2000) Resetting central and peripheral circadian oscillators in transgenic rats. Science 288:682–685

Yang X, Downes M, Yu RT, Bookout AL, He W, Straume M, Mangelsdorf DJ, Evans RM (2006) Nuclear receptor expression links the circadian clock to metabolism. Cell 126:801–810

Yecies JL, Zhang HH, Menon S, Liu S, Yecies D, Lipovsky AI, Gorgun C, Kwiatkowski DJ, Hotamisligil GS, Lee CH et al (2011) Akt stimulates hepatic SREBP1c and lipogenesis through parallel mTORC1-dependent and independent pathways. Cell Metab 14:21–32

Yin L, Wu N, Curtin JC, Qatanani M, Szwergold NR, Reid RA, Waitt GM, Parks DJ, Pearce KH, Wisely GB et al (2007) Rev-erbalpha, a heme sensor that coordinates metabolic and circadian pathways. Science 318:1786–1789

Yoo SH, Yamazaki S, Lowrey PL, Shimomura K, Ko CH, Buhr ED, Siepka SM, Hong HK, Oh WJ, Yoo OJ et al (2004) PERIOD2::LUCIFERASE real-time reporting of circadian dynamics reveals persistent circadian oscillations in mouse peripheral tissues. Proc Natl Acad Sci U S A 101:5339–5346

Yore MM, Syed I, Moraes-Vieira PM, Zhang T, Herman MA, Homan EA, Patel RT, Lee J, Chen S, Peroni OD et al (2014) Discovery of a class of endogenous mammalian lipids with anti-diabetic and anti-inflammatory effects. Cell 159:318–332

Zhang EE, Kay SA (2010) Clocks not winding down: unravelling circadian networks. Nat Rev Mol Cell Biol 11:764–776

Zhang D, Tong X, Arthurs B, Guha A, Rui L, Kamath A, Inoki K, Yin L (2014) Liver clock protein BMAL1 promotes de novo lipogenesis through insulin-mTORC2-AKT signaling. J Biol Chem 289:25925–25935

Chapter 12
Monounsaturated Fatty Acid Mediated Liver-Adipose Tissue Crosstalk and Metabolic Regulation

Maggie S. Burhans and James M. Ntambi

Abstract The maintenance of metabolic health requires complex regulation of metabolic processes in several tissues. The coordination of this metabolic regulation involves extensive crosstalk among tissues. Signaling factors that are secreted into the circulation and impart systemic metabolic effects include molecules such as hepatokines, adipokines and lipokines. Many of these factors regulate lipid metabolism, including de novo lipogenesis. Monounsaturated fatty acids, palmitoleate and oleate, are major products of de novo lipogenesis and exert significant control over metabolic regulation. These fatty acids may circulate as free or esterified forms and subsequently influence metabolic processes in tissues distinct from the ones in which they are synthesized. Examples include the effects of hepatic palmitoleate and oleate in adipose tissue and of adipose tissue-derived palmitoleate in liver. This chapter reviews secreted factors that regulate metabolic processes with a focus on lipids and fatty acids and communication between liver and adipose tissues mediated by these species.

Keywords Oleate • Palmitoleate • Adipose tissue de novo lipogenesis • Metabolic regulation • Monounsaturated fatty acids

---

M.S. Burhans (✉)
Department of Medicine, University of Wisconsin-Madison, Madison, WI 53706, USA
e-mail: mburhans@medicine.wisc.edu

J.M. Ntambi
Department of Biochemistry, University of Wisconsin-Madison, Madison, WI, USA

Department of Nutritional Sciences, University of Wisconsin-Madison, Madison, WI, USA
e-mail: jmntambi@wisc.edu

J.M. Ntambi (ed.), *Hepatic De Novo Lipogenesis and Regulation of Metabolism*,
DOI 10.1007/978-3-319-25065-6_12

1 Introduction

Excessive adiposity now afflicts a majority of adults and an increasing number of children and adolescents worldwide. The global prevalence of overweight and obesity are estimated to be 39 % and 13 %, respectively (World Health Organization 2015). Although it is controversial whether metabolically healthy overweight and obese individuals are at a lower risk for associated morbidity and mortality relative to metabolically abnormal overweight individuals (Hinnouho et al. 2013; Appleton et al. 2013; Kramer et al. 2013), it is generally undisputed that excess adipose tissue increases one's risk for the development of metabolic dysfunction. Indeed, it is estimated that at least 85 % of adults in the United States with type II diabetes are overweight or obese (Centers for Disease Control and Prevention 2004). Additionally, obesity is considered to be a major risk factor for the development for hepatic steatosis, cardiovascular disease, osteoarthritis, and certain types of cancer, among other diseases (Moore 2009; Poirier et al. 2006; Muoio and Newgard 2006). More specifically, visceral adiposity is now recognized as one of the strongest predictors of ectopic fat accumulation in tissues other than in adipose, such as in the liver, and of subsequent development of metabolic disease, including type II diabetes. The mechanisms that connect increased adiposity to the development of other metabolic diseases continue to be actively investigated. Perhaps unsurprisingly, much attention has focused on the role of secreted molecules that regulate metabolism systemically. While many of these secreted factors are peptides with well described target tissues, receptor partners and signaling cascades, recently there has also been an emergence of reports of non-protein factors including specific lipids and fatty acids that control metabolic processes including de novo lipogenesis (DNL), the endogenous synthesis of fatty acids and lipids.

2 De Novo Lipogenesis and Fatty Acid Trafficking

DNL is the metabolic pathway responsible for the endogenous synthesis of cholesterol and fatty acids from acetyl-CoA, with carbohydrates consumed in excess of the body's energy needs and glycogen storage capacity serving as the major acetyl-CoA source (Fig. 12.1). Liver and adipose tissue are the main sites of de novo fatty acid synthesis in humans and rodents. The synthesis of fatty acids occurs through a series of enzymatic reactions that build a sixteen-carbon fatty acyl chain that can be further elongated and/or desaturated to yield monounsaturated fatty acids (MUFA). Transcriptional control of genes in the DNL pathway is regulated predominantly by SREBP-1c, ChREBP, and LXR transcription factors and has been reviewed in detail elsewhere (Strable and Ntambi 2010). Additionally, a complex relationship exists between inflammation and lipid metabolism that is evident in a tissue-specific manner (Glass and Olefsky 2012).

De novo synthesized fatty acids are esterified to a glycerol backbone to form triacylglycerol molecules that are packaged into VLDL particles and secreted from the

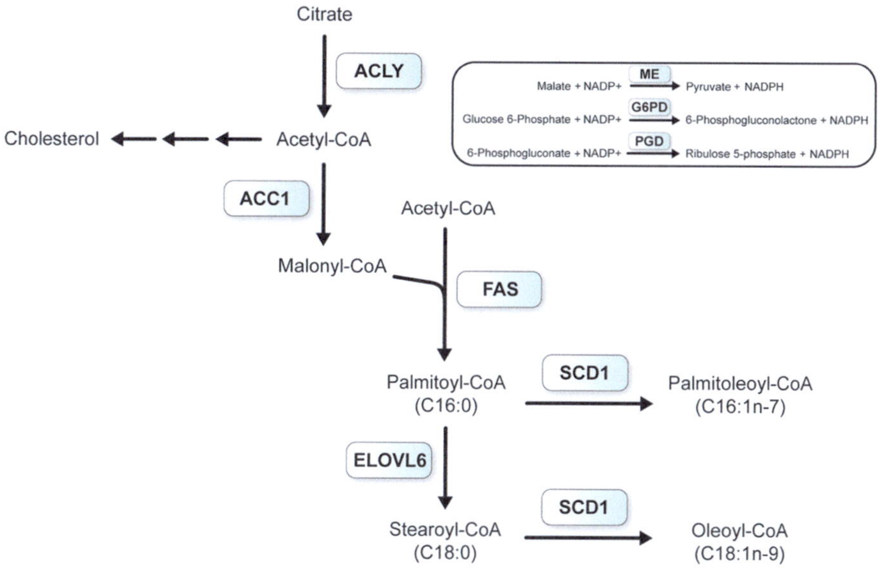

Fig. 12.1 De novo lipogenesis. The de novo synthesis of lipids occurs predominantly in liver and adipose tissues. Acetyl-CoA is carbon donor for de novo synthesis of cholesterol and fatty acids. NADPH is a required cofactor. Three metabolic reactions of the pentose phosphate pathway yield NADPH (*inset*). *ACLY* ATP-citrate lyase, *ACC1* acetyl-CoA carboxylase-1, *FAS* fatty acid synthase, *SCD1* stearoyl-CoA desaturase-1, *ELOVL6* elongation of long-chain fatty acids family member 6, *ME* malic enzyme, *G6PD* glucose 6-phosphate dehydrogenase, *PGD* phosphogluconate dehydrogenase

liver. Secretion of these TG-rich VLDL particles from the liver into the circulation is the predominant pathway by which hepatic de novo synthesized fatty acids are made accessible to extrahepatic tissues. Thus, the fatty acid composition of plasma TG typically reflects that of liver TG fatty acid composition (Cao et al. 2008; Flowers et al. 2012). Studies with stearoyl-CoA desaturase (SCD)-1 liver-specific knockout and the double tissue knockout from liver and adipose demonstrated that the desaturation index, a ratio of a MUFA to its saturated fatty acid precursor, of plasma TG tracks with level of hepatic SCD activity (Flowers et al. 2012). The major products of SCD activity are oleate (18:1n-9) and palmitoleate (16:1n-7) and relative levels of these fatty acids in plasma and liver lipid fractions are significantly reduced in SCD knockout mice as compared to wildtype mice. Studies in mouse models and human subjects with hypertriglyceridemia have also revealed that the plasma TG desaturation index is positively associated with absolute concentration of plasma TGs and can be used as a proxy for hepatic SCD activity (Attie et al. 2002).

Nonesterified fatty acids (NEFA) also contribute to the plasma lipid compartment and fatty acid composition. Plasma NEFA are largely trafficked from adipose tissue, as TG molecules are hydrolyzed and the fatty acids are then released into circulation for use by peripheral tissues. In agreement with this model, lipid flux modeling conducted in fatty acid binding protein (FABP) knockout mice clearly

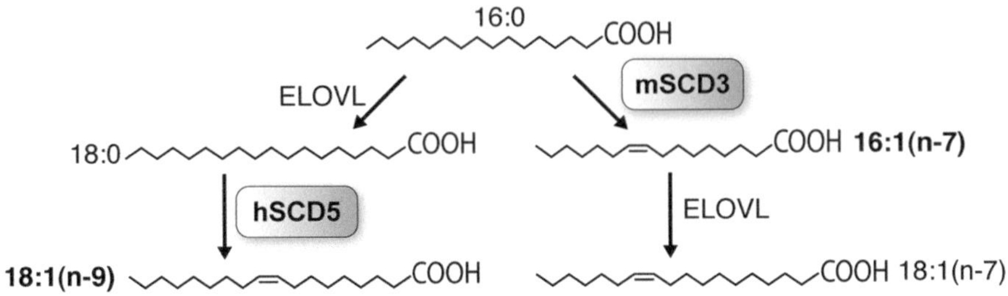

Fig. 12.2 Human and mouse stearoyl-CoA desaturase isoforms synthesize distinct monounsaturated fatty acids. Human SCD5 synthesizes oleate (18:1n-9) from the saturated fatty acid stearate (18:0). Mouse SCD3 synthesizes palmitoleate (16:1n-7) from the saturated fatty acid palmitate (16:0). *hSCD5* human stearoyl-CoA desaturase-5, *mSCD3* mouse stearoyl-CoA desaturase-3, *ELOVL* elongation of long-chain fatty acids family member

demonstrated the role of adipose tissue as the major lipid reservoir that drives the free fatty acid composition profile in plasma (Cao et al. 2008). The relationship between tissue lipids and plasma lipids and fatty acid composition of the distinct fractions has also been demonstrated through dietary studies conducted in genetic mouse models of altered MUFA synthesis. We recently reported on the effects of a high-sucrose lipogenic diet (LD) in two SCD transgenic mouse models (Burhans et al. 2015). In this study, the *SCD5* or *Scd3* allele was crossed into SCD1 global knockout (GKO) mice. Human *SCD5* synthesizes predominantly oleate while mouse *Scd3* expresses predominantly palmitoleate (Fig. 12.2). The ApoE promoter, conferring liver-specificity gene expression, drove the expression of the transgenes. Consistent with the lipid and fatty acid trafficking models, the plasma TG fraction of the LD-fed mice reflected that of the TG fraction in liver as plasma TG from SCD5 transgenic (SCD5Tg) and SCD3 transgenic (SCD3Tg) mice were oleate- and palmitoleate-enriched, respectively. Specifically, both 18:1n-9 and 16:1n-7 were significantly lower in plasma TG of the SCD1 GKO mice compared to WT controls. In contrast, 18:1n-9 was significantly increased in both plasma and liver TG fractions of SCD5Tg mice while 16:1n-7 was significantly increased in plasma and liver TG in SCD3Tg mice. However, the fatty acid composition profile of the plasma NEFA fraction was not significantly different in SCD5Tg and SCD3Tg compared to SCD1 GKO mice; that is, plasma 18:1n-9 and 16:1n-7 free fatty acids were significantly lower in SCD1 GKO, SCD5Tg and SCD3Tg models as compared to WT mice, again consistent with the tissue-plasma lipid trafficking model, as the three models all lack SCD expression in adipose tissue.

3 Hepatokines

Proteins that are synthesized in and secreted from the liver that exert systemic metabolic effects, that is regulating core metabolic pathways in extrahepatic tissues are termed 'hepatokines'. Several hepatokines have been discovered and the list of these liver-secreted factors include fetuin-A, selenoprotein P, and FGF21, among others that have been reviewed recently (Stefan and Haring 2013; Iroz et al. 2015). A major target tissue of FGF21 is white adipose tissue and the beneficial metabolic effects of FGF21 signaling in adipose include an increase in glucose uptake, lowered blood glucose and insulin levels, and lower plasma triglycerides (Kharitonenkov et al. 2005). More recent work demonstrated that FGF21's metabolic effects are mediated through the adipokine adiponectin (Lin et al. 2013). In addition to liver-derived proteins that regulate systemic metabolism, liver-derived lipids also influence metabolic flux and health. Two studies have revealed phosphatidylcholine to be a ligand for PPARα in liver and to integrate hepatic lipogenesis and peripheral fatty acid use in skeletal muscle (Chakravarthy et al. 2009; Liu et al. 2013). Each of these phosphatidylcholine species contains one saturated fatty acid (16:0 or 18:0) and one MUFA (18:1n-9). A diurnal serum lipid integrates hepatic lipogenesis and peripheral fatty acid use.

4 Liver Monounsaturated Fatty Acids and Adipose Tissue De Novo Lipogenesis

Regulation of hepatic DNL is controlled at both the level of transcription and post-translation. Regulation of transcription factors that control DNL can occur via hormones (SREBP-1c), by glucose metabolism (ChREBP), and by fatty acids (SREBP-1c, ChREBP, LXRs). Several studies conducted using whole body and tissue specific knockout mouse models of enzymes in DNL has revealed the dramatic metabolic consequences of altering flux through DNL (Strable and Ntambi 2010). In a series of papers we have demonstrated that hepatic SCD1 activity is required for high-carbohydrate diet-induced hepatic DNL nuclear translocation of the essential transcription factors, lipogenic gene expression, DNL and complex lipid assembly and accumulation (Ntambi et al. 2002; Miyazaki et al. 2004; Miyazaki et al. 2007; Burhans et al. 2015).

We took advantage of the differences in substrate specificity of human SCD5 and mouse SCD3 (Fig. 12.2) to investigate whether oleate and palmitoleate differentially regulate hepatic DNL (Burhans et al. 2015). Our work revealed that restored synthesis of hepatic oleate, but not palmitoleate, was sufficient to induce hepatic lipid accumulation in SCD1 GKO mice under high-carbohydrate lipogenic dietary conditions. Surprisingly, however, hepatic DNL itself was not restored by oleate synthesis and DNL remained low in SCD1 GKO, SCD3Tg and SCD5Tg mice. This suggested that perhaps the extent of oleate restoration was sufficient to induce DNL or perhaps that the regulation of hepatic DNL is regulated by other mechanisms.

In contrast to the lack of differences in hepatic DNL among LD-fed GKO, SCD5Tg and SCD3Tg mice, both fatty acid and cholesterol synthesis rates were significantly induced in gonadal adipose tissue of GKO and SCD3Tg mice. The rate of DNL in SCD5Tg mice was low and not different from WT. The simultaneous increase of both branches of DNL suggests increased activity of an upstream regulator. Although future work is required to probe the mechanism, one possibility is that ATP-citrate lyase (ACLY), a cytosolic enzyme that converts citrate to acetyl-CoA (Fig. 12.1), is repressed by oleate but activated when oleate levels are reduced. In this scenario, ACLY activity in adipose tissue would be increased in both SCD1 GKO and SCD3Tg mice (decreased tissue oleate level), but repressed in SCD5Tg mice (increased tissue oleate), leading to the synthesis of acetyl-CoA and allowing for an increase in both cholesterol and fatty acid synthesis in GKO and SCD3Tg mice only. Additionally, the rate of fatty acid oxidation in adipose tissue was measured in WT, GKO, SCD5Tg and SCD3Tg mice (Burhans et al. 2015). The authors discovered that fatty acid oxidation was significantly elevated in GKO and SCD3Tg, but not SCD5Tg, relative to WT. Although not investigated in detail, oxidation was significantly greater in SCD3Tg mice compared to SCD5Tg. These results suggest differential effects of palmitoleate and oleate in regulation of adipose tissue mitochondrial fatty acid oxidation where palmitoleate stimulates it and oleate represses it.

5 Adipose Tissue De Novo Lipogenesis Is Associated with Metabolic Health

The dramatic induction of DNL in white adipose tissue in GKO and SCD3Tg, 25-fold and 10-fold, respectively, as compared to WT mice, was unexpected. On a per gram of tissue basis, adipose tissue fatty acid synthesis was greater than that in the liver in GKO and SCD3Tg mice. Paradoxically, adipose tissue mass was greatest in WT and SCD5Tg mice despite low rates of DNL in this tissue (Burhans et al. 2015). The upregulation of white adipose tissue fatty acid synthesis in GKO and SCD3Tg may be a compensatory response due to reduced hepatic fatty acid synthesis and the general reduction in body energy stores. Others have demonstrated that reduced hepatic DNL is accompanied by upregulated adipose DNL as a compensatory response (Kuriyama et al. 2005; Beaven et al. 2013). However, in these studies adipose tissue mass was maintained (Kuriyama et al. 2005) or increased (Beaven et al. 2013) relative to control animals. The study from Burhans et al. (2015) may alternatively suggest that the ability to upregulate white adipose tissue DNL is metabolically advantageous in the context of a high-carbohydrate diet and protective against the detrimental effects of such a diet.

Others have also reported upregulation of white adipose tissue DNL in the absence of metabolic derangements. The ability of white adipose tissue to induce the lipogenic program has been proposed to serve important metabolic effects and that this lipogenic flexibility of adipose tissue is indeed metabolically advantageous. Herman et al. (2012) reported that adipose tissue GLUT4 positively regulates ChREBP expression in adipose tissue. Adipose tissue-specific GLUT4 overexpression in mice increased adipose tissue ChREBP expression and DNL, which was associated with enhanced insulin sensitivity. In another study, Kuriyama et al. (2005) demonstrated a reciprocal relationship between liver and adipose tissue DNL rates. Specifically this group showed that in mice with liver-specific deficiency of the SREBP cleavage activating protein (SCAP), a protein required for SREBP-1c nuclear localization, hepatic DNL was reduced whereas adipose tissue DNL was dramatically induced. However, these SCAP liver knockout mice were protected from increased adiposity despite having a higher rate of DNL. Furthermore, the knockout mice exhibited increased glucose clearance and insulin sensitivity. The idea that DNL specifically in adipose tissue maybe be metabolically beneficial may not be surprising given insights obtained from human obesity studies and work from obese animal models. Decreased lipogenic gene expression was reported in obese mouse models (Nadler et al. 2000) and in obese human subjects, which correlated inversely with markers of insulin resistance and hepatic steatosis (Eissing et al. 2013). Eissing et al. also demonstrated increased lipogenic gene expression in white adipose tissue in bariatric surgery patients and the upregulated gene expression coincided with weight loss and improved glucose homeostasis post-surgery. All of these studies taken together provide convincing evidence that unlike hepatic DNL, adipose tissue DNL is associated with metabolic health and resistance to the development of adiposity-associated metabolic dysfunction.

The mechanisms through which hepatic de novo synthesized oleate might regulate specific metabolic pathways in white adipose tissue are unknown. It is possible that oleate or other lipid species that rely on oleate for synthesis or biological function regulate gene expression of metabolic pathways through influence of activity of various transcription factors. It is also possible, as mentioned above, that oleate or a lipid it is incorporated into can directly bind and activate or repress enzymes in metabolic pathways. The differential effects of oleate and palmitoleate from Burhans et al (2015) also suggest that perhaps oleate exerts greater regulatory control over lipid use than does palmitoleate.

6 Adipokines and Cytokines

The role of inter-tissue communication in metabolic regulation and tissue function is becoming more appreciated as an increasingly large number of secreted signaling molecules are discovered. Circulating factors, including secreted adipokines and cytokines from adipose tissue, have significant regulation over systemic metabolism and inflammatory state. Major adipose secreted factors that regulate fat utilization, insulin sensitivity and inflammation in other tissues include leptin, adiponectin, RBP4, TNFα, IL-6, IL-18, NAMPT, resistin, and lipocalin 2, among others. The vast majority of these secreted factors promotes insulin resistance and inflammation in liver and skeletal muscle and has been reviewed extensively elsewhere (Ouchi et al. 2011; Cao 2014). Adiponectin, however, is a notable exception in this list and has garnered attention as a potential therapeutic factor with the potential to restore dysregulated metabolic function as it promotes fat oxidation, increases insulin sensitivity and exerts anti-inflammatory properties (Ohashi et al. 2014; Kadowaki et al. 2006). While most of the secreted factors are peptides with well described target tissues, receptor partners and signaling cascades, recently there has also been an emergence of reports of non-protein factors that also control metabolic processes including the endogenous synthesis of fatty acids and lipids, referred to as DNL.

7 Lipokines

While adipokines and cytokines may be the predominant signaling factors secreted from adipose tissue, much attention has recently turned to role of specific adipose tissue-secreted lipid hormones, referred to as a lipokines. Cao et al. (2008) first proposed that nonesterified palmitoleate, de novo synthesized in and secreted from white adipose tissue, functions as a lipid hormone with systemic metabolic effects. In this initial study conducted in FABP knockout mice, palmitoleate synthesis was increased in adipose tissue, which lead to significantly increased levels in plasma. This pool of palmitoleate repressed hepatic de novo lipogenesis under high-fat dietary conditions. In addition, palmitoleate was shown to improve insulin signaling in skeletal muscle and improve whole-body glucose tolerance. In agreement with these results another study showed that palmitoleate supplementation via gastric tube reduced hepatic lipogenic gene expression and liver TG accumulation and increased insulin sensitivity (Yang et al. 2011). However, a separate study demonstrated that although mice fed a low-fat diet supplemented with palmitoleate were insulin sensitive, hepatic lipid accumulation occurred (Guo et al. 2012). The metabolic effects of palmitoleate are not as clear in humans, although the studies are all association-based. Plasma palmitoleate concentration was not found to correlate significantly with insulin sensitivity in a cohort of obese adults (Fabbrini et al. 2011) or obese children (Okada et al. 2005). In contrast, a separate study demonstrated a significant positive association between plasma nonesterified palmitoleate and insulin sensitivity in a cohort of overweight adults (Arregui et al. 2012).

8 Conclusion

Taken together, the studies discussed in this chapter suggest that endogenously synthesized MUFA exert systemic metabolic effects and influence metabolic function in tissues distinct from which they are synthesized within, with a focus on the liver-adipose tissue axis of tissue crosstalk (Fig. 12.3). Specifically, the model proposes that hepatic de novo synthesized oleate, but not palmitoleate, acts systemically to regulate DNL rate in adipose tissue and total adiposity. In contrast, adipose tissue-derived nonesterified palmitoleate represses DNL in liver, reducing hepatic lipid accumulation. These studies warrant additional studies to investigate the mechanisms through which these two MUFA exert such differential functions and control over metabolic homeostasis.

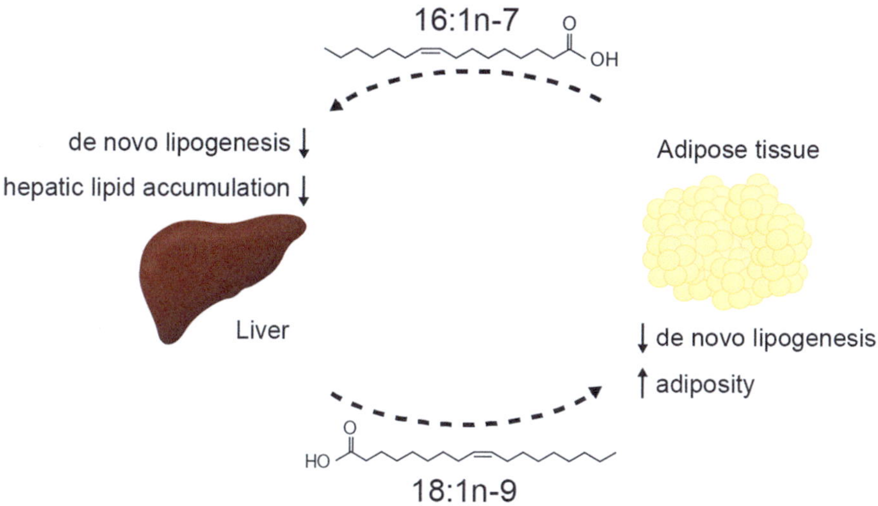

Fig. 12.3 Monounsaturated fatty acids regulate de novo lipogenesis. Nonesterified de novo synthesized palmitoleate (16:1n-7) is released from adipose tissue into circulation and is taken up by the liver where it is proposed to inhibit hepatic de novo lipogenesis and reduce hepatic lipid accumulation. Oleate (18:1n-9) is synthesized in the liver, released into circulation, and taken up by adipose tissue. In adipose tissue oleate is proposed to inhibit de novo lipogenesis but paradoxically increase adipose tissue mass

Acknowledgements This work was supported by National Institutes of Health (NIH) Grant R01 DK062388, ADA 7-13-BS-118, and USDA Hatch W2005 (to J.M.N.) and NIH Predoctoral Training grant T32-DK-007665 (to M.S.B.).

9 References

Appleton SL, Seaborn CJ, Visvanathan R, Hill CL, Gill TK, Taylor AW, Adams RJ (2013) Diabetes and cardiovascular disease outcomes in the metabolically healthy phenotype. Diabetes Care 36:2388–2394

Arregui M et al (2012) Heterogeneity of the stearoyl-CoA desaturase-1 (SCD1) gene and metabolic risk factors in the EPIC-Potsdam study. PLoS One 7, e48338

Attie AD, Krauss RM, Gray-Keller MP, Brownlie A, Miyazaki M, Kastelein JJ, Lusis AJ, Stalenhoef AFH, Stoehr JP, Hayden MR, Ntambi JM (2002) Relationship between stearoyl-CoA desaturase and plasma triglycerides in human and mouse hypertriglyceridemia. J Lipid Res 43:1899–1907

Beaven SW et al (2013) Reciprocal regulation of hepatic and adipose lipogenesis by liver x receptors in obesity and insulin resistance. Cell Metab 18:106–117

Burhans MS, Flowers MT, Harrington KR, Bond LM, Guo C-A, Anderson RM, Ntambi JM (2015) Hepatic oleate regulates adipose tissue lipogenesis and fatty acid oxidation. J Lipid Res 56:304–318

Cao H (2014) Adipocytokines in obesity and metabolic disease. J Endocrinol 220:T47–T59

Cao H et al (2008) Identification of a lipokine, a lipid hormone linking adipose tissue to systemic metabolism. Cell 134:933–944

Centers for Disease Control and Prevention (2004) Prevalence of overweight and obesity among adults with diagnosed diabetes—United States, 1988–1994 and 1999–2002. MMWR Morb Mortal Wkly Rep 53:1066–1068

Chakravarthy MV et al (2009) Identification of a physiologically relevant endogenous ligand for PPARalpha in liver. Cell 138:476–488

Eissing L et al (2013) De novo lipogenesis in human fat and liver is linked to ChREBP-β and metabolic health. Nat Commun 4:1528

Fabbrini E et al (2011) Insulin sensitivity is not associated with palmitoleate availability in obese humans. J Lipid Res 52:808–812

Flowers MT, Ade L, Strable MS, Ntambi JM (2012) Combined deletion of SCD1 from adipose tissue and liver does not protect mice from obesity. J Lipid Res 53:1646–1653

Glass GK, Olefsky JM (2012) Inflammation and lipid signaling in the etiology of insulin resistance. Cell Metab 15:635–645

Guo X et al (2012) Palmitoleate induces hepatic steatosis but suppresses liver inflammatory response in mice. PLoS One 7, e39286

Herman MA et al (2012) A novel ChREBP isoform in adipose tissue regulates systemic glucose metabolism. Nature 484:333–338

Hinnouho GM, Czernichow S, Dugravot A, Batty GD, Kivmaki M, Singh-Manoux A (2013) Metabolically healthy obesity and risk of mortality. Diabetes Care 36:2294–2300

Iroz A, Couty J-P, Postic C (2015) Hepatokines: unlocking the multi-organ network in metabolic diseases. Diabetologia 58(8):1699–1703

Kadowaki T, Yamauchi T, Kubota N, Hara K, Ueki K, Tobe K (2006) Adiponectin and adiponectin receptors in insulin resistance, diabetes, and the metabolic syndrome. J Clin Invest 116:1784–1792

Kharitonenkov A, Shiyanova TL, Koester A, Formd AM, Micanovic R, Galbreath EJ et al (2005) FGF-21 as a novel metabolic regulator. J Clin Invest 115:1627–1635

Kramer CK, Zinman B, Retnakaran R (2013) Are metabolically healthy overweight and obesity benign conditions? Ann Intern Med 159:758–769

Kuriyama H et al (2005) Compensatory increase in fatty acid synthesis in adipose tissue of mice with conditional deficiency of SCAP in liver. Cell Metab 1:41–51

Lin Z, Tian H, Lam KSL, Lin S, Hoo RCL, Konishi M et al (2013) Adiponectin mediates the metabolic effects of FGF21 on glucose homeostasis and insulin sensitivity in mice. Cell Metab 17:779–789

Liu S et al (2013) A diurnal serum lipid integrates hepatic lipogenesis and peripheral fatty acid use. Nature 502:550–554

Miyazaki M et al (2004) Stearoyl-CoA desaturase 1 gene expression is necessary for fructose-mediated induction of lipogenic gene expression by sterol regulatory element-binding protein-1c-dependent and independent mechanisms. J Biol Chem 279:25164–25171

Miyazaki M, Flowers MT, Sampath H, Chu K, Otzelberger C, Liu X, Ntambi JM (2007) Hepatic stearoyl-CoA desaturase-1 deficiency protects mice from carbohydrate-induced adiposity and hepatic steatosis. Cell Metab 6:484–496

Moore JB (2009) Non-alcoholic fatty liver disease: the hepatic consequence of obesity and the metabolic syndrome. Proc Nutr Soc 69:211–220

Muoio DM, Newgard CB (2006) Obesity-related derangements in metabolic regulation. Annu Rev Biochem 75:367–401

Nadler ST et al (2000) The expression of adipogenic genes is decreased in obesity and diabetes mellitus. Proc Natl Acad Sci U S A 97:11371–11376

Ntambi JM et al (2002) Loss of stearoyl-CoA desaturase-1 function protects mice against adiposity. Proc Natl Acad Sci U S A 99:11482–11486

Ohashi K, Shibata R, Murohara T, Ouchi N (2014) Role of anti-inflammatory adipokines in obesity-related diseases. Trends Endocrinol Metab 25:348–355

Okada T et al (2005) Plasma palmitoleic acid content and obesity in children. Am J Clin Nutr 82:747–750

Ouchi N, Parker JL, Lugus JJ, Walsh K (2011) Adipokines in inflammation and metabolic disease. Nat Rev Immunol 11:85–97

Poirier P, Giles TD, Bray GA, Hong Y, Stern JS, Pi-Sunyer FX, Eckel RH (2006) Obesity and cardiovascular disease: pathophysiology, evaluation, and effect of weight loss: an update of the 1997 American Heart Association scientific statement on obesity and heart disease from the obesity committee of the council on nutrition, physical activity, and metabolism. Circulation 113:898–918

Stefan N, Haring H-U (2013) The role of hepatokines in metabolism. Nat Rev Endocrinol 9:144–152

Strable MS, Ntambi JM (2010) Genetic control of de novo lipogenesis: role in diet-induced obesity. Crit Rev Biochem Mol Biol 45:199–214

World Health Organization. Obesity and Overweight. Fact sheet No. 311 (2015). Retrieved from: http://www.who.int/mediacentre/factsheets/fs311/en/

Yang Z-H, Miyahara H, Hatanaka A (2011) Chronic administration of palmitoleic acid reduces insulin resistance and hepatic lipid accumulation in KK-Ay Mice with genetic type 2 diabetes. Lipids Health Dis 10:120

Chapter 13
Adipose Tissue DNL and Its Role in Metabolic Homeostasis

Ludger Scheja and Christoph Buettner

Abstract De novo lipogenesis (DNL), the synthesis of saturated and monounsaturated fatty acids from acetyl-CoA, occurs in most, if not all mammalian cell types. The liver and the adipose tissue are the major sites of DNL where excess nutrients are converted into DNL-derived fatty acids that then become the energy supply of other organs or are stored as triglycerides. DNL in the adipose tissue has been studied to a lesser extent than hepatic DNL. Similar to hepatic DNL, it appears that insulin and a systemic surplus of monosaccharides are powerful stimulants of DNL in white and brown adipose tissue. Although DNL in the liver and adipose tissues share regulatory mechanisms, adipose tissue DNL appears to be a marker of metabolic health whereas hepatic DNL often is increased in unfavorable metabolic states. Here we will discuss mechanisms orchestrating DNL in WAT and BAT and in the liver in physiology and the dysregulation of DNL in obesity and related disorders. We will also review several bioactive lipids that are products of adipose tissue DNL and their role in metabolic health.

Keywords Lipogenesis • Lipokine • Brain • Insulin • Adipose

L. Scheja (✉)
Department of Biochemistry and Molecular Cell Biology, University Medical Center Hamburg-Eppendorf, Hamburg 20246, Germany
e-mail: l.scheja@uke.uni-hamburg.de

C. Buettner
Department of Medicine and Department of Neuroscience, Diabetes, Obesity and Metabolism Institute, Icahn School of Medicine at Mount Sinai, One Gustave L. Levy Place, New York, NY 10029, USA
e-mail: christoph.buettner@mssm.edu

J.M. Ntambi (ed.), *Hepatic De Novo Lipogenesis and Regulation of Metabolism*, DOI 10.1007/978-3-319-25065-6_13

1 Role of Adipose Tissue in Lipid Storage, Synthesis and Homeostasis

Adipose tissue is generally divided into white adipose tissue (WAT) and brown adipose tissue (BAT). The principal role of WAT is to store fatty acids in the form of triglycerides (TGs) during periods of energy surplus, such as the absorptive state after a meal, and to release fatty acids during states of energy deficits such as fasting. In contrast, BAT has a limited ability to store TGs and TGs in BAT are predominantly utilized to fuel adaptive thermogenesis (Cannon and Nedergaard 2004).

DNL in WAT and BAT synthesizefatty acids from non-lipid precursors, in particular glucose (Kashiwagi et al. 1985; Minehira et al. 2003). However, DNL does not account for the majority of fatty acids that are stored in these tissues. Instead both WAT and BAT acquire most of their fatty acids from hydrolysis of circulating TG-rich lipoproteins through lipoprotein lipase (LPL) (Wang and Eckel 2009). Nevertheless, WAT is a qualitatively important sites of DNL as it may be the source for insulin sensitizing fatty acid species and may hence influence systemic metabolism (Cao et al. 2008; Yore et al. 2014).

Rates of DNL in WAT seem to be increased in beneficial metabolic states such as after cold exposure (Goubern and Portet 1986), or caloric restriction (Bruss et al. 2010) and is low in detrimental metabolic conditions such as insulin resistance, ageing or obesity (Scherer et al. 2011). Hence, DNL in adipose tissue is emerging as a metabolic pathway that plays an important role in regulating energy homeostasis and insulin action.

DNL in WAT is regulated by nutrients, such as glucose, and hormones, such as insulin consistent with its major purpose of converting excess calories into fatty acids that are stored in the form of TG. The conversion of excess nutrients into fatty acids, may serve important cellular functions in addition to providing easily storable nutrients. In the fasting state the DNL rate in WAT is low (Palou et al. 2008) while lipolysis is high, resulting in a net release of fatty acids into the circulation. To the contrary, BAT exhibits a high basal DNL rate that is further increased during cold stress (Trayhurn 1979; McCormack 1982; Bianco et al. 1998). The physiological significance of DNL induction in activated BAT is not completely understood. A plausible explanation could be that activated BAT consumes fatty acids at a very high rate and DNL is, therefore, required to preserve lipid membranes and organelles such as lipid droplets (Christoffolete et al. 2004). Hence, the regulation of DNL in BAT reflects the specific role of this organ in adaptive thermogenesis and is not as tightly linked to the nutritional state as DNL in WAT.

Adipose tissue function can be defined as the capacity of the adipose tissue to switch between fatty acid storage and release, depending on metabolic needs. This dynamic adaptation of WAT fatty acid flux warrants metabolic flexibility and is a key aspect of metabolic health. It is therefore not surprising that adipose tissue function is commonly compromised in metabolic disorders like obesity and diabetes, resulting in the spillage of fatty acids during the absorptive state when adipose tissue should take up and retain fatty acids to reduce circulating TGs and maintain lipid homeostasis (Donnelly et al. 2005; Fielding 2011).

Not all fat is equal. There are important differences between visceral and subcutaneous WAT depots. Increased visceral WAT mass is closely linked to the metabolic syndrome and insulin resistance due to impaired adipose tissue functionality (Tchernof and Despres 2013). Subcutaneous WAT can expand with less metabolic detriment as long as visceral adipose tissue mass (used here in a general sense of intraabdominal WAT (Shen et al. 2003)) is low (Kloting et al. 2010). This observation lends credence to the broadly accepted notion that in the context of our current epidemics of obesity and diabetes, subcutaneous WAT is good, while visceral WAT is bad for metabolic health (Tchernof and Despres 2013). They also point towards important physiological and cellular differences between visceral and subcutaneous WAT. One such key difference is that visceral WAT resides within the splanchnic vascular bed and drains to the portal vein (with the exception of retroperitoneal and perigonadal fat, which drains to the inferior vena cava (Shen et al. 2003)). Hence metabolites such as fatty acids and glycerol as well as cytokines released from visceral WAT enter the portal circulation and reach the liver unabated where they exert a major influence on hepatic metabolism. On the other hand, subcutaneous WAT drains into the vena cava and bypasses the liver. Consequently, only a comparatively small fraction of the fatty acid released from the subcutaneous WAT ends up in the liver while the majority of it is distributed through the circulation to peripheral organs, such as muscle that avidly uses fatty acids, in particular during exercise.

Furthermore, compared to subcutaneous adipocytes, visceral adipocytes take up and release fatty acids more rapidly, which is in part due to increased responsiveness of visceral WAT to pro-lipolytic beta-adrenergic signaling (Tchernof and Despres 2013; Richelsen et al. 1991). The advantage of this increased sensitivity to pro-lipolytic beta adrenergic signaling may be that it allows an organism to rapidly increase gluconeogenesis ensuring a steady supply of glucose to the brain which critically depends on glucose as its major energy source, and the production of VLDL by the liver that provides energy to many peripheral organs in particular BAT (Bartelt et al. 2011). The disadvantage is that in obesity increased sympathetic outflow results in unrestrained lipolysis, which is a major detriment for insulin action. These physiological and cellular differences between subcutaneous and visceral WAT may explain the preponderance of visceral WAT in unhealthy obesity and argue for a disproportionately increased influence of visceral WAT on hepatic lipid metabolism.

In sum, the balance between DNL and lipolysis in WAT and their dynamic regulation in response to nutritional availability and energetic demands are major determinants of metabolic homeostasis.

2 Dietary Carbohydrates, Fatty Acids and Insulin Regulate Adipocyte DNL

DNL in adipose tissue is a highly regulated process, which enables WAT and BAT to adapt to the nutritional state and energetic demands. Since the purpose of DNL is the conversion of non-lipids into fatty acids, one might predict that carbohydrates and amino acids induce DNL, and fatty acids to inhibit DNL in WAT. Indeed, to a large extent, that is the case. Cultured adipocytes take up glucose (Collins et al. 2011), acetate (Collins et al. 2011; Morales et al. 1994), acetoacetate (Yoo et al. 2004), glutamate (Yoo et al. 2004), glutamine (Collins et al. 2011; Yoo et al. 2004, 2008), and branched chain amino acids (Burrill et al. 2015) and use these as substrates for DNL. The extent to which substrates other than glucose are used for DNL by WAT has not been accurately determined, and only been indirectly estimated. For example, glutamate is avidly taken up by WAT suggesting that it can be a relevant DNL substrate in vivo (Lackey et al. 2013).

Glucose is not only a DNL substrate but is also a key regulator of DNL enzyme expression. This can be demonstrated in vitro in cultured white and brown rat adipocytes (Foufelle et al. 1992; Saggerson et al. 1988) and in vivo in rodents after the consumption of a high glucose diet (Granneman and Campbell 1984; Baker et al. 1981; Kim and Freake 1996). Glucose stimulates the expression of enzymes involved in DNL such as acetyl-CoA carboxylase (ACC) and fatty acid synthase through the induction of carbohydrate response element binding protein (ChREBP), a transcription factor that is post-translationally activated by glucose metabolites such as glucose-6-phosphate and xylulose-5-phosphate (Filhoulaud et al. 2013). DNL gene expression in adipose tissues of humans and mice correlates with ChREBP activity, and ChREBP knockout mice have reduced DNL enzyme expression in WAT (Herman et al. 2012; Eissing et al. 2013), indicating that ChREBP is important for the transcriptional regulation of DNL. Interestingly, neither the sterol regulatory element-binding protein-1c (SREBP1c) nor the liver X receptors (LXRα/LXRβ), transcription factors in the hepatic transcription of DNL enzymes, have been shown to exert regulatory roles in adipose tissue DNL. Global deletion of SREBP1c or both LXRs reduces expression of DNL enzymes in liver, but not in adipose tissue (Beaven et al. 2013; Shimano et al. 1997; Sekiya et al. 2007; Korach-Andre et al. 2011). In the case of SREBP1c, this could either be due to compensation by SREBP2, as observed in liver (Liang et al. 2002), or because SREBP1c has a more limited role in adipose fatty acid metabolism (Carobbio et al. 2013). The role of LXRs is even more intriguing. Knockout of both LXRs actually leads to higher DNL in WAT while reducing hepatic DNL (Beaven et al. 2013; Korach-Andre et al. 2011; Kalaany et al. 2005). Whether the phenotypes of these lifelong knockout models are accounted for by the cell autonomous versus systemic adaptive responses remains unclear.

Feeding studies indicate that DNL in WAT is only moderately suppressed by dietary fatty acids such as ω3-polyunsaturated fatty acids (Al-Hasani and Joost 2005; Shillabeer et al. 1990; Sun et al. 2011). Endogenously synthesized fatty acids may be more effective, as indicated by suppression of WAT DNL in mice with selective induction of hepatic oleic acid synthesis (Burhans et al. 2015). Mice which lack LPL in the adipose tissue, have reduced fatty acid uptake from the circulation and exhibit higher rates of DNL in WAT (Wagner et al. 2004), providing an example of negative feedback regulation of fatty acids on DNL. Consistent with the notion of negative feedback regulation of DNL by fatty acid is the finding that in humans deficient in LPL, DNL-derived fatty acid species such as palmitoleic acid are increased in WAT (Ullrich et al. 2001).

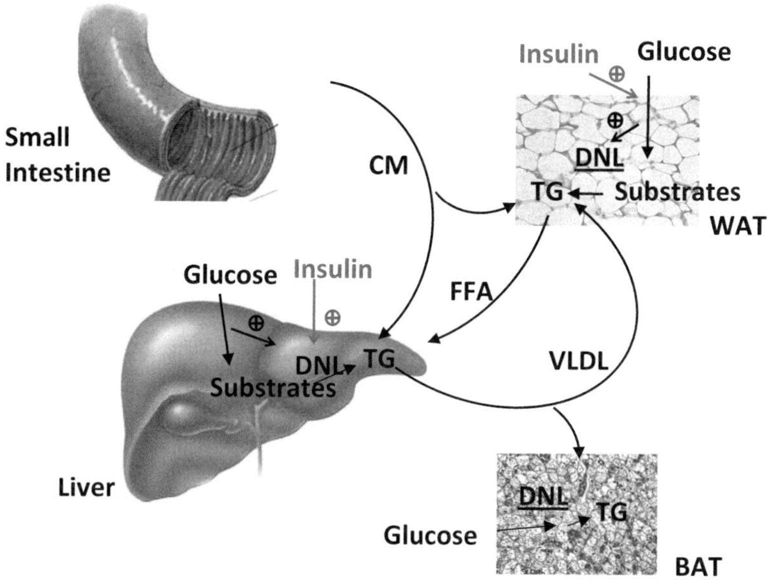

Fig. 13.1 DNL in liver and WAT contributes to a common fatty acid pool. Fatty acids are constantly exchanged between the TG pools in WAT and the liver. The majority of fatty acids in the body are derived from dietary sources and delivered to the organs as chylomicrons (CM) derived from the small intestine. Free fatty acids (FFA) released from WAT TG and not oxidized by liver, muscle or other organs are transiently stored in liver TG and subsequently released as VLDL. The DNL pathway is constitutively active in adipose tissues as well as in liver, and contributes to the respective TG pools. DNL rates are elevated when glucose is abundant and insulin levels are high. In addition to being a substrate, glucose is a major regulator of DNL, as glucose metabolites activate the DNL transcription factor ChREBP. In WAT, expression of DNL enzymes appears to be highly dependent on ChREBP. Hence, WAT DNL is tightly linked to insulin action, as this allows glucose transport through GLUT4, with subsequent ChREBP activation. BAT can take up high amounts of fatty acids and glucose when activated. It exhibits high DNL rates, possibly to maintain integrity of lipid membranes and organelles during high rates of fatty acid oxidation

Insulin is a key regulator of nutrient partitioning and is paramount to lipid storage in WAT (Figure 13.1). Insulin induces DNL in WAT by increasing glucose uptake through the translocation of the glucose transporter GLUT4 to the cell membrane of the adipocyte and enhancing glucose metabolism. An increase in intracellular glucose levels within the adipocytes affects DNL twofold. First, glucose is a key substrate for DNL and secondly, glucose activates ChREBP which induces transcription of DNL enzymes. For example, in rats subjected to a hyperinsulinemic, euglycemic clamp study (a procedure during which insulin can be raised in a controlled fashion while maintaining euglycemia) a fivefold raise in insulin levels induced fivefold increase in DNL rates in WAT (Assimacopoulos-Jeannet et al. 1995). Insulin-dependent regulation of ChREBP accounts in part for the nutritional regulation of DNL in WAT; the lack of insulin-dependent regulation of ChREBP in the insulin resistant state may possibly underpin the reduced DNL observed in obesity and diabetes (see below). Interestingly, insulin can also increase WAT DNL via signaling in the central nervous while suppressing lipolysis as was shown in rats. The latter was demonstrated to be dependent on the suppression of sympathetic outflow by CNS insulin, while the mechanisms through which brain insulin induces DNL in WAT remain only incompletely understood. Of note, while even the constitutive activation of hormone sensitive lipase was completely abrogated after denervation of the epigonadal fat pad, DNL enzyme expression was not markedly altered after denervation (Scherer et al. 2011), although the acute suppression of WAT DNL by brain leptin signaling was abrogated (Buettner et al. 2008). It is possible that DNL in WAT is repressed by even just basal sympathetic input and that brain insulin derepresses WAT DNL by acutely reducing basal sympathetic tone. Hypothalamic insulin action also improves metabolic control which in part could be due to the actions on WAT metabolism and in particular DNL. Short-term overfeeding induces systemic insulin resistance in ratshumans and rodents (Cornier et al. 2006; Brons et al. 2009), which, at least in rodent studies, is a due to impaired brain insulin action (Obici et al. 2002; Pocai et al. 2005). Because Iin human obesity brain insulin action may be impaired (Tschritter et al.2006), brain insulin resistance may be an important cause of decreased WAT DNL. In humans, intranasal insulin, a mode of administration that delivers insulin to the CNS, was also shown to suppress systemic lipolysis while it is unclear if this also increased WAT DNL in visceral WAT as observed in rats. These studies demonstrate that the CNS is an important regulator of DNL although it remains to be better studied if this occurs exclusively via the autonomic nervous system or if there are also circulating factors involved.

Growth hormone inhibits DNL in WAT, presumably via inhibition of insulin signaling (Ng et al. 1990; Rosenbaum et al. 1989). Regulation of DNL by thyroid hormone is more complex. Thyroid hormone has been found to induce DNL enzymes expression or activity in WAT in some (Mariash et al. 1980; Correze et al. 1982; Blennemann et al. 1995; Freake et al. 1989), but not all studies (Roncari and Murthy 1975). Recent evidence indicates that at least part of this induction occurs through thyroid hormone receptor-β induced ChREBP expression in WAT (Gauthier et al. 2010). One important feature of thyroid hormone is that it enhances beta-adrenergic signaling (Rubio et al. 1995; Rubio 1995), which can suppress DNL (Christoffolete et al. 2004), for example through cyclic AMP-mediated phosphorylation and inactivation of ACCα (Holland et al. 1985; Peng et al. 2012).

The role of glucocorticoids in DNL is even less well understood. Glucocorticoids induce insulin resistance (Geer et al. 2014) and hence would be expected to suppress DNL. Yet, chronic glucocorticoid administration, or endogenous glucocorticoid hypersecretion (for example in Cushing's disease) leads to abdominal obesity and glucocorticoids have been demonstrated to selectively increase DNL in visceral adipocytes, while sparing or even reducing subcutaneous fat (Harris et al. 2013; Chimin et al. 2014) (and reviewed in (Geer et al. 2014)). Taken together, it appears that hormones, neurotransmitters, nutrients and local mediators regulate DNL in WAT through modulating the action of the principal regulators of adipose tissue metabolism such as insulin and catecholamines; mechanisms that are distinct from regulation of DNL in other organs such as the liver.

3 How Does WAT DNL Affect Hepatic Lipid Metabolism?

Fatty acids are constantly exchanged between WAT and liver, thus establishing a common pool of fatty acids (Figure 13.1). Fatty acids released from WAT can be utilized by many organs and are particularly important for muscle, liver and BAT as the major energy substrate in the fasting state. Unused circulating fatty acids are re-esterified in the liver and secreted as TGs in the form of very low density lipoproteins (VLDL) into the blood stream (Donnelly et al. 2005; Barrows and Parks 2006; Vedala et al. 2006). In the absorptive state, WAT lipolysis and hepatic VLDL secretion are suppressed (Chirieac et al. 2000; Adiels et al. 2007), while lipoprotein lipase (LPL) is activated in WAT, allowing for rapid clearance of chylomicrons (Wang and Eckel 2009; Kersten 2014). This coordinated response of WAT and liver keeps circulating TGs low and results in the storage of dietary lipids in WAT.

Insulin acting on WAT plays a major role in the metabolic adaptation during the transition from fasting to feeding state by suppressing WAT lipolysis, increasing FA uptake into WAT by activating LPL, and inducing DNL. Consequently, insulin resistance in WAT leads to increased fatty acid flux to the liver, elevation of hepatic lipid content and elevated plasma TG (Adiels et al. 2007). Of note, insulin action in adipose tissue is not only a function of insulin signaling in adipocytes, but in part also occurs via the CNS (Scherer et al. 2011). A corollary of the integrated actions of insulin in orchestrating nutrient flux is that insulin action in adipose tissue in turn is a determinant of hepatic insulin action (Perry et al. 2014, 2015). Fatty acids and glycerol released from WAT are major drivers of hepatic gluconeogenesis, and suppression of lipolytic flux from adipose tissue to the liver by insulin lowers gluconeogenesis which is a pivotal aspect of hepatic insulin action.

Compared to the contribution of dietary lipids to the organismal fatty acid pool, fatty acids derived from DNL in WAT and liver make only a small contribution to the total fatty acid pool. Hepatic DNL is the source of 5 % of all palmitate (the major product of DNL) contained in the VLDL-TG pool of a metabolically healthy animal (Strable and Ntambi 2010). Even in individuals with metabolic syndrome, hepatic DNL has been found to be low in some, but not all studies (Hellerstein 1999; Beysen et al. 2008) and may contribute to up to 30 % of all VLDL-palmitate (Donnelly et al. 2005).

Since DNL enzyme activities in WAT are not higher than those in liver (Hellerstein 1999), the DNL rate in WAT probably does not exceed hepatic DNL, at least in a lean individuals, and hence it seems unlikely that DNL in WAT contributes substantially to the total organismal fatty acid pool. On the other hand, it is plausible that DNL within WAT regulates other lipid pathways. For example, DNL may block fatty acid oxidation through malonyl-CoA mediated inhibition of carnitine-palmitoyl-CoA transferase-1 (Brown et al. 1997), and lead to conservation of essential, polyunsaturated fatty acids such as alpha-linolenic acid which are otherwise readily oxidized (Vermunt et al. 2000). Furthermore, recent evidence suggests that WAT DNL produces bioactive lipids that regulate insulin sensitivity, inflammation and/or adipogenesis. This will be discussed further below.

To better understand the role of adipocyte DNL in systemic lipid homeostasis one should quantitatively assess DNL in WAT versus liver, but this is not as easily done as said. DNL rates can be assessed through isotope tracers of a substrate of DNL that provides carbons (such as glucose or acetate) or hydrogen (in the form of heavy water, D_2O) and the enrichment of these precursors in TGs in a specific organ or in plasma is then used to estimate DNL. Experimentally it is difficult to differentiate hepatic versus WAT DNL due to the rapid exchange of fatty acids between liver and WAT. Fatty acids derived from hepatic DNL are rapidly exported in the form of VLDL and are deposited in WAT via LPL-mediated uptake, and thus the majority of DNL derived fatty acids that are detected in WAT may be derived from hepatic DNL. A second technical challenge is that the total fatty acid pool in WAT is very large compared to that of the liver, and hence, de novo synthesized fatty acids are strongly diluted, which necessitates extended labeling periods to achieve sufficient enrichment in WAT. For these reasons, DNL rates in WAT have not been reliably determined, even though, based on enzyme activities, WAT has been predicted to have a DNL capacity similar to that of liver (Hellerstein 1999), and given the large adipose tissue mass in obese individuals may therefore exceed DNL rates in liver.

4 The Role of Adipose Tissue DNL on Systemic Insulin Sensitivity

In the obese state, DNL in WAT is typically suppressed (Cao et al. 2008; Eissing et al. 2013; Allister et al. 2015) whereas it is increased in the liver. At first sight, reduced DNL may seem counter intuitive since obesity by definition is associated with an increase in lipid storage in WAT. But it is noteworthy that most of the lipids that are stored in WAT are not derived from local DNL, but originate from dietary sources or hepatic DNL.

What is the mechanism underlying the suppression of WAT DNL in obesity? It is possible that diminished glucose transport into adipocytes plays an important role. Both GLUT4 protein downregulation and adipocyte insulin resistance (which further reduces GLUT4 activity by preventing its translocation to the plasma membrane) are prominent features of insulin resistant WAT in obese humans (Eissing et al. 2013) and rodents (Abel et al. 2001) and lead to a reduction in influx of the DNL substrate glucose and prevent ChREBP activation. The suppression of DNL in WAT appears thus to be tightly linked to reduced glucose metabolism and ChREBP activation, although additional mechanisms operating through SREBP1c (Sewter et al. 2002; Boden et al. 2013) and LXRs (Beaven et al. 2013) cannot be ruled out.

Alternatively, elevated hepatic DNL in obesity (Postic and Girard 2008) may suppress WAT DNL. This notion is supported by the observation that transgenic mice with suppression of DNL in the liver exhibit increased WAT DNL (Knowles et al. 2011; Kuriyama et al. 2005; Choi et al. 2010), suggesting at least a tonic suppressive effect of DNL in WAT by DNL in liver, possibly through liver-generated oleic acid (Burhans et al. 2015).

Apart from the mechanisms detailed above which suppress DNL in WAT, it is important to note that insulin resistance does also cause a defect in LPL-mediated fatty acid uptake (Wang and Eckel 2009). Thus, in the setting of deficient lipoprotein hydrolysis, suppression of WAT DNL can be interpreted as a negative feedback mechanism that prevents further WAT expansion, particularly adipocyte hypertrophy which is a hallmark of unhealthy obesity (Tchernof and Despres 2013).

Reduced fat accretion may, however, not be the only consequence of reduced WAT DNL in obese states. Emerging evidence derived from transgenic mouse models supports a role of WAT DNL as an important moderator of inflammation and an enhancer of insulin sensitivity in other organs. In one such mouse model, WAT DNL was induced by increasing glycolysis (and hence DNL (Herman et al. 2012)) through overexpression of inducible phosphofructokinase-2 (iPFK2) which resulted in increased WAT mass and improved insulin sensitivity (Huo et al. 2012). Another line of evidence is based on genetic mouse models in which the expression level of the glucose transporter GLUT4 is altered. Knockout of GLUT4 in adipose tissue using transgenic mice expressing the Cre recombinase through the Fabp4 promoter (aP2-Cre) was found to result in liver and muscle insulin resistance (Abel et al. 2001). Conversely, transgenic overexpression of Glut4 increased glucose disposal (Shepherd et al. 1993), and improved systemic insulin sensitivity in some (Carvalho et al. 2005; Tozzo et al. 1997) but not all (Gnudi et al. 1995) mouse models of insulin resistance studied. Glucose transport through GLUT4 was correlated with activation of ChREBP and hence stimulation of DNL (Herman et al. 2012). In a different transgenic model, mice lacking the fatty acid binding proteins FABP4 and FABP5, adipocytes do not exhibit reduced DNL after high fat diet feeding and the mice remain insulin sensitive (Cao et al. 2008). The observation that experimental elevation of adipose tissue DNL in these models is associated with insulin sensitivity, led to the hypothesis that DNL-derived lipids may confer systemic insulin sensitization (Fig. 13.2).

Two types of adipose tissue DNL-derived "insulin sensitizers" have been identified, palmitoleic acid (C16:1n-7) (Cao et al. 2008) and a branched fatty acid esters of hydroxy fatty acids (FAHFAs) (Yore et al. 2014). C16:1n-7 content was found to be decreased in WAT and plasma of diet-induced obese mice (Cao et al. 2008). The insulin sensitizing and anti-inflammatory effects of C16:1n-7 free fatty acid (FFA) have been demonstrated in vitro in cultured myocytes (Cao et al. 2008; Dimopoulos et al. 2006), and in various cell culture models (Maedler et al. 2003; Talbot et al. 2014), respectively. In vivo also C16:1n-7 dietary supplementation was found to improve whole body insulin sensitivity and exhibited anti-inflammatory properties in mice (Yang et al. 2011; Guo et al. 2012). The cellular mechanisms mediating the anti-inflammatory and insulin sensitizing effects of C16:1n-7 have not been identified so far. It is unknown whether C16:1n-7 needs to be converted to complex lipid mediators (Koeberle et al. 2012) or whether it simply counteracts pro-inflammatory and toxic effects of saturated fatty acids such as palmitate (C16:0) (Talbot et al. 2014; Welters et al. 2004) to achieve its beneficial effects. Interestingly, promoting C16:1n7 elongation to vaccenic acid (C18:1n-7) in the liver by overexpression of fatty acid elongase-5 (ELOVL5) boosts hepatic insulin signaling (Tripathy and

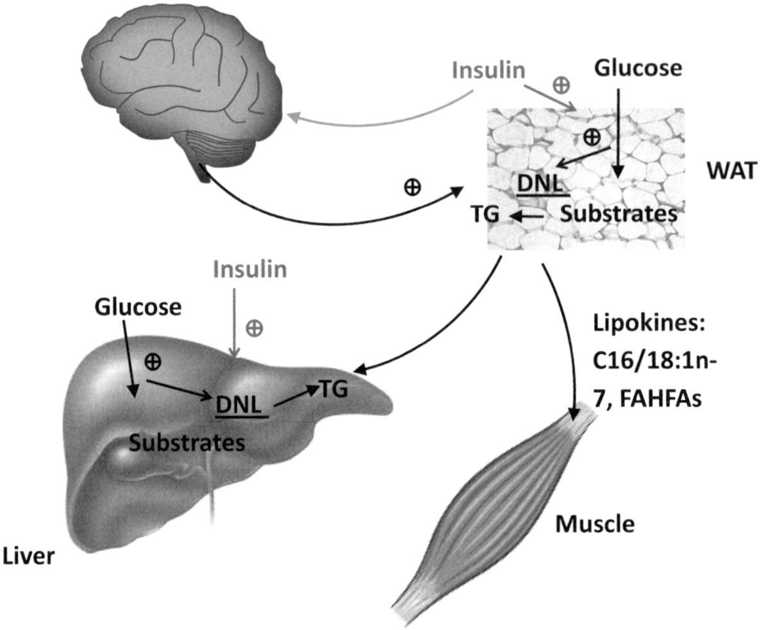

Fig. 13.2 Lipokine hypothesis: WAT DNL confers systemic insulin sensitivity. Recent mouse studies suggest that DNL in WAT regulates systemic metabolism. In the metabolically healthy state, sufficient glucose uptake into adipocytes and intact hypothalamic insulin action result in high DNL rates in WAT which favors the production of systemically acting insulin-sensitizing and anti-inflammatory lipid mediators (lipokines). In obesity and insulin resistance, glucose transport into WAT and hypothalamic insulin action is impaired leading to reduced DNL and diminished lipokine production. This contributes to systemic insulin resistance and metabolic inflammation, together with other WAT-related mechanisms such as increased release of saturated fatty acids and inflammatory cytokines

Jump 2013) and relieves hepatic endoplasmic reticulum stress (Tripathy et al. 2014). It is thus conceivable that conversion of C16:1n-7-FFA derived from WAT to C18:1n-7 in the liver is part of the postulated beneficial effect of WAT-derived C16:1n-7. C16:1n-7 FFA is readily released from subcutaneous adipose tissue in humans (Halliwell et al. 1996), and two studies found a significant association of C16:1n-7 FFA with insulin sensitivity (Stefan et al. 2010; Pinnick et al. 2012), consistent with C16:1n-7 being an insulin-sensitizing lipokine in humans. However, the effect size of these associations was moderate, and could not be confirmed in another study (Fabbrini et al. 2011). Similarly, WAT C16:1n-7 was found to be decreased in WAT of morbidly obese subjects in one study (Garaulet et al. 2011) but not in another (Eissing et al. 2013). Furthermore, hepatic C16:1n-7, as assessed by C16:1n-7 lipoprotein content, is increased in obesity and insulin resistance (Mozaffarian et al. 2010; Paillard et al. 2008; Vessby et al. 1994), probably because hepatic DNL is induced. This indicates that there is no general C16:1n-7 deficiency in human obesity and metabolic syndrome. Taken together, much needs to be learned about the regulatory role and mechanisms of C16:1n-7 in mice and humans (Fig. 13.2).

FAHFAs are adipose tissue-derived signaling molecules that have been identified only very recently (Yore et al. 2014). Little is known about the route of synthesis of FAHFAs. FAHFAs are, however, strongly dependent on glucose-induced DNL, as their concentration is markedly increased in mice with adipose overexpression of GLUT4 and decreased in mice lacking ChREBP (Yore et al. 2014). They constitute a novel lipid class built from a) a DNL-derived fatty acid: C16:0, C16:1n-7, stearic acid (C18:0), and oleic acid (C18:1n-9), esterified to b) a hydroxylated version of the same DNL-derived fatty acids. Although a high number of positional isomers exists, as the hydroxy group can be located at various C-atoms, only a limited number of DNL-dependent. FAHFA species have been identified in adipose tissue. A subset of those, positional isomers of palmitic-acid-hydroxy-stearic acid (PAHSA), have been studied in detail. PAHSA concentration in serum and WAT were found to tightly correlate with insulin sensitivity in both mice and humans. Further, PAHSAs were found to confer a host of metabolic effects, including enhanced secretion of glucagon-like peptide-1, increased adipocyte glucose transport in vitro, reduced inflammation in cultured dendritic cells and adipose tissue and acutely increased insulin sensitivity after oral gavage in mice (Yore et al. 2014). Hence, FAHFAs are strong candidates of a WAT DNL-derived lipid that exerts insulin sensitizing and anti-inflammatory effects. However, the molecular mechanisms that account for these pleiotropic effects need to be better understood. In particular it will be important to identify putative FAHFA receptors, and to differentiate the functions of C16:1n-7-containing FAHFA which were not studied in detail versus other FAHFAs, in order to understand the involvement of FAHFAs in C16:1n-7 insulin-sensitizing effects. In summary, while the concept that DNL-derived lipid mediators such as C16:1n-7 and FAHFAs modulate systemic insulin sensitivity and inflammation by acting on other organs is conceptually attractive, future studies will have to further test this hypothesis.

In addition to being systemically active, lipokines may also have local effects in WAT or BAT through modulating the secretion of pro- and anti-inflammatory peptide mediators (adipokines) which in turn can influence metabolism in other organs. For example, adiponectin, an anti-inflammatory adipokine secreted by WAT has been demonstrated in mouse studies to suppress hepatic DNL through AMP-dependent kinase signaling (Ma and Liu 2013). Conversely, classical, pro-inflammatory cytokines such as tumor necrosis factor-α and Interleukin-1α can increase hepatic DNL (Negrin et al. 2014; Grunfeld et al. 1990). It is thus conceivable that WAT DNL-derived lipid mediators influence liver DNL and other aspects of metabolism indirectly through modulation of adipokines, especially those that derived from visceral WAT. One recently described example is Nrg4, which is produced in WAT and BAT, that diminishes hepatic DNL through ErbB3/ErbB4 signaling (Wang et al. 2014). If Nrg4 production controls adipose tissue DNL adipose tissue DNL, as suggested by the observation that its induction coincides with increased DNL after cold induction, needs to be shown.

In conclusion, strong experimental evidence points towards a causal connection of DNL in WAT and BAT to metabolism in liver and other organs, which is beyond mere shifts in fatty acid fluxes and fuel metabolism. It appears that a multitude of paracrine and endocrine factors are dependent on DNL produced in adipose tissue. Future research will need to be directed at dissecting the regulatory effects of these lipid and peptide mediators. This knowledge will help our understanding of the role of WAT and BAT DNL in metabolic disease, and may lead to novel therapeutic approaches to the treatment or the prevention of obesity, type 2 diabetes and other diseases linked to the metabolic syndrome.

Acknowledgment This study was supported by ADA 7-11-CD-02, NIH grants DK AA023416 and DK0836581 to C.B. and to L.S.

The authors declare that no competing financial interests exist.

References

Abel ED et al (2001) Adipose-selective targeting of the GLUT4 gene impairs insulin action in muscle and liver. Nature 409:729–733

Adiels M et al (2007) Acute suppression of VLDL1 secretion rate by insulin is associated with hepatic fat content and insulin resistance. Diabetologia 50:2356–2365

Al-Hasani H, Joost HG (2005) Nutrition-/diet-induced changes in gene expression in white adipose tissue. Best Pract Res Clin Endocrinol Metab 19:589–603

Allister CA et al (2015) In vivo $2H_2O$ administration reveals impaired triglyceride storage in adipose tissue of insulin-resistant humans. J Lipid Res 56:435–439

Assimacopoulos-Jeannet F, Brichard S, Rencurel F, Cusin I, Jeanrenaud B (1995) In vivo effects of hyperinsulinemia on lipogenic enzymes and glucose transporter expression in rat liver and adipose tissues. Metabolism 44:228–233

Baker N, Mead J Jr, Kannan R (1981) Hepatic contribution to newly made fatty acids in adipose tissue in rats and inhibition of hepatic and extrahepatic lipogenesis from glucose by dietary corn oil. Lipids 16:568–576

Barrows BR, Parks EJ (2006) Contributions of different fatty acid sources to very low-density lipoprotein-triacylglycerol in the fasted and fed states. J Clin Endocrinol Metab 91:1446–1452

Bartelt A et al (2011) Brown adipose tissue activity controls triglyceride clearance. Nat Med 17:200–205

Beaven SW et al (2013) Reciprocal regulation of hepatic and adipose lipogenesis by liver X receptors in obesity and insulin resistance. Cell Metab 18:106–117

Beysen C et al (2008) A pilot study of the effects of pioglitazone and rosiglitazone on de novo lipogenesis in type 2 diabetes. J Lipid Res 49:2657–2663

Bianco AC, Carvalho SD, Carvalho CR, Rabelo R, Moriscot AS (1998) Thyroxine 5′-deiodination mediates norepinephrine-induced lipogenesis in dispersed brown adipocytes. Endocrinology 139:571–578

Blennemann B, Leahy P, Kim TS, Freake HC (1995) Tissue-specific regulation of lipogenic mRNAs by thyroid hormone. Mol Cell Endocrinol 110:1–8

Boden G et al (2013) Comparison of in vivo effects of insulin on SREBP-1c activation and INSIG-1/2 in rat liver and human and rat adipose tissue. Obesity 21:1208–1214

Brons C et al (2009) Impact of short-term high-fat feeding on glucose and insulin metabolism in young healthy men. J Physiol 587:2387–2397

Brown NF et al (1997) Mouse white adipocytes and 3 T3-L1 cells display an anomalous pattern of carnitine palmitoyltransferase (CPT) I isoform expression during differentiation. Inter-tissue and inter-species expression of CPT I and CPT II enzymes. Biochem J 327(Pt 1):225–231

Bruss MD, Khambatta CF, Ruby MA, Aggarwal I, Hellerstein MK (2010) Calorie restriction increases fatty acid synthesis and whole body fat oxidation rates. Am J Physiol Endocrinol Metab 298:E108–E116

Burhans MS et al (2015) Hepatic oleate regulates adipose tissue lipogenesis and fatty acid oxidation. J Lipid Res 56:304–318

Burrill JS et al (2015) Inflammation and ER stress regulate branched-chain amino acid uptake and metabolism in adipocytes. Mol Endocrinol 29:411–420

Cannon B, Nedergaard J (2004) Brown adipose tissue: function and physiological significance. Physiol Rev 84:277–359

Cao H et al (2008) Identification of a lipokine, a lipid hormone linking adipose tissue to systemic metabolism. Cell 134:933–944

Carobbio S et al (2013) Adaptive changes of the Insig1/SREBP1/SCD1 set point help adipose tissue to cope with increased storage demands of obesity. Diabetes 62:3697–3708

Carvalho E, Kotani K, Peroni OD, Kahn BB (2005) Adipose-specific overexpression of GLUT4 reverses insulin resistance and diabetes in mice lacking GLUT4 selectively in muscle. Am J Physiol Endocrinol Metab 289:E551–E561

Chimin P et al (2014) Chronic glucocorticoid treatment enhances lipogenic activity in visceral adipocytes of male Wistar rats. Acta Physiol (Oxf) 211:409–420

Chirieac DV et al (2000) Glucose-stimulated insulin secretion suppresses hepatic triglyceride-rich lipoprotein and apoB production. Am J Physiol Endocrinol Metab 279:E1003–E1011

Choi CS et al (2010) Liver-specific pyruvate dehydrogenase complex deficiency upregulates lipogenesis in adipose tissue and improves peripheral insulin sensitivity. Lipids 45:987–995

Christoffolete MA et al (2004) Mice with targeted disruption of the Dio2 gene have cold-induced overexpression of the uncoupling protein 1 gene but fail to increase brown adipose tissue lipogenesis and adaptive thermogenesis. Diabetes 53:577–584

Collins JM et al (2011) De novo lipogenesis in the differentiating human adipocyte can provide all fatty acids necessary for maturation. J Lipid Res 52:1683–1692

Cornier MA, Bergman BC, Bessesen DH (2006) The effects of short-term overfeeding on insulin action in lean and reduced-obese individuals. Metabolism 55:1207–1214

Correze C, Berriche S, Tamayo L, Nunez J (1982) Effect of thyroid hormones and cyclic AMP on some lipogenic enzymes of the fat cell. Eur J Biochem 122:387–392

Dimopoulos N, Watson M, Sakamoto K, Hundal HS (2006) Differential effects of palmitate and palmitoleate on insulin action and glucose utilization in rat L6 skeletal muscle cells. Biochem J 399:473–481

Donnelly KL et al (2005) Sources of fatty acids stored in liver and secreted via lipoproteins in patients with nonalcoholic fatty liver disease. J Clin Invest 115:1343–1351

Eissing L et al (2013) De novo lipogenesis in human fat and liver is linked to ChREBP-beta and metabolic health. Nat Commun 4:1528

Fabbrini E et al (2011) Insulin sensitivity is not associated with palmitoleate availability in obese humans. J Lipid Res 52:808–812

Fielding B (2011) Tracing the fate of dietary fatty acids: metabolic studies of postprandial lipaemia in human subjects. Proc Nutr Soc 70:342–350

Filhoulaud G, Guilmeau S, Dentin R, Girard J, Postic C (2013) Novel insights into ChREBP regulation and function. Trends Endocrinol Metab 24:257–268

Foufelle F et al (1992) Glucose stimulation of lipogenic enzyme gene expression in cultured white adipose tissue. A role for glucose 6-phosphate. J Biol Chem 267:20543–20546

Freake HC, Schwartz HL, Oppenheimer JH (1989) The regulation of lipogenesis by thyroid hormone and its contribution to thermogenesis. Endocrinology 125:2868–2874

Garaulet M, Hernandez-Morante JJ, Tebar FJ, Zamora S (2011) Relation between degree of obesity and site-specific adipose tissue fatty acid composition in a Mediterranean population. Nutrition 27:170–176

Gauthier K et al (2010) Thyroid hormone receptor beta (TRbeta) and liver X receptor (LXR) regulate carbohydrate-response element-binding protein (ChREBP) expression in a tissue-selective manner. J Biol Chem 285:28156–28163
Geer EB, Islam J, Buettner C (2014) Mechanisms of glucocorticoid-induced insulin resistance: focus on adipose tissue function and lipid metabolism. Endocrinol Metab Clin North Am 43:75–102
Gnudi L, Tozzo E, Shepherd PR, Bliss JL, Kahn BB (1995) High level overexpression of glucose transporter-4 driven by an adipose-specific promoter is maintained in transgenic mice on a high fat diet, but does not prevent impaired glucose tolerance. Endocrinology 136:995–1002
Goubern M, Portet R (1986) Modulation of fatty acid synthesis in vivo in brown adipose tissue, liver and white adipose tissue of cold-acclimated rats. Ann Nutr Metab 30:380–385
Granneman JG, Campbell RG (1984) Effects of sucrose feeding and denervation on lipogenesis in brown adipose tissue. Metabolism 33:257–261
Grunfeld C et al (1990) Evidence for two classes of cytokines that stimulate hepatic lipogenesis: relationships among tumor necrosis factor, interleukin-1 and interferon-alpha. Endocrinology 127:46–54
Guo X et al (2012) Palmitoleate induces hepatic steatosis but suppresses liver inflammatory response in mice. PLoS One 7, e39286
Halliwell KJ, Fielding BA, Samra JS, Humphreys SM, Frayn KN (1996) Release of individual fatty acids from human adipose tissue in vivo after an overnight fast. J Lipid Res 37:1842–1848
Harris C et al (2013) Large increases in adipose triacylglycerol flux in Cushingoid CRH-Tg mice are explained by futile cycling. Am J Physiol Endocrinol Metab 304:E282–E293
Hellerstein MK (1999) De novo lipogenesis in humans: metabolic and regulatory aspects. Eur J Clin Nutr 53(Suppl 1):S53–S65
Herman MA et al (2012) A novel ChREBP isoform in adipose tissue regulates systemic glucose metabolism. Nature 484:333–338
Holland R, Hardie DG, Clegg RA, Zammit VA (1985) Evidence that glucagon-mediated inhibition of acetyl-CoA carboxylase in isolated adipocytes involves increased phosphorylation of the enzyme by cyclic AMP-dependent protein kinase. Biochem J 226:139–145
Huo Y et al (2012) Targeted overexpression of inducible 6-phosphofructo-2-kinase in adipose tissue increases fat deposition but protects against diet-induced insulin resistance and inflammatory responses. J Biol Chem 287:21492–21500
Kalaany NY et al (2005) LXRs regulate the balance between fat storage and oxidation. Cell Metab 1:231–244
Kashiwagi A et al (1985) The effects of short-term overfeeding on adipocyte metabolism in Pima Indians. Metabolism 34:364–370
Kersten S (2014) Physiological regulation of lipoprotein lipase. Biochim Biophys Acta 1841:919–933
Kim TS, Freake HC (1996) High carbohydrate diet and starvation regulate lipogenic mRNA in rats in a tissue-specific manner. J Nutr 126:611–617
Kloting N et al (2010) Insulin-sensitive obesity. Am J Physiol Endocrinol Metab 299:E506–E515
Knowles C, Liu ZM, Yang J (2011) Compensatory increase in lipogenic gene expression in adipose tissue of transgenic mice expressing constitutively active AMP-activated protein kinase-alpha1 in liver. Biochem Biophys Res Commun 412:249–252
Koeberle A, Shindou H, Harayama T, Shimizu T (2012) Palmitoleate is a mitogen, formed upon stimulation with growth factors, and converted to palmitoleoyl-phosphatidylinositol. J Biol Chem 287:27244–27254
Korach-Andre M et al (2011) Liver X receptors regulate de novo lipogenesis in a tissue-specific manner in C57BL/6 female mice. Am J Physiol Endocrinol Metab 301:E210–E222
Kuriyama H et al (2005) Compensatory increase in fatty acid synthesis in adipose tissue of mice with conditional deficiency of SCAP in liver. Cell Metab 1:41–51
Lackey DE et al (2013) Regulation of adipose branched-chain amino acid catabolism enzyme expression and cross-adipose amino acid flux in human obesity. Am J Physiol Endocrinol Metab 304:E1175–E1187

Liang G et al (2002) Diminished hepatic response to fasting/refeeding and liver X receptor agonists in mice with selective deficiency of sterol regulatory element-binding protein-1c. J Biol Chem 277:9520–9528
Ma Y, Liu D (2013) Hydrodynamic delivery of adiponectin and adiponectin receptor 2 gene blocks high-fat diet-induced obesity and insulin resistance. Gene Ther 20:846–852
Maedler K, Oberholzer J, Bucher P, Spinas GA, Donath MY (2003) Monounsaturated fatty acids prevent the deleterious effects of palmitate and high glucose on human pancreatic beta-cell turnover and function. Diabetes 52:726–733
Mariash CN, Kaiser FE, Schwartz HL, Towle HC, Oppenheimer JH (1980) Synergism of thyroid hormone and high carbohydrate diet in the induction of lipogenic enzymes in the rat. Mechanisms and implications. J Clin Invest 65:1126–1134
McCormack JG (1982) The regulation of fatty acid synthesis in brown adipose tissue by insulin. Prog Lipid Res 21:195–223
Minehira K et al (2003) Effect of carbohydrate overfeeding on whole body and adipose tissue metabolism in humans. Obes Res 11:1096–1103
Morales LM, Campos G, Ryder E, Casanova A (1994) Insulin and lipogenesis in rat adipocytes. II. Effect of high levels of insulin and glucose on lipid synthesis in isolated rat adipocytes. Biochem Biophys Res Commun 199:1151–1157
Mozaffarian D et al (2010) Circulating palmitoleic acid and risk of metabolic abnormalities and new-onset diabetes. Am J Clin Nutr 92:1350–1358
Negrin KA et al (2014) IL-1 signaling in obesity-induced hepatic lipogenesis and steatosis. PLoS One 9, e107265
Ng FM, Adamafio NA, Graystone JE (1990) Effects of exogenous growth hormone on lipid metabolism in the isolated epididymal fat pad of the growth hormone-deficient little mouse. J Mol Endocrinol 4:43–49
Obici S, Zhang BB, Karkanias G, Rossetti L (2002) Hypothalamic insulin signaling is required for inhibition of glucose production. Nat Med 8:1376–1382
Paillard F et al (2008) Plasma palmitoleic acid, a product of stearoyl-coA desaturase activity, is an independent marker of triglyceridemia and abdominal adiposity. Nutr Metab Cardiovasc Dis 18:436–440
Palou M et al (2008) Sequential changes in the expression of genes involved in lipid metabolism in adipose tissue and liver in response to fasting. Pflugers Arch 456:825–836
Peng IC et al (2012) Glucagon regulates ACC activity in adipocytes through the CAMKKbeta/AMPK pathway. Am J Physiol Endocrinol Metab 302:E1560–E1568
Perry RJ, Samuel VT, Petersen KF, Shulman GI (2014) The role of hepatic lipids in hepatic insulin resistance and type 2 diabetes. Nature 510:84–91
Perry RJ et al (2015) Hepatic acetyl CoA links adipose tissue inflammation to hepatic insulin resistance and type 2 diabetes. Cell 160:745–758
Pinnick KE et al (2012) Gluteofemoral adipose tissue plays a major role in production of the lipokine palmitoleate in humans. Diabetes 61:1399–1403
Pocai A et al (2005) Hypothalamic K(ATP) channels control hepatic glucose production. Nature 434:1026–1031
Postic C, Girard J (2008) Contribution of de novo fatty acid synthesis to hepatic steatosis and insulin resistance: lessons from genetically engineered mice. J Clin Invest 118:829–838
Richelsen B, Pedersen SB, Moller-Pedersen T, Bak JF (1991) Regional differences in triglyceride breakdown in human adipose tissue: effects of catecholamines, insulin, and prostaglandin E2. Metabolism 40:990–996
Roncari DA, Murthy VK (1975) Effects of thyroid hormones on enzymes involved in fatty acid and glycerolipid synthesis. J Biol Chem 250:4134–4138
Rosenbaum M, Gertner JM, Leibel RL (1989) Effects of systemic growth hormone (GH) administration on regional adipose tissue distribution and metabolism in GH-deficient children. J Clin Endocrinol Metab 69:1274–1281
Rubio A (1995) Raasmaja, A. and Silva, J.E. Thyroid hormone and norepinephrine signaling in brown adipose tissue. II: Differential effects of thyroid hormone on beta 3-adrenergic receptors in brown and white adipose tissue. Endocrinology 136:3277–3284

Rubio A, Raasmaja A, Maia AL (1995) Kim, K.R. and Silva, J.E. Effects of thyroid hormone on norepinephrine signaling in brown adipose tissue. I. Beta 1- and beta 2-adrenergic receptors and cyclic adenosine 3′,5′-monophosphate generation. Endocrinology 136:3267–3276
Saggerson ED, McAllister TW, Baht HS (1988) Lipogenesis in rat brown adipocytes. Effects of insulin and noradrenaline, contributions from glucose and lactate as precursors and comparisons with white adipocytes. Biochem J 251:701–709
Scherer T et al (2011) Brain insulin controls adipose tissue lipolysis and lipogenesis. Cell Metab 13:183–194
Sekiya M et al (2007) SREBP-1-independent regulation of lipogenic gene expression in adipocytes. J Lipid Res 48:1581–1591
Sewter C et al (2002) Human obesity and type 2 diabetes are associated with alterations in SREBP1 isoform expression that are reproduced ex vivo by tumor necrosis factor-alpha. Diabetes 51:1035–1041
Shen W et al (2003) Adipose tissue quantification by imaging methods: a proposed classification. Obes Res 11:5–16
Shepherd PR et al (1993) Adipose cell hyperplasia and enhanced glucose disposal in transgenic mice overexpressing GLUT4 selectively in adipose tissue. J Biol Chem 268:22243–22246
Shillabeer G, Hornford J, Forden JM, Wong NC, Lau DC (1990) Hepatic and adipose tissue lipogenic enzyme mRNA levels are suppressed by high fat diets in the rat. J Lipid Res 31:623–631
Shimano H et al (1997) Elevated levels of SREBP-2 and cholesterol synthesis in livers of mice homozygous for a targeted disruption of the SREBP-1 gene. J Clin Invest 100:2115–2124
Stefan N et al (2010) Circulating palmitoleate strongly and independently predicts insulin sensitivity in humans. Diabetes Care 33:405–407
Strable MS, Ntambi JM (2010) Genetic control of de novo lipogenesis: role in diet-induced obesity. Crit Rev Biochem Mol Biol 45:199–214
Sun C, Wei ZW, Li Y (2011) DHA regulates lipogenesis and lipolysis genes in mice adipose and liver. Mol Biol Rep 38:731–737
Talbot NA, Wheeler-Jones CP, Cleasby ME (2014) Palmitoleic acid prevents palmitic acid-induced macrophage activation and consequent p38 MAPK-mediated skeletal muscle insulin resistance. Mol Cell Endocrinol 393:129–142
Tchernof A, Despres JP (2013) Pathophysiology of human visceral obesity: an update. Physiol Rev 93:359–404
Tozzo E, Gnudi L, Kahn BB (1997) Amelioration of insulin resistance in streptozotocin diabetic mice by transgenic overexpression of GLUT4 driven by an adipose-specific promoter. Endocrinology 138:1604–1611
Trayhurn P (1979) Fatty acid synthesis in vivo in brown adipose tissue, liver and white adipose tissue of the cold-acclimated rat. FEBS Lett 104:13–16
Tripathy S, Jump DB (2013) Elovl5 regulates the mTORC2-Akt-FOXO1 pathway by controlling hepatic cis-vaccenic acid synthesis in diet-induced obese mice. J Lipid Res 54:71–84
Tripathy S et al (2014) Fatty acid elongase-5 (Elovl5) regulates hepatic triglyceride catabolism in obese C57BL/6 J mice. J Lipid Res 55:1448–1464
Tschritter O et al (2006) The cerebrocortical response to hyperinsulinemia is reduced in overweight humans: a magnetoencephalographic study. Proc Natl Acad Sci U S A 103:12103–12108
Ullrich NF, Purnell JQ, Brunzell JD (2001) Adipose tissue fatty acid composition in humans with lipoprotein lipase deficiency. J Invest Med 49:273–275
Vedala A, Wang W, Neese RA, Christiansen MP, Hellerstein MK (2006) Delayed secretory pathway contributions to VLDL-triglycerides from plasma NEFA, diet, and de novo lipogenesis in humans. J Lipid Res 47:2562–2574
Vermunt SH, Mensink RP, Simonis MM, Hornstra G (2000) Effects of dietary alpha-linolenic acid on the conversion and oxidation of 13C-alpha-linolenic acid. Lipids 35:137–142
Vessby B, Tengblad S, Lithell H (1994) Insulin sensitivity is related to the fatty acid composition of serum lipids and skeletal muscle phospholipids in 70-year-old men. Diabetologia 37:1044–1050

Wagner EM et al (2004) Defective uptake of triglyceride-associated fatty acids in adipose tissue causes the SREBP-1c-mediated induction of lipogenesis. J Lipid Res 45:356–365
Wang H, Eckel RH (2009) Lipoprotein lipase: from gene to obesity. Am J Physiol Endocrinol Metab 297:E271–E288
Wang GX et al (2014) The brown fat-enriched secreted factor Nrg4 preserves metabolic homeostasis through attenuation of hepatic lipogenesis. Nat Med 20:1436–1443
Welters HJ, Tadayyon M, Scarpello JH, Smith SA, Morgan NG (2004) Mono-unsaturated fatty acids protect against beta-cell apoptosis induced by saturated fatty acids, serum withdrawal or cytokine exposure. FEBS Lett 560:103–108
Yang ZH, Miyahara H, Hatanaka A (2011) Chronic administration of palmitoleic acid reduces insulin resistance and hepatic lipid accumulation in KK-Ay Mice with genetic type 2 diabetes. Lipids Health Dis 10:120
Yoo H, Stephanopoulos G, Kelleher JK (2004) Quantifying carbon sources for de novo lipogenesis in wild-type and IRS-1 knockout brown adipocytes. J Lipid Res 45:1324–1332
Yoo H, Antoniewicz MR, Stephanopoulos G, Kelleher JK (2008) Quantifying reductive carboxylation flux of glutamine to lipid in a brown adipocyte cell line. J Biol Chem 283:20621–20627
Yore MM et al (2014) Discovery of a class of endogenous mammalian lipids with anti-diabetic and anti-inflammatory effects. Cell 159:318–332

Chapter 14
The Lipogenic Effect of Insulin Revisited

Sonia M. Najjar

Abstract Insulin promotes lipogenesis and suppresses hepatic gluconeogenesis. Paradoxically, the positive effect of insulin on lipid production in liver is maintained under conditions of insulin resistance, unlike its defective action on gluconeogenesis. In this chapter, we revisit this gluconeogenesis-selectivity notion by reviewing evidence that in fact, under normo-insulinemic conditions, insulin acutely reduces de novo lipogenesis in liver. This is mediated by the ability of insulin pulses in portal vein to phosphorylate the Carcino-Embryonic Antigen-related Cell Adhesion Molecule 1 (CEACAM1), an abundant protein in liver that promotes hepatic insulin clearance in a phosphorylation-dependent manner. In the presence of sustained hyperinsulinemia, pulsatility of insulin release diminishes together with its signaling, giving way to its chronic positive effect on lipogenic genes transcription. This would set the stage to redefine hepatic insulin resistance as a comprehensive process that includes defective insulin action in glucose as well as lipid production.

Keywords De novo Lipogenesis • Hepatic insulin clearance • Hyperinsulinemia • Acute insulin effect • Chronic insulin effect • Insulin resistance • Non-alcoholic fatty liver disease (NAFLD)

1 Introduction

Metabolic syndrome, characterized by insulin resistance (Takahashi and Fukusato 2010), is a constellation of metabolic abnormalities that comprise visceral abdominal obesity, dyslipidemia and hypertension (Eckel et al. 2010). The disease is increasingly associated with non-alcoholic fatty liver disease

S.M. Najjar, Ph.D. (✉)
Center for Diabetes and Endocrine Research, University of Toledo College of Medicine and Life Sciences, Health Science Campus, 3000 Arlington Ave., Mail Stop 1009, Block Health Science Bldg., CeDER, Toledo, OH 43614-2598, USA

Department of Physiology and Pharmacology, University of Toledo College of Medicine and Life Sciences, Toledo, OH USA
e-mail: Sonia.Najjar@utoledo.edu

J.M. Ntambi (ed.), *Hepatic De Novo Lipogenesis and Regulation of Metabolism*,
DOI 10.1007/978-3-319-25065-6_14

(NAFLD), a rising epidemic that affects most patients with type 2 diabetes and insulin resistance (Adams et al. 2005; Farrell and Larter 2006; Marchesini et al. 2003). This tight association has provided the impetus to identify NAFLD as the hepatic manifestation of metabolic syndrome (Marchesini et al. 2003; Marra et al. 2008).

In this chapter, we will revisit the mechanistic basis of fat accumulation in liver and its link to insulin resistance. We recognize that several mechanisms are involved, in particular the adverse effect of obesity-associated activation of pro-inflammatory pathways in blunting insulin action, as originally proposed by Hotamisligil et al. (Hotamisligil et al. 1993). Because the interaction between inflammation and insulin resistance has been extensively reviewed (Ferrante 2007; Gregor and Hotamisligil 2011; Lumeng and Saltiel 2011; Najjar and Russo 2014; Odegaard and Chawla 2013; Osborn and Olefsky 2012; Saltiel 2012; Shoelson et al. 2006), we will limit our current discussion to the role of hepatic lipid accumulation in overall insulin action. In particular, we will use our findings concerning the regulatory role of carcinoembryonic antigen-related cell adhesion molecule 1 (CEACAM1)-dependent pathways in hepatic lipid and insulin metabolism as a tool to shed novel insight on the molecular mechanisms linking insulin resistance to altered lipid homeostasis in metabolic syndrome.

2 CEACAM1 and Its Role in Insulin Metabolism

2.1 CEACAM1: Gene and Protein

CEACAM1 (previously called biliary glycoprotein [BGP], C-CAM1, CD66a or pp120/HA4/ectoATPase) is a type I membrane glycoprotein with a single transmembrane domain (Beauchemin et al. 1999). It is highly conserved (Zebhauser et al. 2005) and is ubiquitously expressed in all cell types (Öbrink 1997). In liver, CEACAM1 is very abundant, in particular in hepatocytes, but it is not detected at the protein level in skeletal muscle or white adipose tissue (Najjar 2002).

Najjar et al. (Najjar et al. 1993) initially reported that *Ceacam1* gene consists of nine exons, the seventh of which is alternative spliced to yield a long (71–73 a.a in the rat) or a short (10–12 a.a) intracellular tail that differ by the presence or absence of tyrosine and serine phosphorylation sites in their cytoplasmic tail. The full-length protein consists of a membrane-distal Ig variable domain (IgV)-like amino (N)-terminal domain region that is highly conserved among the carcinoembryonic antigen family of proteins, followed by 1–3 membrane-proximal C2-type Ig constant domain-related sequences. Also highly conserved, is the cytoplasmic tail of the long isoform of CEACAM proteins. Because it contains tyrosine phosphorylation sites that harbor CEACAM1 regulatory effect on insulin and lipid metabolism, we will focus this review on the function of the long isoform.

2.2 *CEACAM1: A Key Role in Hepatic Insulin Clearance*

Insulin is released from pancreatic β-cells into the portal vein in a pulsatile fashion (Matveyenko et al. 2012), and is rapidly cleared in the hepatocyte (circulating half-life of insulin is ~3–5 min, and ~50 % insulin is cleared during its first pass through the liver). Acute pulses of insulin reaching the hepatocyte trigger insulin binding to its receptor, followed by its activation and simultaneous receptor-mediated intracellular uptake into clathrin-coated vesicles (Carpentier 1994). Subsequently, insulin dissociates from its receptor in the endosomal compartment to be degraded by the insulin-degrading enzyme while the receptor recycles to the membrane [reviewed in (Najjar 2002)]. This internalization-degradation process constitutes the basic mechanism of insulin clearance in liver, a dominant site of systemic insulin extraction relative to kidney.

The activated receptor phosphorylates several signaling molecules, such as insulin receptor substrates 1 and 2 (IRS-1, IRS-2), Shc and CEACAM1 (Kitamura and Accili 2004; Najjar et al. 1995; Siddle 2012; White 2003). Of these, CEACAM1 has been extensively shown to enhance the rate of intracellular insulin transport in a phosphorylation-dependent manner (Formisano et al. 1995; Najjar 2002). These findings are buttressed by impaired hepatic insulin clearance and resulting hyperinsulinemia in mice expressing a liver-specific dominant-negative phosphorylation-defective CEACAM1 mutant (L-SACC1) (Poy et al. 2002), null mutation of *Ceacam1* (*Cc1*$^{-/-}$) (Deangelis et al. 2008), or an activated SH-containing phosphatase-1 (SHP-1) that dephosphorylates CEACAM1 (Dubois et al. 2006).

Hyperinsulinemic clamp studies reveal that hyperinsulinemia causes secondary insulin resistance in *Ceacam1* mutant mice (Deangelis et al. 2008; Dubois et al. 2006; Park et al. 2006). Moreover, transgenic hepatic overexpression of CEACAM1 curbs dietary induced insulin resistance mainly by maintaining hepatic insulin clearance and lowering insulin levels (Al-Share et al. 2015). This lends further support to the notion that impaired insulin clearance and its resulting hyperinsulinemia mediate systemic insulin resistance. Thus, studies on the role of CEACAM1 in insulin clearance provide more convincing evidence that hyperinsulinemia causes insulin resistance (Poy et al. 2002). Whereas other substrates of the insulin receptor mediate the metabolic action of insulin by participating in its signaling transduction, CEACAM1 indirectly regulates it by promoting insulin clearance in liver. In this respect, studies on CEACAM1-dependent insulin clearance pathways have advanced our understanding of how hyperinsulinemia itself causes systemic insulin resistance, as originally proposed by McGarry (Mcgarry 1992), and reevaluated by other investigators in recent years (Corkey 2012; Dankner et al. 2009; Pories and Dohm 2012).

3 Hepatic Lipid Metabolism and Insulin Action

Insulin resistance in liver is linked to increased fat accumulation and NAFLD pathogenesis. Elevated fat deposition in liver induces the intracellular content of fatty acids and their metabolites. This, in turn, adversely affects insulin signaling

(Freedland 2004), at least partly, by activating classical and novel protein kinase C (PKC) pathways (Kim et al. 2003; Pereira et al. 2014; Perry et al. 2014), and IKKβ inflammatory signaling (Nguyen et al. 2005). Together, these metabolic and inflammatory effectors could hamper insulin action in liver.

3.1 *Visceral Adipose Tissue and Hepatic Steatosis*

Several distinct but intertwined mechanisms lead to hepatic steatosis: increased free fatty acid (FFA) transport from white adipose tissue and elevated de novo lipogenesis with compromised output and reduced hepatic fatty acid β-oxidation.

Under conditions of positive nutrient supply and visceral obesity, the permissive effect of insulin on lipid formation and storage in white adipose tissue is defected, leading to lipolysis and FFA release (Bergman and Ader 2000; Defronzo 1988; Lewis et al. 2002; Ostman et al. 1979). Mobilized FFA are then transported to the liver (and other extra-adipocytic tissues) to cause systemic glucose intolerance and insulin resistance, which could progress to type 2 diabetes (Baldeweg et al. 2000; Charles et al. 1997; Knowler et al. 2002), owing to the adverse effect of chronic FFA transport into pancreatic β-cells (Unger and Zhou 2001). Mobilized FFA reaching the liver primarily via the portal vein, contribute to insulin resistance (the portal hypothesis) (Bays et al. 2004; Bergman and Ader 2000; Kabir et al. 2005; Rebrin et al. 1995) by promoting lipogenesis (Oakes et al. 1997; Sidossis et al. 1998) and gluconeogenesis (Massillon et al. 1997; Williamson et al. 1966), and curbing the suppressive effect of insulin on glycogenolysis (Boden et al. 2002).

Transported FFA to the liver can also impair hepatic insulin clearance to cause chronic hyperinsulinemia (Mittelman et al. 2000), followed by peripheral insulin resistance. We have shown that this is, at least in part, mediated by reducing hepatic CEACAM1 levels (Al-Share et al. 2015).

Elevated FFA uptake into liver triggers hepatic fatty acid β-oxidation to limit fat accumulation (Chakravarthy et al. 2005). Unchecked fatty acid β-oxidation and ω-peroxidation could cause oxidative stress (Chavin et al. 1999) in addition to activating NF-κB inflammatory pathways, thereby resulting in insulin resistance and mitochondrial dysfunction (Arkan et al. 2005).

Taken together, this highlights the well-documented notion that elevated visceral obesity plays an important role in hepatic insulin resistance by regulating lipid homeostasis along the white adipose tissue-liver axis.

3.2 De Novo *Lipogenesis in Liver*

During feeding, carbohydrates are converted to fatty acids to ultimately produce triglycerides, which could constitute a source of energy at times of fasting. Insulin secretion during feeding regulates transcriptionally lipogenic genes, including fatty acid synthase (FAS) that plays a critical role in de novo lipogenesis by catalyzing the conversion of malonyl-CoA to palmitic acid. Recent studies by the Sul laboratory (Wong et al. 2009) showed that insulin's induction of FAS transcription is regulated by the coordinated action of the nuclear sterol regulatory element-binding protein 1c (nSREBP-1c) and Upstream Stimulatory Factor (USF-1), and that this requires USF-1 phosphorylation by DNA-dependent protein kinase (DNA-PK) in immediate response to insulin release. Induction of FAS transcription by insulin during feeding is consistent with the abundance of this enzyme in hepatocytes, and with the two-to-three fold higher insulin level in the portal relative to systemic circulation (Ward et al. 1990).

De novo lipogenesis under conditions of hepatic insulin resistance: In contrast to blunting the suppressive effect of insulin on hepatic glucose production via Foxo-dependent mechanisms, hyperinsulinemia promotes the positive effect of insulin on de novo lipogenesis in liver (Brown and Goldstein 2008; Haeusler and Accili 2008). Thus, hepatic insulin resistance paradoxically favors the lipogenic effect of insulin. In this respect, hepatic insulin resistance has been regarded as selective by virtue of the fact that it comprises a defective ability of insulin to suppress glycogenolysis and gluconeogenesis, but not lipogenesis.

Is hepatic insulin resistance really selective? Given the abundance of FAS in liver, why is fatty acid synthesis in liver barely detectable even under stimulating conditions (feeding), and why is elevated de novo lipogenesis associated with the clinico-pathologic state of NAFLD?

These questions prompted us to investigate**:** (1) whether FAS activity in liver is counter-regulated in response to insulin, and (2) whether this is regulated by a phosphorylation cascade that is initiated by the acute release of insulin pulses in the portal circulation during refeeding (Najjar et al. 2005). Overnight-fasted wild-type mice manifested an acute rise of insulin after 1, 4 and 6–8 h of refeeding with a regular chow diet. The transient surge of insulin after 4 and 7 h induced phosphorylation of the insulin receptor and CEACAM1, and was associated with elevated CEACAM1/FAS interaction in liver, but not white adipose tissue, leading to reduced hepatic FAS activity (Najjar et al. 2005), as summarized in Fig. 14.1. Together, this suggests that physiologically low FAS activity in liver stems from the acute negative effect of insulin, and that this requires pulsatility of insulin release into the portal vein (Matveyenko et al. 2012) to trigger CEACAM1 phosphorylation in hepatocytes under normo-insulinemic conditions. Failure of insulin to reduce FAS activity in white adipose tissue is consistent with absence of insulin pulsatility in the systemic circulation, and with the physiologically low levels of CEACAM1 in this tissue.

Under conditions of chronic hyperinsulinemia (as in *Ob/Ob* and L-SACC1 phosphorylation-defective CEACAM1 mutant mice), the levels of insulin induced during refeeding are persistently high (Najjar et al. 2005). Consistent with insulin resistance, CEACAM1 phosphorylation and interaction with FAS are ablated (Najjar et al. 2005). This paves the way for insulin-induced activation of FAS transcription to dominate, thereby causing more lipogenesis. Because this positive effect of chronically elevated levels of insulin on FAS transcription is in contrast to its acute down-regulatory effect on FAS activity, we propose to identify hepatic

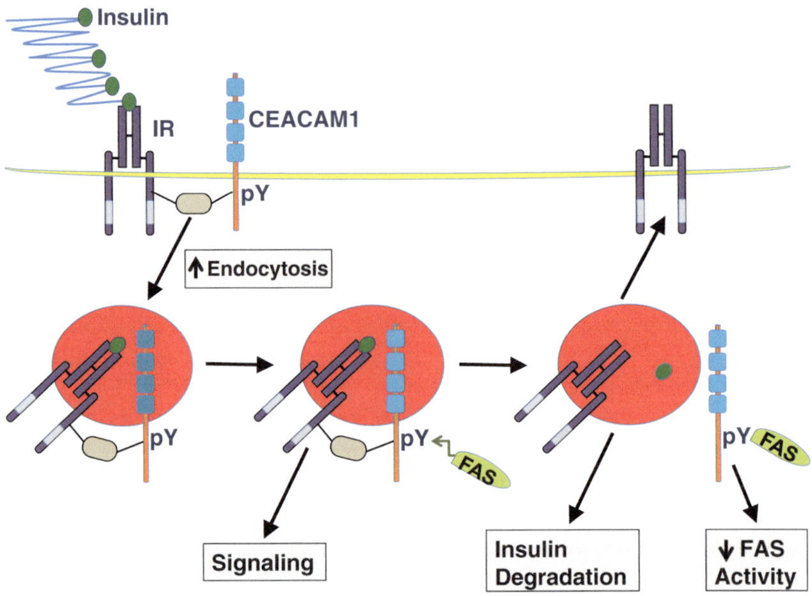

Fig. 14.1 Model of the dual effect of CEACAM1 on insulin and fat metabolism. In response to the pulsatile release of insulin from pancreatic β-cells, the insulin receptor (IR) is activated to phosphorylate CEACAM1 on a conserved tyrosine residue (pY). This leads to the indirect association of CEACAM1 with IR and rapid increase in the rate of the insulin-receptor endocytosis complex, followed by its uptake into clathrin-coated pits and vesicles. Subsequently, fatty acid synthase (FAS) binds to the same phosphorylated tyrosine residue in CEACAM1, causing its dissociation from the insulin-receptor complex. This destabilizes the complex and promotes insulin detachment and degradation by insulin degrading enzyme, while the receptor recycles to the plasma membrane. CEACAM1 binding to FAS causes a decrease in its enzymatic activity

insulin resistance as a comprehensive process comprising not only a defect of insulin to suppress glucose production, but also a defect in its downregulatory action on FAS activity, and subsequently, de novo lipogenesis.

In summary, our studies (Najjar et al. 2005) propose that CEACAM1 phosphorylation in response to acute rise of insulin pulses protects the liver against physiologically high portal insulin level (Ward et al. 1990). Under conditions of insulin resistance and hyperinsulinemia, when insulin pulsatility is masked and CEACAM1 phosphorylation is reduced, the negative acute effect of insulin gives way to its positive chronic effect on FAS transcription to induce de novo lipogenesis in liver.

Impaired insulin clearance regulates de novo lipogenesis: In addition to regulating de novo lipogenesis by mediating the acute negative effect of insulin on FAS enzymatic activity, CEACAM1 also regulates lipogenesis indirectly by promoting hepatic insulin clearance and reducing circulating insulin levels. This notion is buttressed by our recent findings that forced overexpression of CEACAM1 in liver curbs hyperinsulinemia, systemic insulin resistance and the rise in hepatic FAS protein content induced by high-fat diet (Al-Share et al. 2015). Together with increased fatty acid β-oxidation, in part mediated by induction of FGF21, this restricts significantly hepatic steatosis in response to sustained high-fat feeding (Al-Share et al. 2015). Consequently, it limits triglycerides' redistribution to the white adipose tissue, together with the development of diet-induced inflammation and fibrosis in this tissue (Lester et al. 2015).

The phenotype of *Ceacam1* mutant mice further demonstrates the tight connection between insulin clearance and insulin action in lipid metabolism (Bergman 2000). Reduction of CEACAM1 expression and phosphorylation in liver impairs insulin clearance to cause hyperinsulinemia with resulting hepatic lipogenesis (due in part to unchecked FAS activity) (Najjar et al. 2005). This increases triglyceride output and redistribution to white adipose tissue, as reflected by elevated visceral obesity in these mice. Together, this demonstrates how impaired hepatic insulin clearance in liver regulates insulin action by affecting lipid metabolism along the liver-adipose tissue axis. Moreover, it is important to note that mice with liver-specific insulin receptor null mutation (LIRKO) exhibit impairment of insulin clearance with exaggerated insulin resistance, but without modified triglyceride synthesis (Biddinger et al. 2008). The difference between these mouse models of impaired insulin clearance is likely due to the maintenance of a certain level of insulin receptor signaling in *Ceacam1* mutant mice, leading to a milder degree of hyperinsulinemia and lipogenesis, as observed in obese subjects with NAFLD and metabolic syndrome. In support of this notion, hepatic CEACAM1 levels are markedly reduced in obese insulin resistant patients that also develop fatty liver disease (Lee 2011).

Mechanistically, the effect of CEACAM1 on FAS is tied to its effect on insulin uptake. As summarized in Fig. 14.1, binding of FAS to CEACAM1 requires its phosphorylation and internalization as part of the insulin-insulin receptor complex (Najjar et al. 2005). Binding of FAS causes CEACAM1 detachment, an event that promotes insulin dissociation from its receptor to facilitate its degradation. Thus, CEACAM1 is well positioned to act as a unifying mechanism for the intertwined regulation of insulin and lipid metabolism in liver.

4 Insulin Resistance in NAFLD

Patients with metabolic syndrome universally manifest features of NAFLD, and it is widely accepted that treating metabolic syndrome improves liver function in these patients. Yet, a clear-cut molecular mechanism linking hepatic insulin resistance to hepatic steatosis in NAFLD has not been clearly identified (Farrell 2009). The universally accepted divergence in insulin regulation of hepatic glucose and lipid homeostasis during insulin resistance, with a selective defect in insulin action on gluconeogenesis, but not lipogenesis (Brown and Goldstein 2008; Haeusler and Accili 2008) likely contributes to this ambiguity. Our studies showing that insulin resistance also includes a defect in insulin's ability to suppress FAS enzymatic activity in liver and subsequently, hepatic de novo lipogenesis, may resolve some of the questions concerning the manifestation of "insulin resistance" in fatty liver disease.

5 Conclusion

That impaired insulin clearance is associated with insulin resistance in type 2 diabetes was first proposed by Mirsky and Broh-Kahn (Mirsky and Broh-Kahn 1949). Several investigations have followed to explore this connection in various animal models (Ader et al. 2014), and in individuals with obesity (Jones et al. 2000; Meistas et al. 1983) and predisposition to metabolic syndrome and type 2 diabetes (Lee et al. 2013; Pivovarova et al. 2013). The insulin-degrading enzyme has been identified as a diabetes susceptibility gene (Sladek et al. 2007; Zeggini et al. 2007). Our studies demonstrate that CEACAM1 is an important promoter of receptor-mediated insulin entry into the hepatocyte, an important first step in its clearance process.

In addition to insulin resistance, chronic hyperinsulinemia caused by impaired insulin clearance, in turn results in hepatic steatosis in mice with *Ceacam1* mutations. This phenotype is consistent with the mechanistically linked metabolic effects of CEACAM1 in liver: promoting insulin clearance to maintain insulin sensitivity and mediating an inhibitory physiologic acute downregulatory effect of insulin on de novo lipogenesis. While more clinical studies are needed, the important role of reduced CEACAM1 levels in liver as a molecular link between impaired insulin clearance and hepatic steatosis in metabolic syndrome has been highlighted by a recent clinical study demonstrating a marked reduction of hepatic CEACAM1 levels in obese insulin resistant obese subjects with fatty liver disease (Lee 2011). Thus, *Ceacam1* mutant mice are indisputably reliable replicates of the human disease. Importantly, they provide a valuable tool to understand the molecular underpinning of the intertwined regulation of insulin action by insulin and lipid metabolism. Their phenotype identifies low hepatic level of CEACAM1 as a missing link between insulin resistance (hyperinsulinemia) and hepatic steatosis, and promotes this plasma membrane protein as a therapeutic target against NAFLD and metabolic syndrome.

Acknowledgements The author thanks all the past and current trainees, and collaborators who participated in the studies of the effect of CEACAM1. This work was supported by grants from the NIH R01 DK054254, R01 DK083850, R01 HL112248 and 5P01 HL036573 to SMN.

Disclosures: None.

References

Adams LA, Lymp JF, St Sauver J et al (2005) The natural history of nonalcoholic fatty liver disease: a population-based cohort study. Gastroenterology 129:113–121

Ader M, Stefanovski D, Kim SP et al (2014) Hepatic insulin clearance is the primary determinant of insulin sensitivity in the normal dog. Obesity (Silver Spring) 22:1238–1245

Al-Share QY, Deangelis AM, Lester SG et al (2015) Forced hepatic overexpression of CEACAM1 curtails diet-induced insulin resistance. Diabetes 64:2780–2790

Arkan MC, Hevener AL, Greten FR et al (2005) IKK-beta links inflammation to obesity-induced insulin resistance. Nat Med 11:191–198

Baldeweg SE, Golay A, Natali A et al (2000) Insulin resistance, lipid and fatty acid concentrations in 867 healthy Europeans. European Group for the Study of Insulin Resistance (EGIR). Eur J Clin Invest 30:45–52

Bays H, Mandarino L, Defronzo RA (2004) Role of the adipocyte, free fatty acids, and ectopic fat in pathogenesis of type 2 diabetes mellitus: peroxisomal proliferator-activated receptor agonists provide a rational therapeutic approach. J Clin Endocrinol Metab 89:463–478, Review

Beauchemin N, Draber P, Dveksler G et al (1999) Redefined nomenclature for members of the carcinoembryonic antigen family. Exp Cell Res 252:243–249

Bergman RN (2000) Non-esterified fatty acids and the liver: why is insulin secreted into the portal vein? Diabetologia 43:946–952

Bergman RN, Ader M (2000) Free fatty acids and pathogenesis of type 2 diabetes mellitus. Trends Endocrinol Metab 11:351–356

Biddinger SB, Hernandez-Ono A, Rask-Madsen C et al (2008) Hepatic insulin resistance is sufficient to produce dyslipidemia and susceptibility to atherosclerosis. Cell Metab 7:125–134

Boden G, Cheung P, Stein TP et al (2002) FFA cause hepatic insulin resistance by inhibiting insulin suppression of glycogenolysis. Am J Physiol Endocrinol Metab 283:E12–E19

Brown MS, Goldstein JL (2008) Selective versus total insulin resistance: a pathogenic paradox. Cell Metab 7:95–96

Carpentier JL (1994) Insulin receptor internalization: molecular mechanisms and physiopathological implications. Diabetologia 37(Suppl 2):S117–S124

Chakravarthy MV, Pan Z, Zhu Y et al (2005) "New" hepatic fat activates PPARalpha to maintain glucose, lipid, and cholesterol homeostasis. Cell Metab 1:309–322

Charles MA, Eschwege E, Thibult N et al (1997) The role of non-esterified fatty acids in the deterioration of glucose tolerance in Caucasian subjects: results of the Paris Prospective Study. Diabetologia 40:1101–1106

Chavin KD, Yang S, Lin HZ et al (1999) Obesity induces expression of uncoupling protein-2 in hepatocytes and promotes liver ATP depletion. J Biol Chem 274:5692–5700

Corkey BE (2012) Banting lecture 2011: hyperinsulinemia: cause or consequence? Diabetes 61: 4–13

Dankner R, Chetrit A, Shanik MH et al (2009) Basal-state hyperinsulinemia in healthy normoglycemic adults is predictive of type 2 diabetes over a 24-year follow-up: a preliminary report. Diabetes Care 32:1464–1466

Deangelis AM, Heinrich G, Dai T et al (2008) Carcinoembryonic antigen-related cell adhesion molecule 1: a link between insulin and lipid metabolism. Diabetes 57:2296–2303

Defronzo RA (1988) Lilly lecture 1987. The triumvirate: beta-cell, muscle, liver. A collusion responsible for NIDDM. A collusion responsible for NIDDM. Diabetes 37:667–687

Dubois MJ, Bergeron S, Kim HJ et al (2006) The SHP-1 protein tyrosine phosphatase negatively modulates glucose homeostasis. Nat Med 12:549–556

Eckel RH, Alberti KG, Grundy SM et al (2010) The metabolic syndrome. Lancet 375:181–183

Farrell GC (2009) The liver and the waistline: Fifty years of growth. J Gastroenterol Hepatol 24(Suppl 3):S105–S118

Farrell GC, Larter CZ (2006) Nonalcoholic fatty liver disease: from steatosis to cirrhosis. Hepatology 43:S99–S112

Ferrante AW Jr (2007) Obesity-induced inflammation: a metabolic dialogue in the language of inflammation. J Intern Med 262:408–414

Formisano P, Najjar SM, Gross CN et al (1995) Receptor-mediated internalization of insulin. Potential role of pp 120/HA4, a substrate of the insulin receptor kinase. J Biol Chem 270:24073–24077

Freedland ES (2004) Role of a critical visceral adipose tissue threshold (CVATT) in metabolic syndrome: implications for controlling dietary carbohydrates: a review. Nutr Metab (Lond) 1:12

Gregor MF, Hotamisligil GS (2011) Inflammatory mechanisms in obesity. Annu Rev Immunol 29: 415–445

Haeusler RA, Accili D (2008) The double life of Irs. Cell Metab 8:7–9

Hotamisligil GS, Shargill NS, Spiegelman BM (1993) Adipose expression of tumor necrosis factor-alpha: direct role in obesity-linked insulin resistance. Science 259:87–91
Jones CN, Abbasi F, Carantoni M et al (2000) Roles of insulin resistance and obesity in regulation of plasma insulin concentrations. Am J Physiol Endocrinol Metab 278:E501–E508
Kabir M, Catalano KJ, Ananthnarayan S et al (2005) Molecular evidence supporting the portal theory: a causative link between visceral adiposity and hepatic insulin resistance. Am J Physiol Endocrinol Metab 288:E454–E461
Kim YB, Kotani K, Ciaraldi TP et al (2003) Insulin-stimulated protein kinase C lambda/zeta activity is reduced in skeletal muscle of humans with obesity and type 2 diabetes: reversal with weight reduction. Diabetes 52:1935–1942
Kitamura Y, Accili D (2004) New insights into the integrated physiology of insulin action. Rev Endocr Metab Disord 5:143–149
Knowler WC, Barrett-Connor E, Fowler SE et al (2002) Reduction in the incidence of type 2 diabetes with lifestyle intervention or metformin. N Engl J Med 346:393–403
Lee W (2011) The CEACAM1 expression is decreased in the liver of severely obese patients with or without diabetes. Diagn Pathol 6:40
Lee CC, Haffner SM, Wagenknecht LE et al (2013) Insulin clearance and the incidence of type 2 diabetes in Hispanics and African Americans: the IRAS Family Study. Diabetes Care 36:901–907
Lester SG, Russo L, Ghanem SS et al. (2015) Hepatic CEACAM1 overexpression protects against diet-induced fibrosis and inflammation in white adipose tissue. Front Endocrinol (Lausanne). 6:116
Lewis GF, Carpentier A, Adeli K et al (2002) Disordered fat storage and mobilization in the pathogenesis of insulin resistance and type 2 diabetes. Endocr Rev 23:201–229
Lumeng CN, Saltiel AR (2011) Inflammatory links between obesity and metabolic disease. J Clin Invest 121:2111–2117
Marchesini G, Bugianesi E, Forlani G et al (2003) Nonalcoholic fatty liver, steatohepatitis, and the metabolic syndrome. Hepatology 37:917–923
Marra F, Gastaldelli A, Svegliati Baroni G et al (2008) Molecular basis and mechanisms of progression of non-alcoholic steatohepatitis. Trends Mol Med 14:72–81
Massillon D, Barzilai N, Hawkins M et al (1997) Induction of hepatic glucose-6-phosphatase gene expression by lipid infusion. Diabetes 46:153–157
Matveyenko AV, Liuwantara D, Gurlo T et al (2012) Pulsatile portal vein insulin delivery enhances hepatic insulin action and signaling. Diabetes 61:2269–2279
Mcgarry JD (1992) What if Minkowski had been ageusic? An alternative angle on diabetes. Science 258:766–770
Meistas MT, Margolis S, Kowarski AA (1983) Hyperinsulinemia of obesity is due to decreased clearance of insulin. Am J Physiol Endocrinol Metab 245:E155–E159
Mirsky IA, Broh-Kahn RH (1949) The inactivation of insulin by tissue extracts; the distribution and properties of insulin inactivating extracts. Arch Biochem 20:1–9
Mittelman SD, Van Citters GW, Kim SP et al (2000) Longitudinal compensation for fat-induced insulin resistance includes reduced insulin clearance and enhanced beta-cell response. Diabetes 49:2116–2125
Najjar SM (2002) Regulation of insulin action by CEACAM1. Trends Endocrinol Metab 13:240–245
Najjar SM, Russo L (2014) CEACAM1 loss links inflammation to insulin resistance in obesity and non-alcoholic steatohepatitis (NASH). Semin Immunopathol 36:55–71
Najjar SM, Accili D, Philippe N et al (1993) pp 120/ecto-ATPase, an endogenous substrate of the insulinreceptortyrosinekinase,isexpressedastwovariablysplicedisoforms.JBiolChem268:1201–1206
Najjar SM, Philippe N, Suzuki Y et al (1995) Insulin-stimulated phosphorylation of recombinant pp 120/HA4, an endogenous substrate of the insulin receptor tyrosine kinase. Biochemistry 34: 9341–9349
Najjar SM, Yang Y, Fernstrom MA et al (2005) Insulin acutely decreases hepatic fatty acid synthase activity. Cell Metab 2:43–53

Nguyen MT, Satoh H, Favelyukis S et al (2005) JNK and tumor necrosis factor-alpha mediate free fatty acid-induced insulin resistance in 3 T3-L1 adipocytes. J Biol Chem 280:35361–35371
Oakes ND, Cooney GJ, Camilleri S et al (1997) Mechanisms of liver and muscle insulin resistance induced by chronic high-fat feeding. Diabetes 46:1768–1774
Öbrink B (1997) CEA adhesion molecules: multifunctional proteins with signal-regulatory properties. Curr Opin Cell Biol 9:616–626
Odegaard JI, Chawla A (2013) Pleiotropic actions of insulin resistance and inflammation in metabolic homeostasis. Science 339:172–177
Osborn O, Olefsky JM (2012) The cellular and signaling networks linking the immune system and metabolism in disease. Nat Med 18:363–374
Ostman J, Arner P, Engfeldt P et al (1979) Regional differences in the control of lipolysis in human adipose tissue. Metabolism 28:1198–1205
Park SY, Cho YR, Kim HJ et al (2006) Mechanism of glucose intolerance in mice with dominant negative mutation of CEACAM1. Am J Physiol Endocrinol Metab 291:E517–E524
Pereira S, Park E, Mori Y et al (2014) FFA-induced hepatic insulin resistance in vivo is mediated by PKCδ, NADPH oxidase, and oxidative stress. Am J Physiol Endocrinol Metab 307:E34–E46
Perry RJ, Samuel VT, Petersen KF et al (2014) The role of hepatic lipids in hepatic insulin resistance and type 2 diabetes. Nature 510:84–91
Pivovarova O, Bernigau W, Bobbert T et al (2013) Hepatic insulin clearance is closely related to metabolic syndrome components. Diabetes Care 36:3779–3785
Pories WJ, Dohm GL (2012) Diabetes: have we got it all wrong? Hyperinsulinism as the culprit: surgery provides the evidence. Diabetes Care 35:2438–2442
Poy MN, Yang Y, Rezaei K et al (2002) CEACAM1 regulates insulin clearance in liver. Nat Genet 30:270–276
Rebrin K, Steil GM, Getty L et al (1995) Free fatty acid as a link in the regulation of hepatic glucose output by peripheral insulin. Diabetes 44:1038–1045
Saltiel AR (2012) Insulin resistance in the defense against obesity. Cell Metab 15:798–804
Shoelson SE, Lee J, Goldfine AB (2006) Inflammation and insulin resistance. J Clin Invest 116:1793–1801
Siddle K (2012) Molecular basis of signaling specificity of insulin and IGF receptors: neglected corners and recent advances. Front Endocrinol 3:34
Sidossis LS, Mittendorfer B, Walser E et al (1998) Hyperglycemia-induced inhibition of splanchnic fatty acid oxidation increases hepatic triacylglycerol secretion. Am J Physiol Endocrinol Metab 275:E798–E805
Sladek R, Rocheleau G, Rung J et al (2007) A genome-wide association study identifies novel risk loci for type 2 diabetes. Nature 445:881–885
Takahashi Y, Fukusato T (2010) Pediatric nonalcoholic fatty liver disease: overview with emphasis on histology. World J Gastroenterol 16:5280–5285
Unger RH, Zhou YT (2001) Lipotoxicity of beta-cells in obesity and in other causes of fatty acid spillover. Diabetes 50(Suppl 1):S118–S121
Ward GM, Walters JM, Aitken PM et al (1990) Effects of prolonged pulsatile hyperinsulinemia in humans. Enhancement of insulin sensitivity. Diabetes 39:501–507
White MF (2003) Insulin signaling in health and disease. Science 302:1710–1711
Williamson JR, Kreisberg RA, Felts PW (1966) Mechanism for the stimulation of gluconeogenesis by fatty acids in perfused rat liver. Proc Natl Acad Sci U S A 56:247–254
Wong RH, Chang I, Hudak CS et al (2009) A role of DNA-PK for the metabolic gene regulation in response to insulin. Cell 136:1056–1072
Zebhauser R, Kammerer R, Eisenried A et al (2005) Identification of a novel group of evolutionarily conserved members within the rapidly diverging murine Cea family. Genomics 86:566–580
Zeggini E, Weedon MN, Lindgren CM et al (2007) Replication of genome-wide association signals in UK samples reveals risk loci for type 2 diabetes. Science 316:1336–1341

Index

A

ACC. *See* Acetyl-CoA carboxylase (ACC)
ACC1. *See* Acetyl-CoA carboxylase 1 (ACC 1)
Acetate, 192, 193
Acetyl-coA carboxylase (ACC), 2
Acetyl-coA carboxylase 1 (ACC1), 36
5′ Adenosine monophosphate (AMP), 15
Adipokines, 261
Adipose tissue de novo lipogenesis, 260, 261, 270, 271, 273
  advantages, 265
  BAT, 264
  beneficial metabolic states, 264
  dietary carbohydrates, 265, 267, 268
  disadvantage, 265
  fatty acids, 264, 265, 267, 268
  function, 264
  hepatic lipid metabolism, 269, 270
  insulin regulate adipocyte, 265, 267, 268
  lipokine hypothesis, 272, 273
  liver monounsaturated fatty acids, 259, 260
  metabolic health
    dramatic induction, 260
    lipogenic program, 260
    mechanisms, 261
  nutrients, 264
  synthesize, 264
  systemic insulin sensitivity
    anti-inflammatory effects, 271
    FAHFAs, 273
    glucose transport, 271
    lipid storage, 270
    mechanism, 270
    synthesis, 273
    transgenic mice, 270, 271
  WAT, 264
AFLD. *See* Alcoholic fatty liver disease (AFLD)
Akt
  Akt phosphorylation, mSin1, 60
  lipogenesis, 64
  tuberous sclerosis complex, 60
Alcoholic fatty liver disease (AFLD)
  acetaldehyde dehydrogenase, 220
  characteristics, 220
  factors, 220
  lipogenic effects, 220
  mechanisms, 220
AMP. *See* 5′ Adenosine monophosphate (AMP)
AMP-activated protein kinase (AMPK), 35
  chronic diseases, 35
  exercise and adiponectin, 34
  glucose and lipid homeostasis, 33
  hypothalamus, 34
  LKB1/AMPK pathway, 35
  master energy sensor and metabolic switch, 34
  metformin (*see* Metformin)
  obesity and type 2 diabetes, 35
  phosphorylation, 33
  polyphenols, 37, 38
  protein kinase LKB1, 34
  and xenobiotics, 34
ATP citrate lyase (Acly), 247

J.M. Ntambi (ed.), *Hepatic De Novo Lipogenesis and Regulation of Metabolism*,
DOI 10.1007/978-3-319-25065-6

B

Bacterial-transformed secondary bile acids, 194
Basic helix-loop-helix–leucine zipper (bHLH-Zip), 38
BH3-only proteins, 107, 110, 112, 113
bHLH-Zip. *See* Basic helix-loop-helix–leucine zipper (bHLH-Zip)
Bile acids, 194
Brown adipose tissue (BAT). *See* White adipose tissue (WAT)
Butyrate, 193

C

Cancer, 19
Carbohydrate responsive element-binding protein (ChREBP)
  acetylation and O-GlcNAcylation, 13
  ChREBPβ, 13–14
  cytosol/nucleus, 12
  DNA-binding activity, 12
  high-fat diet and high-fructose diet, 14
  histone acetyltransferase (HAT), 13
  liver steatosis, 14
  phosphorylation/dephosphorylation, 12
  PKA phosphorylation, 12
Carcino-embryonic antigen-related cell adhesion molecule 1 (CEACAM1)
  gene and protein, 282
  hepatic insulin clearance, 283
  insulin and fat metabolism, 285, 286
Cardiovascular disease (CVD)
  characterization, 221
  cluster, risk factors, 221
  components, dyslipidemia, 221
  metabolic syndrome, 221
  VLDL, 221, 222
Carnitine palmitoyltransferase 1 α (CPT-1α), 36
CB1 receptor
  adiponectin synthesis, 131, 135
  CB_1-deficient and ApoE-deficient mice, 133
  chemical approach, 132
  chronic alcohol, 133
  FAAH, 132
  hepatic lipogenesis, 134
  hepatic mRNA expression levels, 131
  hepatic stellate cells, 133
  obesity, 132
  paracrine mechanism, 133
  rimonabant, 133
  triacylglycerols, 132
  zebrafish larvae, 132
CB_2 receptor, 135
CCAAT/enhancer-binding homologous protein (CHOP), 110, 111
CDI. *See* Clostridium difficile infection (CDI)
CEACAM1. *See* Carcino-embryonic antigen-related cell adhesion molecule 1 (CEACAM1)
Choline, gut microbiota
  cardiovascular diseases, 195
  FMO3, 195
  phosphatidylcholine (lecithin), 195
  requirement, 195
  SREBPs, 196
  toxic methylamines, 195
Chromosomal maintenance 1 (CRM-1), 79
Circadian rhythm, 233, 240, 242
c-Jun N-terminal kinase (JNK), 112
Clostridium difficile infection (CDI), 200
CPT-1α. *See* Carnitine palmitoyltransferase 1 α (CPT-1α)
CRM-1. *See* Chromosomal maintenance 1 (CRM-1)
CVD. *See* Cardiovascular disease (CVD)
Cytochrome P450 2E1 (CYP2E1), 115
Cytokines, 261

D

De novo lipogenesis (DNL) pathway, 146–148
  acetyl-CoA source, 256, 257
  AMP and AMPK, 15
  antisense oligonucleotide (ASO), 19
  Bmal1-knockout mice, 16
  calorie supply, 1
  cancer, 19
  circadian rhythms and metabolism, 16
  ER stress and UPR, 17
  fatty acid (FA), 1, 20, 256
  lipid and fatty acid trafficking models, 258
  liver, 2–3, 17
  metabolic syndromes, 17–19, 256
  NEFA, 257
  obesity and diabetes, 16
  polymorphisms, 16
  PPARγ and PPARδ regulation, 15
  Rev-erbα and Rev-erbβ, 16
  SIRT1, 15–16
  synthesis, 256
  transcriptional control, 256
  triglyceride secretion and β-oxidation, 3–4
Death receptors, 114
Delta 5 (Δ5) desaturase
  fatty acids, 151
  hepatic pathology, 152
  LCPUFA, 151

molecular mechanisms, 151
plasmatic concentration, 152
SCD1, 151
Delta 6 (Δ6) desaturase, 151, 152
Dietary carbohydrates, 265–268
Dietary fiber, 189
DNL pathway. *See* De novo lipogenesis (DNL) pathway
Dyslipidemia, 222

E
Endocannabinoids (eCBs)
adipose tissue, 130
arachidonoyl ethanolamide (AEA), 130
Cannabis sativa, 129
G protein-coupled receptors, 129
hepatocytes, 131
immune and hematopoietic systems, 130
lipophilicity, 130
liver, 130
Endoplasmic reticulum (ER), 5, 40, 108–110, 166, 167
ER. *See* Endoplasmic reticulum (ER)
ER-associated degradation (ERAD) pathway, 108
Extra-hepatic effect
immune function, 193
intestinal macrophages, 194
lipogenesis, 193

F
FAAH. *See* Fatty-acid amide hydrolase (FAAH)
FAHFAs. *See* Fatty acid esters of hydroxy fatty acids (FAHFAs)
Farnesoid X receptor (FXR), 243
FAS. *See* Fatty acid synthase (FAS)
Fasn. *See* Fatty acid synthase (Fasn)
Fatty acid esters of hydroxy fatty acids (FAHFAs), 249
Fatty acid synthase (FAS), 32
Fatty acid synthase (Fasn), 32, 36, 42, 43, 50, 247
Fatty liver, 131, 132, 134, 136
Fatty-acid amide hydrolase (FAAH), 130
FC. *See* Free cholesterol (FC)
Fecal microbiota transplantation (FMT), 200
FGF21. *See* Fibroblast growth factor 21 (FGF21)
Fibrosis
DGAT2, 102
FXR activation, 106, 107
liver, 116, 117
FMT. *See* Fecal microbiota transplantation (FMT)
Forkhead box O1 (FoxO1)
gluconeogenesis, 82, 83
hyperglycemia, 84
hypertriglyceridemia, 87, 90
independent inhibition, 83
insulin signaling, 79, 82, 88
lipogenesis, 88, 89
NAFLD, 88, 89
transcription factor, 78, 79
Free cholesterol (FC), 105
FXR. *See* Farnesoid X receptor (FXR)

G
GK. *See* Glucokinase (GK)
Glucocorticoids, 268
Glucokinase (GK), 10
Gluconeogenesis
independent inhibition, 83, 84
insulin action, 82
insulin signaling, 83
kidney, 82
PEPCK, 82
pivotal metabolic pathway, 82
purposes, 82
Glucose-6-phosphatase (G6Pase), 82, 83
Glucose-ChREBP pathway
carbohydrate level, 10
chromatography and mass spectrometry, 11
DNA-binding activity, 11
GK, 10
glucose promotes lipogenic gene transcription, 10
glycolysis and lipogenesis, 10
glucose-sensing module (GSM), 11
hepatocytes, 10
lipogenic gene transcription, 10
Mlx, 11
NLS, 11
nuclear export signals, 11
promoter-mapping analysis, 10
Glucose-response activation conserved element (GRACE), 12
Glycogen synthase (Gys2), 65
G6Pase. *See* Glucose-6-phosphatase (G6Pase)
GRACE. *See* Glucose-response activation conserved element (GRACE)
Gut microbiota
bacterial-transformed secondary bile acids, 194
bile acids, 194
choline, 195, 196

Gut microbiota (*cont.*)
cyanobacteria, 188
development, 189
dietary fiber, 189
factors, 188
feeding mice, 189
functions, 190
hepatic lipogenesis, 190, 191
liver metabolism, 189, 190
microbiota-mediated metabolic diseases, 198, 199
microbiotal metabolites, 196–198
plethora, 190
potential treatment, 199, 200
prominent producers, 190
SCFAs, 190
triglycerides accumulation, 189
versatile function, 188
Gys2. *See* Glycogen synthase (Gys2)

H
HCC. *See* Hepatocellular carcinoma (HCC)
Hepatic d*e novo* lipogenesis, 217–222, 243, 244, 246, 247
biochemical pathways, 208
blood glucose and lipids, 231
cancer cells, 51
cellular metabolism, 32
circadian clock transcription factors, 231
citrate, 32
clock regulators and metabolic homeostasis, 239–240
crosstalk tissue, 248–249
dietary nutrient composition, 208, 209
endogenous ligands, 242
enzymes, 210
exercise, 216, 217
feeding/fasting responses, 232–233
gain-/loss-of function mouse models
Acaca/b, 247
Acly, 247
Fasn, 247–248
Gck, 247
Pdha1, 247
Scd1, 248
glucose and lipid metabolism, 32
hormonal fluctation, 208, 209
hormonal regulation, 213–216
hepatic lipid metabolism, 240
hepatic lipid synthesis and catabolism, 241
insulin and glucose, hepatocytes, 32, 33
isoforms, 208
lipogenic pathway, 32
carbohydrates, 243
complex lipids, 243, 244
mitochondrial synthesis, 243
mouse genetic models, 244, 245
liver metabolism, 233
long-term regulation, 209
LXR and FXR, 242, 243
metabolic diseases, 223
metabolic ligands, 240
molecular clock, 233–235
mouse genetic models, 231
negative feedback loop, 242
nuclear receptors, 240
nutrient-sensing mechanisms, 51
nutritional regulation, 210–213
palmitic acid, 208
pathophysiological role
AFLD, 220, 221
CVD, 221
homeostasis, 217
insulin resistance, 218–219
NAFLD, 217, 218, 220, 221
VLDL, 222
PPARs, 242
Rev-erb α and β, 240
RORs and REV-ERBs, 240
short-term regulation, 209
Hepatic insulin clearance
hepatocyte, 283
hyperinsulinemia, 283
mechanism, 283
phosphorylates, 283
Hepatic insulin resistance
diacylglycerol, 219
gluconeogenesis, 218
lipogenesis, 218
NAFLD, 219
steatosis, 219
Hepatic lipid droplets
characteristics, 144
fusion/coalescence, 146
mechanism, 145
phospholipid monolayer, 144
Hepatic lipid metabolism
adipocyte DNL, 270
dietary lipids, 269
fatty acids, 269
hepatosteatosis, 174
liver, 175, 284–287
mechanism, 174
metabolic adaptation, 269
steatosis, 284

visceral adipose tissue, 284
VLDL secretion, 174
Hepatic lipid synthesis
cholesterol biosynthetic enzymes, 44
hepatocytes, 45
human HepG2 hepatoma cells, 45
inhibiting SREBP-1c-dependent lipogenesis, 44
insulin resistance and type 2 diabetes, 44
lipid-lowering effect, 44
lipogenic genes, 45
liver, 44
luciferase reporters, 45
mimic hyperglycemia and insulin resistance, 45
phosphorylation, 44
polyphenols and metformin, 45
refeeding, 44
Hepatic lipogenesis, 208
bacteroidetes, 191
DNL (*see* Hepatic de novo lipogenesis (DNL))
environmental factor, 190
fatty acid, 208
SCFAs, 191
Hepatic selective insulin resistance, 48
Hepatic steatosis, 284
Hepatocellular carcinoma (HCC), 117, 194, 195
Hepatokines, 258–259
High fat diet (HFD), 131, 135
HMC coA reductase (HMGCR), 106
Hormonal regulation
glucagon, 215, 216
insulin, 214, 215
Hormones, 69
Hyperglycemia, 84
Hypertriglyceridemia
characterization, 87, 90
chromosomal translocations, 90
insulin signaling, 87
pathogenesis, 87

I
IGF binding protein (IGFBP), 68
Inflammation. *See* Nonalcoholic fatty liver disease (NAFLD)
Ingenuity pathway analysis (IPA), 65
Insulin-dependent inhibition, 81
Insulin, FoxO1
gluconeogenesis, 82, 83
inhibitory action, 79, 82
TG metabolism, 84, 85
VLDL-TG production, 85, 86
Insulin-like growth factor 1 (IGF-1)
Akt phosphorylation, 67
AMPK, 68
glycogenesis, 68
hormones, 68
IGFBP, 68
lipogenesis, 67, 68
steatosis and hepatocellular carcinoma, 69
Insulin-receptor substrates (IRS), 6
Insulin resistance (IR), 143, 200, 232
liver, 283
metabolic syndrome, 281
NAFLD, 287
Inter-organ communication, 248
IPA. *See* Ingenuity pathway analysis (IPA)
IR. *See* Insulin resistance (IR)
IRS. *See* Insulin-receptor substrates (IRS)

J
JNK. *See* c-Jun N-terminal kinase (JNK)

L
LCPUFA. *See* Long-chain poly-unsaturated fatty acids (LCPUFA)
LDLR. *See* Low density lipoprotein receptor (LDLR)
LDs. *See* Lipid droplets (LDs)
Lipid biosynthesis, 40–43
fatty acids, triacylglycerol, cholesterol and phospholipid, 39
glucose and insulin, 39
intestine and spleen, 39
posttranscriptional regulation
cholesterogenic gene expression, 40
cholesterol, 40
GSK3 activity, 42
nutrient regulation, 40, 41
proteolytic activation, 40
PUFA, 41
SCAP, 40
sterols, 40
ubiquitin-proteasome pathway, 41
sterol-sensitive feedback inhibition process, 39
transcriptional regulation
hyperglycemia, 42
hypertriglyceridemia and obesity, 43
lipid metabolism, 43
liver, 42
nuclear receptors, 43
nutrients and hormones, 42
nutrient sensing pathways, 42

Lipid biosynthesis (*cont.*)
obesity-induced insulin-resistant mice, 42
plasma glucagon levels, 42
triglycerides, 43
type 2 diabetes, 43
Lipid droplets (LDs)
abnormal liver function tests, 164
electron microscopy, 164
endoplasmic reticulum, 166, 167
expansion, 167
hepatocytes, 164–166
hepatosteatosis, 164
proteomic studies, 164
Lipid mediator, 244
Lipid metabolism
cholesterol and bile acids, 105–107
free fatty acids and triglycerides, 102–103
hepatosteatosis, 175
LPC, 103
microarray analyses, 175
SL, 103, 105
triglycerides (TG), 101
Lipoapoptosis
apoptosis, 107
Bcl-2 family proteins and mitochondrial-lysosomal pathway, 112, 113
CHOP-dependent apoptosis, 110–111
c-Jun NH2-terminal kinase signaling pathway, 111, 112
death receptors, 114
ER stress, 108–110
fibrosis, 107
Lipogenesis
adiponectin and leptin, 66
Akt2, 64
clinical and preclinical data, 88
DNL pathway, 146–148
elongation, 146
ELOVL6 expression, 148
FGF21, 69
FOXO1 and FOXA2, 65
IGF-1, 68
lipin1, 64
metabolic diseases, 149
mTORC1, 63
numerous studies, 146
potential mechanism, 89
PPARα ligands, 149
pyruvate, 146
steatosis, 149
transcription factors, 148
Lipogenic effect, insulin
CEACAM1, 282, 283
hepatic lipid metabolism, 283–287
NAFLD, 287
Lipokine hypothesis, 262, 272, 273
Lipolysis
adipocytes, 169–170
non-adipose tissues, 170, 171
Liver
chronic inflammation, 196
functions, 195, 196
gluconeogenesis, 65
hepatic lipid metabolism, 174, 175
human polymorphism, 173
hyperlipidemia, 63
independent inhibition, 83, 84
insulin-AKT pathway, 8–9
insulin resistance, 9–10
integrates insulin signaling, 82–83
IPA, 65
lipid homeostasis, 63, 64
lipid synthesis, 5, 64
lipin1, 63
lipogenesis, 4–5, 64
lipogenic gene expression and fatty acid synthesis, 5
PLIN2, 173
PPARα, 65
SREBPs, 63
Liver, DNL
hepatic insulin clearance, 286, 287
hepatic insulin resistance, 285, 286
insulin secretion, 284
Liver metabolism, 233
Liver monounsaturated fatty acids
advantage, 259
hepatic DNL, 259
mechanism, 259
palmitoleate and oleate, regulation, 260
regulation, 259
Liver X receptors (LXRs), 7, 283
Long-chain poly-unsaturated fatty acids (LCPUFA), 151
Low density lipoprotein receptor (LDLR), 38
LPC. *See* Lysophosphatidyl choline (LPC)
LXR-responsive elements (LXREs), 41
LXRs. *See* Liver X receptors (LXRs)
Lysophosphatidyl choline (LPC), 101, 103

M

MAGL. *See* Monoglyceride lipase (MAGL)
Mammalian target of rapamycin complex 1 (mTORC1), 48, 232
Max-like factor X (Mlx), 11
MCD. *See* Methionine choline deficient (MCD)

Mechanistic target of rapamycin (mTOR)
anti-proliferative effects, 58
cytoplasm and nucleus, 58
hepatic lipid metabolism, 66–67
hormones, 69
metabolism and physiology, 61
molecular regulation, 59–60
mTORC1, 58
mTORC2, 61–62
protein kinase C (PKC), 58
Metabolic endotoxemia, 189, 198
Metabolic syndromes, 17–19
Metformin
ACC1, 36
glucose metabolism, 35
glucose production, liver, 37
hepatocellular lipids, 36
high glucose-induced insulin resistance and triglyceride, 35
in vitro insulin-resistant state, 35
PKA, 37
rat hepatocytes, 36
type 2 diabetes, 35
tyrosine phosphorylation, 35
Methionine choline deficient (MCD), 101
Microbiotal metabolites
component, 197
endogenous alcohol, 197
inflammasomes, 197
liver diseases, 196
LPS, 197
NAFLD, 196, 198
translocation, 196
Microbiota-mediated metabolic diseases
bacteroidetes, 198
components, 199
host immunity, 198
IgA deficiency, 199
immune receptor, 198
Microsomal triglyceride transfer protein (MTP), 239
Mlx. *See* Max-like factor X (Mlx)
Molecular clock, 238–239
Monoglyceride lipase (MAGL), 130
Monounsaturated fatty acids (MUFAs), 132, 262, 263
mTOR complex 1 (mTORC1)
adipogenesis, 66
adipose-derived hormones leptin and adiponectin, 68
cell growth and metabolism, 58
cytoplasm and nucleus, 58
definition, 58
lipogenesis, 63
lysosome, 60
Rheb, 60
SREBPs, 63
mTOR complex 2 (mTORC2)
Igfbp1 transcription, 68
insulin/IGF-1 signaling pathway, 58, 59
lipogenesis, 64
mTORC1, 58
MTP. *See* Microsomal triglyceride transfer protein (MTP)
MUFAs. *See* Monounsaturated fatty acids (MUFAs)

N
NAFLD. *See* Non-alcoholic fatty liver disease (NAFLD)
NASH. *See* Nonalcoholic steatohepatitis (NASH)
NES. *See* Nuclear export signal (NES)
Neutral lipid synthesis, 166
'New Age' metabolic disorders
CDI, 200
FMT, 200
insulin resistance, 200
probiotic Lactobacillus strains, 200
widespread consumption, 199
NLS. *See* Nuclear localization signal (NLS)
Non-alcoholic fatty liver disease (NAFLD), 88, 89
accumulation, lipids, 143
characterization, 142
chronic liver disease, 100
clinical-histological phenotypes, 100
delta 5 ($\Delta 5$) and delta 6 ($\Delta 6$), 151, 152
DNL dysregulation, 217–218
gut metabolites, 196
hepatic inflammatory responses, 116–118
hepatic lipid droplets, 144–146
hepatic steatosis, 101, 196
hepatocyte apoptosis, 101
insulin resistance, 101, 287
lensing model, 145
lipogenesis, 146, 148–150
liver histology, 142
metabolic syndrome, 101
oxidative stress, 115–116, 144
public health issue, 142
SCD1, 153, 154
steatohepatitis, 115
steatosis, 142
T2DM, 101
TNFα receptor, 144
two-hit hypothesis, 101

Nonalcoholic steatohepatitis (NASH), 100
Nuclear export signal (NES), 79
Nuclear localization signal (NLS), 11
Nutrient sensors
  diet-induced hepatic steatosis and hypercholesterolemia, 49
  gluconeogenesis, 48
  hepatic lipogenic and cholesterol synthetic processes, 50
  hepatocytes, 50
  lipogenic gene expression, 50
  lipotoxicity, 50
  mTORC1, 48
  protein kinase, 50
  tandem mass spectrometry and mutagenesis studies, 50
  triglyceride and cholesterol metabolism, 50
Nutritional regulation
  carbohydrate, 210
  fatty acid composition, 212, 213
  fructose, 211, 212
  glucose, 211
  lipogenic effect, 210
  palmitoleate, 213

O
Obesity
  AMPK-mediated inhibition, 44
  hypertriglyceridemia, 43
  metformin, 35
  SREBP-1c and SREBP-2, 42
Oxidative stress. *See* Nonalcoholic fatty liver disease (NAFLD)

P
PC. *See* Phosphatidylcholine (PC)
pdha1. *See* Pyruvate dehydrogenase (pdha1)
PEPCK. *See* Phosphoenolpyruvate carboxykinase (PEPCK)
Perilipins (PLIN)
  lipolysis, 169, 170
  proteins, 168, 169
  structures, 168, 171
Perilipins 1 (PLIN1)
  172, adipocytes
  172, chronic hepatic steatosis
  172, fatty liver
  172, steroidogenic cells
Perilipins 2 (PLIN2)
  lipid metabolism, 175
  liver, 173
Perilipins 3 (PLIN3)
  cytoplasmic sorting factor, 175
  factors, 176
  histologic analyses, 176
  liver pathology, 177
  protein, 177
Perilipins 5 (PLIN5)
  cardiomyocytes, 178
  hepatocytes, 178
  human pancreatic islets, 178
  human/mouse obesity, 177
  insulin sensitivity, 179
  mouse primary hepatocytes, 178
Peroxisome proliferator-activated receptors (PPARs), 242
Phosphatidylcholine (PC), 195, 248, 249
Phosphoenolpyruvate carboxykinase (PEPCK), 82
Phosphorylation
  acetyl-CoA carboxylase and HMG-CoA reductase, 48
  AMP/ADP, 46
  AMPKα subunit, 46
  bioinformatics tools, 46, 47
  cholesterol metabolism, 48
  nutrient signaling crosstalk, 48
  protein-protein interaction, 46
  Ser-327, 47
  SREBP-1/SREBP-2, 47
  zebrafish, 47
PKA. *See* Protein kinase A (PKA)
PLIN. *See* Perilipins (PLIN)
Polyunsaturated fatty acids (PUFA), 41
PPARs. *See* Peroxisome proliferator-activated receptors (PPARs)
Propionate, 193
Protein kinase A (PKA), 37
PUFA. *See* Polyunsaturated fatty acids (PUFA)
Pyruvate dehydrogenase (pdha1), 247

R
RAR-related orphan receptors (RORs), 240
Retinoic acid receptor-gamma (RARγ), 133

S
SCAP. *See* SREBP-cleavage activating protein (SCAP)
Scd1. *See* Stearoyl-CoA dehydrogynase (Scd1)
SCD1. *See* Stearoyl-CoA desaturase 1 (SCD1)
SCN. *See* Supra chiasmic nucleus (SCN)
Short-chain fatty acids (SCFAs), 191–193

extra-hepatic effect, 193, 194
metabolic functions
acetate, 192, 193
butyrate, 193
colonocytes, 192
dietary fiber, 191
propionate, 193
SHP. *See* Small heterodimer partner (SHP)
Sphingolipid (SL), 103–105
SREBP-1c. *See* Sterol regulatory element binding protein-1c (SREBP-1c)
SREBP-cleavage activating protein (SCAP), 6
SREBPs. *See* Sterol regulatory element-binding proteins (SREBPs)
Stearoyl-coA dehydrogynase (Scd1), 248
Stearoyl-coA desaturase 1 (SCD1)
hepatic fat metabolism, 153
mechanism, 153
numerous studies, 153
pathogenesis, 154
steatosis, 154
synthesis, 153
Sterol regulatory element binding protein-1c (SREBP-1c), 32
Sterol-regulatory element binding proteins (SREBPs), 39, 63, 196
adult animals, 39
amino acids, 39
bHLH-Zip, 38
ER, 5
intracellular lipid metabolism, 38
IRS, 6
lipid biosynthesis (*see* Lipid biosynthesis)
LXRs, 7
mTORC1 protein kinase, 6
positive feedback loop, 7
protein stability, 7–8
SCAP-SREBP-1c, 6
site-1 protease (SIP1), 6
site-2 protease (S2P), 6
SREBP-1a and SREBP-1c, 5
TSC1–TSC2 complex, 6
Supra chiasmic nucleus (SCN), 233

T
T2DM. *See* Type 2 diabetes mellitus (T2DM)
TCA cycle. *See* Tricarboxylic acid (TCA) cycle
Tetrahydrocannabinol (THC), 129
TGs. *See* Triglycerides (TGs)
THC. *See* Tetrahydrocannabinol (THC)
Thyroid hormone, 268
TNF receptor. *See* Tumor necrosis factor (TNF) receptor
TNFα. *See* Tumor necrosis factor-α (TNFα) receptor
Tricarboxylic acid (TCA) cycle, 2, 32
Triglycerides (TGs), 1, 2
Tumor necrosis factor (TNF) receptor, 114
Tumor necrosis factor-α (TNFα) receptor, 144
Type 2 diabetes
berberine, 38
biguanide drugs, 35
fatty liver and cardiovascular complications, 35
hepatic steatosis and dyslipidemia, 43
hyperglycemia, 37
hyperinsulinemia, 48
metformin, 35
obesity, 35

V
Very long chain fatty acids (VLCFA), 3
Very-low density lipoproteins (VLDLs), 232
cytosolic triglyceride pool, 222
dyslipidemia, 222
fatty acids, 222
metabolism, 221
Visceral adipose tissue, 284
Visceral adiposity, 259
VLCFA. *See* Very long chain fatty acids (VLCFA)
VLDLs. *See* Very low-density lipoproteins (VLDLs)
VLDL-TG production, 85, 86

W
White adipose tissue (WAT)
dynamic adaptation, 264
dynamic regulation, 265
fatty acids, 264
glucose transporter, 267
growth hormone inhibits, 268
insulin resistance, 265
metabolic syndrome, 265
triglycerides (TGs), 264

Z
Zeitgeber time zero (ZT0), 240
ZT0. *See* Zeitgeber time zero (ZT0)

MIX
Papier aus verantwortungsvollen Quellen
Paper from responsible sources
FSC® C105338

If you have any concerns about our products,
you can contact us on
ProductSafety@springernature.com

In case Publisher is established outside the EU,
the EU authorized representative is:
Springer Nature Customer Service Center GmbH
Europaplatz 3, 69115 Heidelberg, Germany

Printed by Libri Plureos GmbH
in Hamburg, Germany